Vectorial

Fundamentals and Applications

Optical Fields

Vectorial
Fundamentals and Applications
Optical Fields

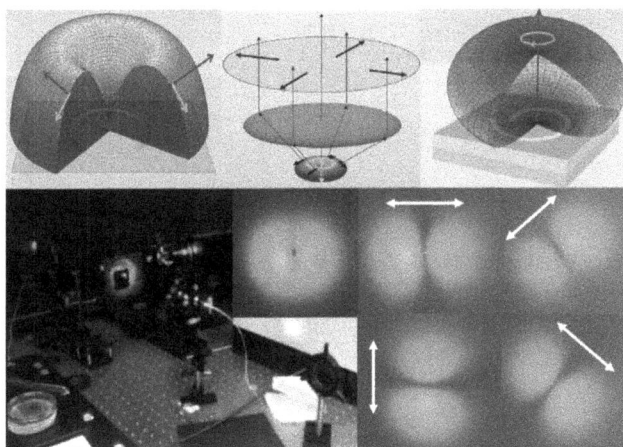

Editor

Qiwen Zhan
University of Dayton, USA

World Scientific

NEW JERSEY · LONDON · SINGAPORE · BEIJING · SHANGHAI · HONG KONG · TAIPEI · CHENNAI

Published by

World Scientific Publishing Co. Pte. Ltd.

5 Toh Tuck Link, Singapore 596224

USA office: 27 Warren Street, Suite 401-402, Hackensack, NJ 07601

UK office: 57 Shelton Street, Covent Garden, London WC2H 9HE

Library of Congress Cataloging-in-Publication Data
Zhan, Qiwen.
    Vectorial optical fields : fundamentals and applications / Qiwen Zhan, University of Dayton, USA.
    pages cm
  Includes bibliographical references and index.
  ISBN 978-9814449885 (hardcover : alk. paper)
  1. Polarization (Light) 2. Electrooptics. I. Title.
  TA1750.Z43 2013
  535.5'2--dc23

              2013026562

British Library Cataloguing-in-Publication Data
A catalogue record for this book is available from the British Library.

In-house Editor: Song Yu

PREFACE

Polarization of light plays important roles in optical sciences and engineering. This vector nature of light has been extensively exploited in numerous optical and photonic designs. Most previous research dealt with light beams with spatially homogeneous states of polarization. There has been a recent increasing interest in vectorial optical fields with spatially inhomogeneous state of polarizations. The early interests started with the studies of optical imaging such as optimal concentration of electromagnetic radiation in the focal region (Sheppard and Larkin, *J. Mod. Opt.* **41**, 1495, (1994); *Optik,* **107**, 79, (1997)). Many techniques of generating and manipulating vectorial optical fields and their applications have been reported ever since. New effects and phenomena have been predicted and observed for optical fields with these unconventional polarization states. Overwhelming attentions were given to the high numerical aperture focusing of those so-called cylindrical vector (CV) beams named and studied by Brown *et al.* (*Opt. Express* **7**, 77 (2000)). Sharper focusing with radial polarization was experimentally demonstrated by Leuchs *et al.* (*Phys. Rev. Lett.* **91**, 233901 (2003)). Zhan and Leger explored focal field engineering with CV beams (*Opt. Express* **10**, 324 (2002)). Their peculiar physics and broad potential applications have drawn increasing studies in CV beams and other more generalized optical vectorial fields. Many of the earlier developments were reviewed in several recent monographs (Hasman *et al.*, *Progress in Optics*, **47**, 215, (2005); Zhan, *Adv. Opt. Photon.,* **1**, 1, (2009); Brown, *Progress in Optics,* **56**, 81, (2011)).

Researchers around the world continue to come up with new vectorial optical fields concepts and develop techniques to generate, characterize and utilize them. To capture the latest development in this important emerging field of optics, it is my pleasure to introduce you this review volume book, *Vectorial Optical Fields: Fundamentals and Applications*, which is the first of its kind to my best knowledge. This review volume consists of chapters contributed by scientists around the world who are active in this field with balanced depth and breadth. In Chapter 1, I will provide a review of the fundamentals and recent developments in CV beams, which also serves the foundations for further discussions in the following chapters. Beyond CV beams, vectorial optical fields with more complicated spatial polarization distributions exhibit many novel properties and physics that attracted rapidly increasing interests. Professor Hui-Tian Wang (Nankai University, China) will present the latest progress of this field in Chapter 2. In Chapter 3, Professor Alfred Meixner and his team (University of Tübingen, Germany) will review the latest work in the use of CV beams for imaging and spectroscopy of single nanoparticles and single quantum systems. The use of vectorial optical fields to engineer optical focal field with specific desired characteristics, such as optical needle, optical bubble, flattop focusing, and 3D optical polarization etc., will be reviewed in Chapter 4 by Dr. Weibin Chen (Seagate Technology, USA). In Chapter 5, recent progresses on plasmonic interactions with vectorial optical fields will be summarized by Dr. Guanghao Rui (University of Dayton, USA). The polarization diversity within the cross-section of vectorial optical fields offers interesting opportunities in optical measurement. This will be illustrated by several examples of optical measurement techniques including microellipsometry, polarimetry and interferometry in Chapter 6. In Chapter 7, Professor Yangjian Cai and his group (Soochow University, China) will provide a comprehensive review of partially coherent vector beams and their propagation.

With simple and efficient generation methods becoming more and more accessible, the research into vectorial optical fields experienced exponential increase recently and the scope of their applications continued to expand. Considering the rapid growth in this field, omission of many important aspects that also deserve to be in this review volume is inevitable. For example, attentions should be paid to the recent developments of vectorial fiber lasers that can produce vectorial optical outputs owing to their relatively high gain, compactness and flexibility. Diverse applications of vectorial optical fields, including laser micro-machining and material processing, novel extremely high-density data storage, three-dimensional micro- and nano-fabrication, are very exciting as well. Nevertheless, it is my intention for this review volume to serve as an introduction to this nascent and promising field for researchers who are interested in this area, especially young graduate students and scientists, and encourage them to work in this field. I hope that you will enjoy reading this book and find it useful and stimulating for your research.

Finally, I would like to thank Professors Yangjian Cai, Alfred Meixner, Fei Wang, Hui-Tian Wang, Chengliang Zhao, and Drs. Weibin Chen, Anna M. Chizhik, Alexey I. Chizhik, Yiming Dong, Regina Jäger, Guanghao Rui, Frank Wackenhut, Gaofeng Wu, and Shijun Zhu for their important contributions to this book. I am also grateful to my current and former students and colleagues who have contributed to the understanding of the subject of vectorial otpical fields, participated in stimulating discussions, conducted difficult experiments and collected excellent experimental data, and co-authored papers in the past: Dr. Weibin Chen, Dr. Guanghao Rui, Mr. Shuangyang Yang, Dr. Alain Tschimwangan, Mr. Renjie Zhou, Dr. Wen Cheng, Dr. Wei Han, Prof. Jiming Wang, Prof. Yanfang Yang, Dr. Yiqiong Zhao, Dr. Jianing Dai, Dr. Liangcheng Zhou, Prof. James R. Leger, Prof. Daniel Ou-yang, Prof. Yongping Li, Prof. Hai Ming, Prof. Lixin Xu, Dr. Don C. Abeysinghe, and Dr. Robert L. Nelson. I would also like to acknowledge financial supports from the Fraunhofer Society, the State of Ohio, the US Air Force Research Laboratory, and the National Science Foundation in the past.

Qiwen Zhan, Editor
Professor of Electro-Optics, Electrical and Computer Engineering
University of Dayton

CONTENTS

Preface...v

Chapter 1 : Cylindrical Vector Beams 1
   Qiwen Zhan
  1. Introduction 1
  2. Mathematical description of cylindrical vector beams......2
  3. Graphical representation of cylindrical vector beams 4
  4. Generation of cylindrical vector beams...........................7
   4.1. Passive generation methods in free space 7
   4.2. Passive generation methods using optical fiber 11
   4.3. Active generation methods..13
  5. Cylindrical vector beams under high NA focusing 16
  6. Summary 23
  References..24

Chapter 2 : Vector Optical Fields and their Novel Effects 27
   Hui-Tian Wang
  1. Introduction 27
  2. Generation of vector optical fields 29
   2.1. Local linearly polarized vector fields.....................33
   2.2. Hybridly polarized vector fields 40
  3. Novel effects 45
   3.1. Optical cages..45
   3.2. Axial-symmetry broken vector fields 51
   3.3. Young's two-slit interference..................................57
   3.4. Optical orbital angular momentum (OAM) 63
  4. Summary 68
  References..69

Chapter 3 : Cylindrical Vector Beams for Spectroscopic Imaging of 73
   Single Molecules and Nanoparticles
   Regina Jäger, Anna M. Chizhik, Alexey I. Chizhik,
   Frank Wackenhut, Alfred J. Meixner
  1. Introduction 73

2. Theoretical background 75

3. Instrumentation 86

4. Fluorescence spheres to probe the quality of CVBs 88

5. Single molecules 89

6. Single nanoparticles 103

   6.1 Single SiO_2 nanoparticles 103

   6.2. Single silicon nanocrystals 106

   6.3. Excitation isotropy of single CdSe/ZnS quantum .. 108
     dots

   6.4. Optical characterization of single gold nanorods ... 111

7. Orientation and position determination of a single 113
   quantum emitter inside an optical microcavity

8. Conclusions 118

References 118

Chapter 4 : Comprehensive Focal Field Engineering with Vectorial ... 125
   Optical Fields
   Weibin Chen and Qiwen Zhan

1. Introduction 126

2. Three-dimensional focus shaping with CV beams 127

3. Three-dimensional polarization control within focal 133
   volume

4. Spherical spot with controllable 3D polarization 141

5. Focus shaping through inverse dipole array 146
   radiation

   5.1. High purity optical needle field 147

   5.2. 3D optical tube, flattop focus and optical chain 153

6. Conclusions 157

References 157

Chapter 5 : Plasmonics with Vectorial Optical Fields 161
   Guanghao Rui and Qiwen Zhan

1. Surface plasmon polaritons 161

2. Interaction of vectorial fields with plasmonic 164
   structures

   2.1. Planar metallic thin film 164

2.2. Bull's eye structures................................ 169

2.3. Extraordinary optical transmission with radial..... 177
    polarization

2.4. Polarization mode matching and optimal 181
    plasmonic focusing

2.5. Archimedes' spiral plasmonic lens 182

2.6. Applications in near-field optical probe designs... 191

3. Conclusions 198

References.. 199

Chapter 6 : Optical Measurement Techniques Utilizing Vectorial 201
    Optical Fields
    Qiwen Zhan

1. Introduction 201

2. Manipulation techniques for vectorial optical fields 202

3. Microellipsometer with rotational symmetry 206

    3.1. Microellipsometer with rotational symmetry......... 206

    3.2. Nulling microellipsometer with rotational............. 211
        symmetry

4. Radial polarization interferometer 214

5. Rapid mueller matrix polarimetry 216

6. Atomic spin analyzer...................................... 217

7. Summary 218

References.. 219

Chapter 7 : Partially Coherent Vector Beams: From Theory to........... 221
    Experiment
    *Yangjian Cai, Fei Wang, Chengliang Zhao, Shijun
    Zhu, Gaofeng Wu, and Yiming Dong*

1. Introduction 221

2. Characterizations of partially coherent vector beams ... 223

3. Partially coherent vector beams with uniform state 227
    of polarization: theory

    3.1. Partially coherent electromagnetic Gaussian 227
        Schell-model beam

3.2. Tensor method for treating the paraxial 229
propagation of partially coherent electromagnetic
Gaussian Schell-model beam

3.3. Statistics properties of a partially coherent 232
electromagnetic Gaussian Schell-model beam
in a Gaussian cavity

3.4. Propagation of a partially coherent 235
electromagnetic Gaussian Schell-model beam in
turbulent atmosphere

3.5. Coincidence fractional Fourier transform with a ... 240
partially coherent electromagnetic Gaussian Schell-
model beam

3.6. Degree of paraxiality of a partially coherent.......... 244
electromagnetic Gaussian Schell-model beam

4. Partially coherent vector beams with uniform state 247
of polarization: experiment

4.1. Experimental generation and measurement of a 247
partially coherent electromagnetic beam

4.2. Experimental coupling of a partially coherent 251
electromagnetic Gaussian Schell-model beam into a
single-mode optical fiber

5. Partially coherent vector beams with non-uniform 254
state of polarization: theory

5.1. Cylindrical vector partially coherent beam and 255
its paraxial propagation

5.2. Tight focusing properties of a partially coherent .. 259
azimuthally polarized beam

6. Partially coherent vector beams with non-uniform 265
state of polarization: experiment

7. Summary 268

References... 269

Index 275

CHAPTER 1

CYLINDRICAL VECTOR BEAMS

Qiwen Zhan

Electro-Optics Program, University of Dayton
300 College Park, Dayton, Ohio 45469, USA
E-mail: qzhan1@udayton.edu

Cylindrical vector beams are considered one special class of optical vector fields as the axially symmetric beam solution to the full vector electromagnetic wave equation. In this chapter, a brief introduction of the mathematical and graphical descriptions of the cylindrical vector beams will be provided. Various active and passive methods to generate these cylindrical vector beams are reviewed. Expressions for the focal field distributions of the cylindrical vector beams focused by high numerical aperture objective lens are derived. The cylindrical polarization symmetry of this special class of optical vector beams gives rise to unique focusing properties under high numerical aperture condition with important applications including optical imaging and manipulation.

1. Introduction

Cylindrical vector beams, also known as CV beams, are vector-beam solutions of Maxwell's equations that obey axial symmetry in both amplitude and phase.[1, 2, 3] Laser beams with such axial polarization symmetry has been investigated many decades ago.[4] However, due to the difficulties associated with the generation and manipulation of these beams as well as the lack of practical applications, little attention was paid to this field until recently. In the past decade, researchers discovered unique focusing properties of CV beams under high numerical aperture

(NA) condition with potential applications in imaging, machining, particle trapping, data storage, and sensing, etc. This field has seen rapidly growing interests driven by these potential applications. In this chapter, we will first derive the mathematical description of CV beams by solving the full vector electromagnetic wave equation. Graphical representation of CV beams will be given to illustrate the characteristics of their polarization distribution within their cross sections. We will also review both active and passive methods for the generation of CV beams. Expressions for the focal field under high NA focusing will be derived and examples will be given to illustrate some of the unique properties of CV beams when they are strongly focused. Applications of these focusing properties are further explored and discussed in later chapters.

2. Mathematical description of cylindrical vector beams

In free space, the paraxial harmonic beam-like solutions are obtained by solving the scalar Helmholtz equation:

$$(\nabla^2 + k^2)E = 0 . \tag{1}$$

In Cartesian coordinates, the general solution of a beam-like paraxial solution at a specific frequency takes the form of:

$$E(x, y, z, t) = u(x, y, z)\exp[i(kz - \omega t)] . \tag{2}$$

Utilizing the slow varying envelop approximation:

$$\begin{cases} \dfrac{\partial^2 u}{\partial z^2} << k^2 u \\ \dfrac{\partial^2 u}{\partial z^2} << k\dfrac{\partial u}{\partial z}, \end{cases}$$

one can derive the Hermite-Gauss (HG) solution HG_{mn} mode as:

$$u(x, y, z)$$
$$= AH_m\left(\sqrt{2}\frac{x}{w(z)}\right)H_n\left(\sqrt{2}\frac{y}{w(z)}\right)\frac{w_0}{w(z)}\exp[-i\phi_{mn}(z)]\exp\left[-\frac{k}{2q(z)}r^2\right], \tag{3}$$

where $H_m(x)$ is the Hermite polynomials, $w(z)$ is the beam size, w_0 is the beam size at beam waist, $z_0 = \pi w_0^2 / \lambda$ is the Rayleigh range, $q(z) = z + jz_0$ is the complex beam parameter and $\varphi_{mn}(z) = (m+n+1)$ $\tan^{-1}(z/z_0)$ is the Gouy phase shift. For $m=n=0$, this solution reduces to the fundamental Gaussian beam solution:

$$u(r,z) = A\frac{w_0}{w(z)}\exp[-i\varphi(z)]\exp\left[-\frac{k}{2q(z)}r^2\right]. \qquad (4)$$

In cylindrical coordinates, the general form of a beam-like paraxial solution takes the following formula:

$$E(r,\phi,z,t) = u(r,\phi,z)\exp[i(kz - \omega t)], \qquad (5)$$

By inserting this trial solution into the scalar Helmholtz equation (1) and applying the slow varying envelop approximation, one can get:

$$\frac{1}{r}\frac{\partial}{\partial r}\left(r\frac{\partial u}{\partial r}\right) + \frac{1}{r^2}\frac{\partial^2 u}{\partial \phi^2} + 2ik\frac{\partial u}{\partial z} = 0. \qquad (6)$$

From this equation, the Laguerre-Gauss solution LG_{pl} modes can be obtained by using the separation of variables in r and φ.

$$u(r,\phi,z) = A(\sqrt{2}\frac{r}{\omega})^l L_p^l\left(2\frac{r^2}{\omega^2}\right)\frac{w_0}{w(z)}\exp[-i\varphi_{pl}(z)]\exp\left[-\frac{k}{2q(z)}r^2\right], \qquad (7)$$
$$\times\exp(-jl\phi)$$

where $L_p^l(x)$ is the associated Laguerre polynomials and $\varphi_{pl}(z) = (2p+l+1)\tan^{-1}(z/z_0)$ is the Gouy phase shift. For $l=p=0$, the solution also reduces to the fundamental Gaussian beam solution.

Another solution to Eq. (6) that obeys rotational symmetry (independent of azimuthal angle φ) has also been found. These solutions take the following general formula:

$$u(r,z) = A\frac{w_0}{w(z)}\exp[-i\varphi(z)]\exp\left[-\frac{k}{2q(z)}r^2\right]J_0\left(\frac{\beta r}{1+iz/z_0}\right), \qquad (8)$$
$$\times\exp\left[-\frac{\beta^2 z/(2k)}{1+iz/z_0}\right]$$

where β is a scaling parameter, $\phi(z)$ is the Gouy phase shift, and $J_0(x)$ is the 0th order Bessel function of the first kind. This is the co-called scalar Bessel-Gauss beam solution. When $\beta = 0$, the solution again reduces to the fundamental Gaussian beam solution.

The solutions derived above (Hermite-Gauss, Laguerre-Gauss and Bessel-Gauss) are the paraxial beam-like solutions that correspond to spatially homogeneous polarization or scalar beams. For these beams, the local state of polarization (SOP) is independent on the location of observation points within the beam cross section. However, if the full vector wave equation for the electric field is considered, we have:

$$\nabla \times \nabla \times \vec{E} - k^2 \vec{E} = 0. \tag{9}$$

A axially symmetric beamlike vector solution with the electric field aligned in the azimuthal direction should have the form of:

$$\vec{E}(r,z) = U(r,z)\exp[i(kz - \omega t)]\vec{e}_\phi, \tag{10}$$

where $U(r,z)$ satisfies the following equation under paraxial and slow varying envelope approximation:

$$\frac{1}{r}\frac{\partial}{\partial r}\left(r\frac{\partial U}{\partial r}\right) - \frac{U}{r^2} + 2ik\frac{\partial U}{\partial z} = 0. \tag{11}$$

The solution that obeys azimuthal polarization symmetry has trial solution as:

$$U(r,z) = AJ_1\left(\frac{\beta r}{1+iz/z_0}\right)\exp\left[-\frac{i\beta^2 z/(2k)}{1+iz/z_0}\right]u(r,z), \tag{12}$$

with $u(r, z)$ being the fundamental Gaussian solution given in Eq. (4), $J_1(x)$ is the 1st order Bessel function of the first kind. This is the azimuthally polarized vector Bessel-Gauss beam solution. Similarly the transverse magnetic field solution can be found as:

$$\vec{H}(r,z) = -BJ_1\left(\frac{\beta r}{1+iz/z_0}\right)\exp\left[-\frac{i\beta^2 z/(2k)}{1+iz/z_0}\right]u(r,z)\exp[i(kz - \omega t)]\vec{h}_\phi. \tag{13}$$

The transverse component of the electric field of this solution is aligned in the radial direction. Hence it represents the radial polarization for the electric field. It should be noted that there is also a z-component of the

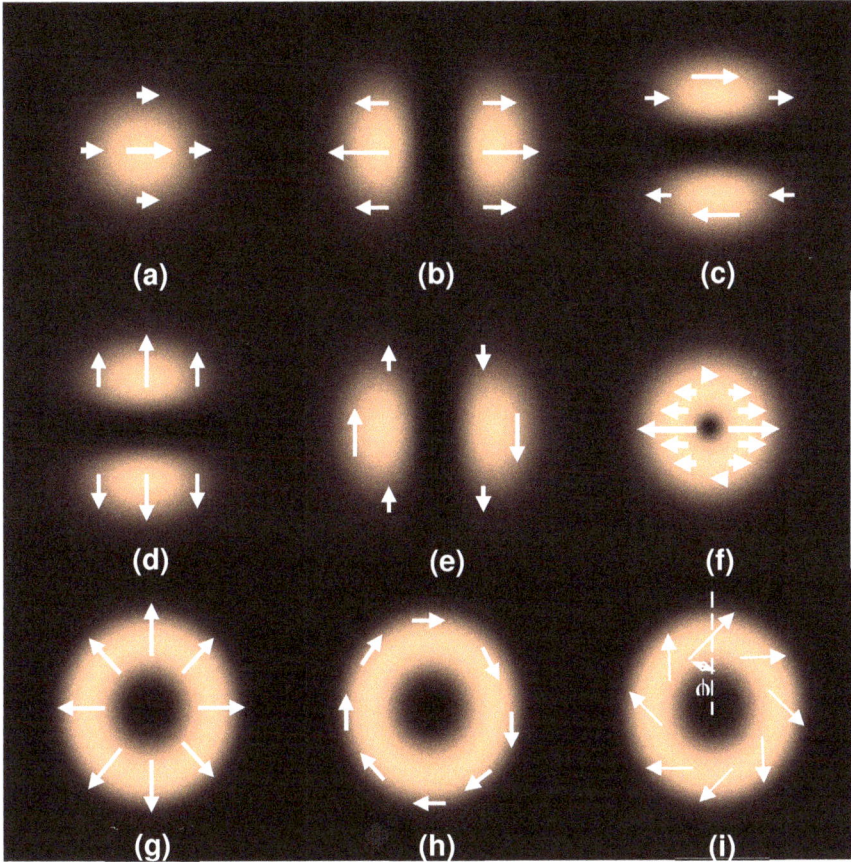

Fig. 1. Illustration of the instantaneous electric field spatial distribution for different modes. (a) x-polarized fundamental Gaussian mode; (b) x-polarized HG_{10} mode; (c) x-polarized HG_{01} mode; (d) y-polarized HG_{01} mode; (e) y-polarized HG_{01} mode; (f) x-polarized LG_{01} mode; (g) radial polarization; (h) azimuthal polarization; and (i) generalized CV beams.[2]

electric field as well due to the finite size of the beam. However, this z-component is weak and can be ignored under paraxial conditions.

3. Graphical representation of cylindrical vector beams

Graphical representation of the spatial distributions of the *instantaneous* electric field vector for linearly polarized Hermite-Gauss, Laguerre-Gauss modes and the CV modes are shown in Fig. 1 to illustrate the characteristics of the spatial polarization distribution of these modes. The SOPs of the modes shown in Fig. 1(a)-(f) are regarded to be spatially homogeneous despite of the electric field may have opposite instantaneous direction caused by the inhomogeneous phase distribution across the beam. The field illustrated in Fig. 1(g) has polarization aligned in the radial direction, which is called the radial polarization. Similarly, the polarization pattern shown in Fig. 1(h) is termed azimuthal polarization. The generalized CV beam as a linear superposition of the radial and azimuthal polarizations is shown in Fig. 1(i). Due to the transverse field continuity, CV modes feature a singularity point in the center of the beam.

In many applications, simplified distributions have been used to describe the amplitude profile of CV beams, especially for CV beams with large cross section. For very small β, the vector Bessel-Gauss beam at the beam waist can be approximated as:

$$\vec{E}(r,z) = A r \exp(-\frac{r^2}{w^2})\vec{e}_i, i = r, \phi \; . \tag{14}$$

This turns out to be the LG_{01} modes without the vortex phase term $\exp(i\varphi)$. Using Eqs. (3) and (7), it can be shown that CV beams can also be expressed as superposition of orthogonally polarized Hermite-Gauss HG_{01} and HG_{10} modes:

$$\vec{E}_r = HG_{10}\vec{e}_x + HG_{01}\vec{e}_y \; , \tag{15}$$

$$\vec{E}_\phi = HG_{01}\vec{e}_x + HG_{10}\vec{e}_y \; , \tag{16}$$

where E_r and E_φ denotes radial and azimuthal polarization respectively. This is graphically illustrated in Fig. 2. Such a linear superposition principle has been explored in the generation of CV beams through interferometric means.

In some cases, particularly for CV beams generated with collimated beams through passive devices for high NA focusing studies, annular distribution with an opaque center block can be used to describe the field after these devices or in the pupil plane:

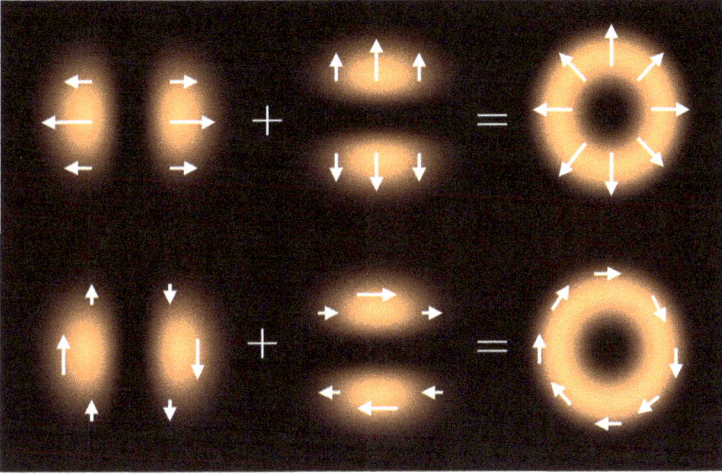

Fig. 2. Formation of radial and azimuthal polarizations through interferometric superposition of orthogonally polarized HG modes.[2]

$$\vec{E}(r) = P(r)\vec{e}_i, i = r, \phi \ , \tag{17}$$

where $P(r)$ is the beam cross section or pupil function. For example, for a uniform annular illumination, $P(r)$ can be written as:

$$P(r) = \begin{cases} 1, & r_1 < r < r_2 \\ 0, & 0 < r < r_1 \end{cases} . \tag{18}$$

4. Generation of cylindrical vector beams

In general, the techniques used to generate cylindrical vector beams can be broadly divided into two main categories: active and passive, depending on whether gain medium is involved. A detailed overview of a variety of these methods can be found in a previous review monograph.[2] A variety of passive methods have been demonstrated to generate CV beams in free space. For example, interferometric methods, such as Mach-Zehnder interferometer, combined with spiral phase element (SPE) have been demonstrated.[5] More often, passive generation methods convert commonly used spatially homogeneous polarizations (e.g. linear or circular polarization) into the spatially inhomogeneous cylindrical vector polarizations, which entail the use of devices with spatially variant polarization properties.

4.1. *Passive generation methods in free space*

CV beams can be generated in free space by using devices with spatially arranged retardation axis. For example, λ/2-plates with spatially variant axis directions can be used to convert a linear polarization into CV polarizations. This type of device can be realized by carefully taping or gluing together several segmented λ/2-plates with different discrete crystal angles (Fig. 3). Due to the discrete nature of this approach, this type of device provides quasi spatial alignment of the polarization. Mode selector, such as near confocal Fabry-Perot interferometer, can be used to further clean up the polarization distribution pattern.[6] Continuous rotation of linear polarization input using stress-induced space-variant plate[7] has been reported, which eliminates the needs for such a mode selector.

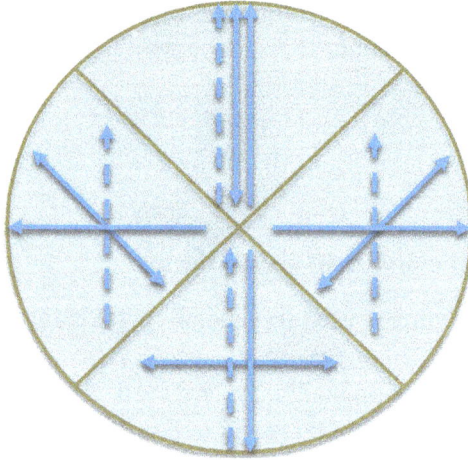

Fig. 3. Illustration of a segmented λ/2-plate that can convert a linearly polarized incident polarization into a quasi-radial polarization. In this case, four segments of λ/2-plates with their fast axis (denoted as double arrows in the diagram) aligned at 45°, 90°, 135° and 180° with respect to the horizontal directions are assembled together. For a vertically polarized input (denoted as the dashed arrows), each of the λ/2-plate segments rotates the input linear polarization to the quasi-radial directions (denoted as the solid arrows).

Devices with continuous axial birefringence and dichroism have been applied to generate CV beam in free space. Simple setups with radial polarizer made from either birefringent or dichroic materials can be used to generate the CV beams. When a circularly polarized collimated beam is used as the input to the radial polarizer, the beam after the radial polarizer will be polarized either radially or azimuthally, depending on the type of radial analyzer used. However, one needs to pay attention to an additional geometrical phase (also known as the Berry's phase or Pancharatnam's phase) that arises from this operation. For a circularly polarized input:

$$E_{in} = e_x + je_y = (\cos\varphi e_r - \sin\varphi e_\varphi) + j(\sin\varphi e_r + \cos\varphi e_\varphi) = e^{j\varphi}(e_r + je_\varphi),$$

(19)

where e_x and e_y are the unit vectors in the Cartesian coordinates, e_r and e_φ are the unit vectors in the cylindrical coordinates. After the beam passes through the radial polarizer with the local transmission axes aligned along the radial direction, the output can be expressed as:

$$E_{out} = e^{j\varphi}e_r.$$

(20)

Apparently, although the electrical field is aligned along the radial direction, there is a spiral geometrical phase with topological charge of 1 on top of it. This geometric spiral phase has been confirmed by interferometric measurement.[8] In order to obtain a true CV beam, a spiral phase element (SPE) with the opposite topological charge is necessary to compensate the geometric phase.

Spatially variant birefringence and dichroism produced by subwavelength dielectric and metallic structures have been extensively exploited to generate CV beams passively. For subwavelength dielectric structures, the form birefringent effect is typically utilized. Direct laser writing technique produced radial polarization converters (known as z-polarizer, z-plate, q-plate or s-waveplate) that converts linear polarized input into CV beams are available commercially.[9] Subwavelength metallic structure such as concentric rings (also known as bulls eye pattern) etched into metallic thin film (Fig. 4) is another popular choice for creating CV beams. Utilizing the high contrast in the transmittance for TE and TM polarization, metallic bulls eye structures can convert

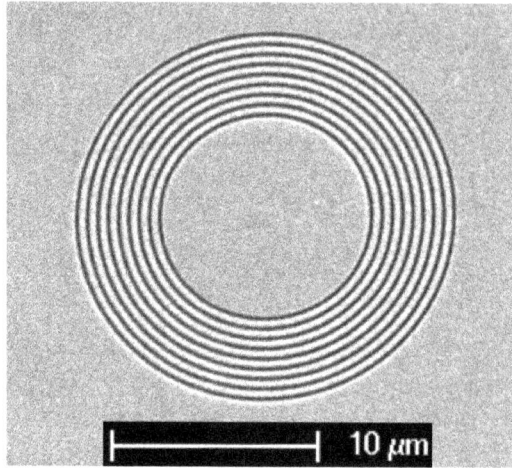

Fig. 4. An example of bulls eye structure made of 9 concentric rings fabricated in silver film with focused ion beam (FIB) milling.

circular polarization into radial or azimuthal polarizations. However, extra care needs to be taken to compensate the spiral geometric phase term described above when this type of structures is illuminated with circularly polarized beam.

A very popular and powerful passive generation method involves the use of liquid crystal (LC) materials. Spatially variant polarization rotation introduced by purposefully aligned LC molecules can be exploited to produce CV beams. Linear polarization is typically used as input and then locally rotated to the desired spatial polarization pattern. One example is devices with Twisted Nematic (TN) LC molecules sandwiched between a linearly and a circularly rubbed plates.[10, 11] Due to the circular rubbing of the second plate, the TN LC molecules continuously rotates from the initial linear rubbing direction to the corresponding spatially distributed rubbing direction on the other plate. An incident beam that is linearly polarized perpendicular or parallel to the linear rubbing direction will follow the molecule rotation, creating a radial or azimuthal polarization on the exiting side. A π-step phase plate is necessary to correct a geometric phase similar to what we mentioned previously. In a similar approach, continuous rotation of linear polarization input using photo-aligned liquid crystal polymers has been reported.[12] LC based devices that are specifically designed for CV beam generation with lower cost have also been developed and are commercially available nowadays.[13]

Despite its relative high cost, LC spatial light modulators (SLM) offer the flexibility and capability to generate almost arbitrary complex field distributions. One such example is illustrated in Fig. 5. The combination of the $\lambda/4$-plate and the LC-SLM essentially forms a polarization rotator where the amount of rotation is determined by the phase retardation of each pixel on the SLM[14]. By designing the phase pattern on the second SLM, the input linear polarization can be converted into any arbitrary polarization distribution, including the CV beams. The use of four sets of LC-SLM has also been proposed to generate arbitrarily complex optical vector fields.[15]

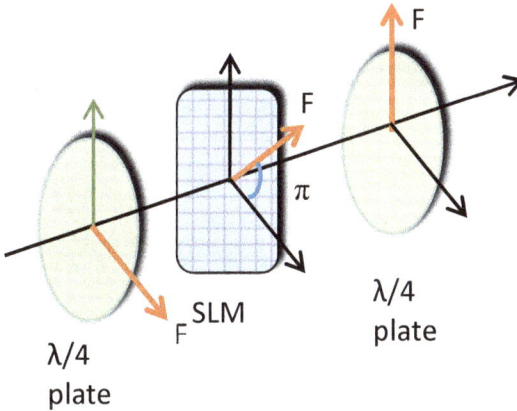

Fig. 5. Illustration of a pixelated polarization rotator (PR) through sandwiching a liquid crystal SLM between two $\lambda/4$-plates. Although a transmission type of design is shown here, reflection type of setup that uses one $\lambda/4$-plate has also been demonstrated in Ref.14.

4.2. *Passive generation methods using optical fiber*

Generation of CV beams with few-mode fiber is another technique that deserves special attention. If the parameters are chosen correctly, a multi-mode step index optical fiber can support the TE_{01} and TM_{01} modes possessing cylindrical polarization symmetry, with the TE_{01} mode being azimuthally polarized and the TM_{01} mode being radially polarized (Fig. 6). Under weakly-guiding approximation, these modes have the same cut-off parameter that is lower than all the other modes except the HE_{11} fundamental mode. In order the achieve high excitation efficiency, the V-number of the optical fiber is typically chosen to be between 2.41 and 3.9 for the working wavelength such that the fiber supports the 2nd order modes while avoids the excitation of higher order modes.

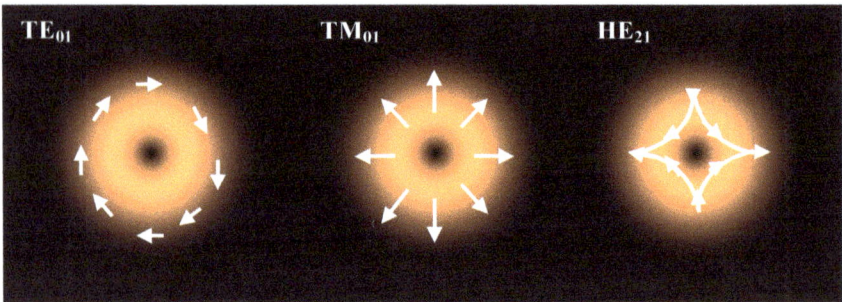

Fig. 6. Intensity patterns of LP_{11} modes of optic fiber superimposed with polarization distributions. The TE_{01} and TM_{01} modes have azimuthal and radial polarization symmetry respectively.

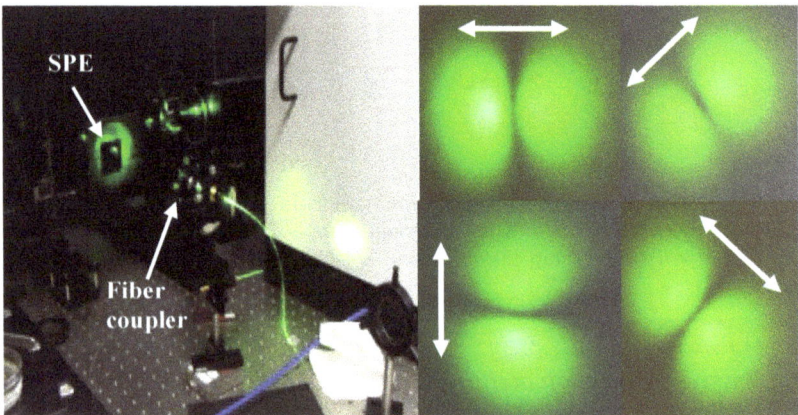

Fig. 7. Picture of a laboratory setup that generates CV beams using few-mode optic fiber. The intensity patterns of radial polarization generated with this setup after passing through a linear polarizer are shown to the right.[2]

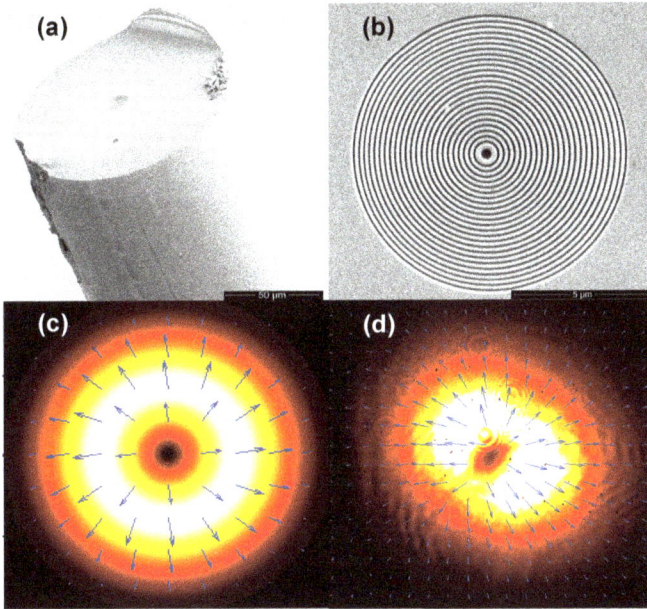

Fig. 8. (a) SEM image of a few-mode fiber with bulls eye pattern integrated at the core region; (b) Zoom-in of the bulls eye structure; (c) Calculated polarization pattern transmitted through the bulls eye structure; (d) Experimental measurement of the polarization pattern.[16]

Through choosing an optical fiber with proper parameters and careful alignment with an effort to minimize the excitation of the fundamental mode and other higher order modes, CV mode excitation in fiber can be achieved. High efficiency can be obtained by preforming the incident polarization either in phase or polarization. A laboratory picture for CV mode excitation using a charge-1 spiral phase element (SPE) is shown in Fig. 7. Few-mode fiber with metallic bulls eye pattern fabricated on the fiber end (Fig. 8) has also been demonstrated recently to facilitate the launching of CV beams into optical fiber.[16]

4.3. *Active generation methods*

Typical active CV beam generation methods use intracavity devices to force the laser to oscillate in CV modes. The intracavity devices can be axial birefringent (intrinsic, form or induced) or axial dichroic

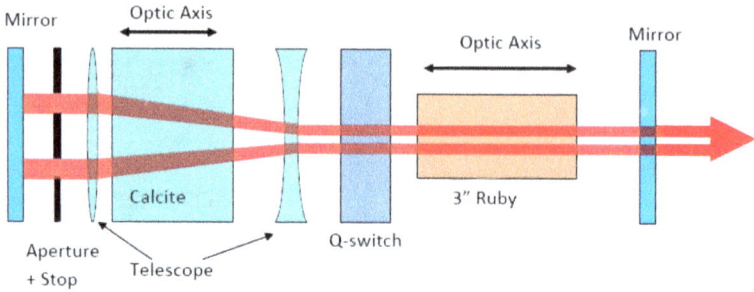

Fig. 9. Diagram of a Ruby laser that generates CV beam output.[4] The optics axis of the calcite crystal is parallel to the resonator axis. The combination of the calcite crystal, the telescope and the aperture/stop provides the mode discrimination to force the laser operation in CV polarization mode (azimuthal polarization in this case).

(concentrically aligned molecules or metallic structures, conical axicon, or conical Brewster angle reflectors) to provide the polarization mode selection. One of the earliest experiments utilized intracavity axial birefringence from a calcite crystal placed in a telescope setup with its crystal axis parallel to optical axis of the cavity (Fig. 9).[4] Due to double refraction, the e-polarization and o-polarization experience slightly different magnification. A central aperture stop is used to discriminate one polarization by creating higher loss. The cylindrical symmetry of the entire system ensures the oscillation mode to have cylindrical polarization symmetry. Since calcite is negative birefringent, the azimuthal polarization was generated directly in this setup. Radial polarization was generated with optical active materials (quartz) to rotate the electric field by 90°.

CV beams can also be generated with intracavity interferometric methods using folded mirrors or prisms based on the linear superposition principle given in Eqs. (2.15) and (2.16). For example, an intracavity Sagnac interferometer setup[17] is illustrated in Fig. 10. In this method, linearly polarized HG_{01} modes are created by placing a thin wire across the center of the cavity. A dove prism provides the necessary rotation to create the orthogonally polarized HG_{10} mode. The Sagnac interferometer combines the two modes and creates the CV output.

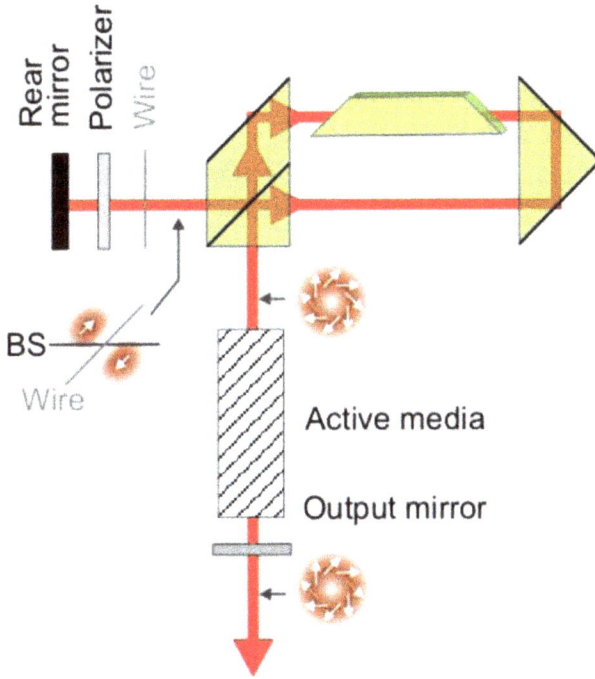

Fig. 10. A CV beam laser design with Sagnac interferometer as an end mirror (from Ref. 17).

The techniques mentioned above used bulk intracavity devices for creating axial birefringence or dichroism. Particular interests were shown in the developing compact CV beam sources recently. Modern micro- and nano-fabrication tools enable the creation of diffractive phase plate or polarization selective end mirror devices for CV beam generation. This type of devices allows a much more compact laser cavity design and can be exploited to generate high output power. For example, GaAs VCSEL (vertical-cavity surface-emitting laser) operated at 850 nm that can generate radially polarized output has been demonstrated.[18] In this work, a gold bulls eye structure is integrated within the output aperture on the top of the VCSEL. Similar structure has been used on InGaAsP semiconductor laser operated at 1.6 μm.[19] Depending on the metallic bulls eye design, the laser could operate either in the radial polarization mode or azimuthal polarization mode. In another effort, a dye-doped nematic liquid crystal microlaser has been demonstrated to produce azimuthally polarized output.[20] The polarization selection was achieved through concentric alignment of liquid crystal molecules by patterned rubbing on one of the high reflectivity surface of the cavity. This simple structure also allows the lasing wavelength to be tuned electrically without losing the polarization properties. Over 20 nm tuning range center around 630 nm has been reported at a 1.6-volt driving voltage.

The fact that a few-mode optical fiber can support both radially and azimuthally polarized modes enables the design of CV fiber lasers. Fiber lasers have attracted increasing interest owing to their relatively high gain, compactness and flexibility. Various kinds of fiber laser designs have been explored to produce CV beams. The basic underlying principle is to use appropriate elements in the cavity, such as dual conical prism, axicon, polarization mode converter, fiber optic polarization controller, and spatially variable retarder etc., to select the desired polarization. For example, a simple design that can generate both radially and azimuthally polarized beams using a c-cut calcite crystal with a three-lens telescope in an Erbium doped fiber laser cavity operated near 1.6 μm has been demonstrated.[21] The development of vectorial fiber laser techniques deserves special attention due to their unique characteristics

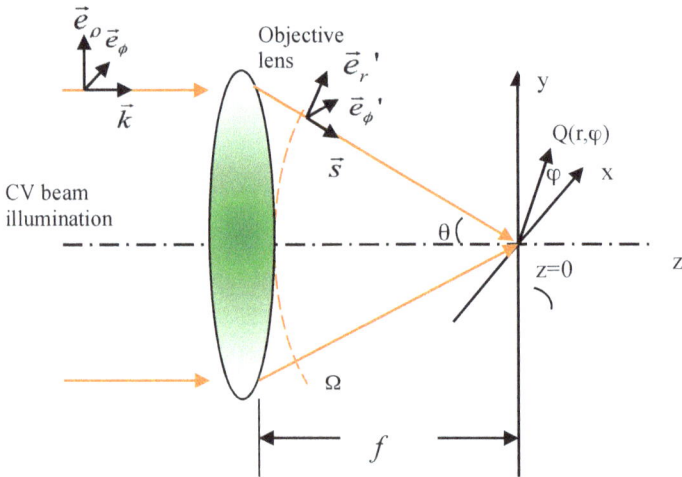

Fig. 11. Focusing of a cylindrical vector beam. In the diagram, f is the focal length of the objective lens. $Q(r, \varphi)$ is an observation point in the focal plane.

5. Cylindrical vector beams under high NA focusing

The strong interests in CV beams are largely driven by the unique properties under high NAfocusing and the potential applications of these focusing properties. Thus it is very important to be able to numerically calculate and predict the focal field characteristics of CV beams under various conditions. Highly focused polarized beams can be numerically analyzed with the Richards-Wolf vectorial diffraction method. The geometry of the problem is shown in Fig. 11. The illumination is a generalized CV beam with a planar wavefront over the pupil. An aplanatic objective lens produces a spherical wave converging to the focal point. For a generalized CV beam illumination, the field near the focus can be expressed as a linear combination of the focal fields arising from the radial and azimuthal polarization components. The focal field of a generalized CV beam can be written as

$$\vec{E}(r,\phi,z) = E_r \vec{e}_r + E_z \vec{e}_z + E_\phi \vec{e}_\phi , \qquad (21)$$

where $\vec{e}_r, \vec{e}_z, \vec{e}_\phi$ are the unit vectors in the radial, longitudinal and azimuthal directions, respectively. E_r, E_z and E_ϕ are the amplitudes of the three orthogonal components that can be derived as:[22]

$$E_r(r,\phi,z) = 2A\cos\phi_0 \int_0^{\theta_{max}} P(\theta)\sin\theta\cos\theta J_1(kr\sin\theta)e^{ikz\cos\theta}d\theta , \quad (22)$$

$$E_z(r,\phi,z) = i2A\cos\phi_0 \int_0^{\theta_{max}} P(\theta)\sin^2\theta J_0(kr\sin\theta)e^{ikz\cos\theta}d\theta , \quad (23)$$

$$E_\phi(r,\phi,z) = 2A\sin\phi_0 \int_0^{\theta_{max}} P(\theta)\sin\theta J_1(kr\sin\theta)e^{ikz\cos\theta}d\theta , \quad (24)$$

where θ_{max} is the maximal angle corresponding to the NA of the objective lens, k is the wave number and $J_n(x)$ is the Bessel function of the first kind with order n. A is a constant that is given by the objective lens focal length f and wavelength λ as:

$$A = \frac{\pi f l_0}{\lambda} \ , \tag{25}$$

with l_0 being the peak field amplitude at the pupil plane. $P(\theta)$ is the pupil apodization function that strongly depends on the objective lens design. Assuming the incident field at the pupil plane has amplitude distribution of $P(r_p)$ normalized to l_0, with r_p being the radial position in the pupil plane, this amplitude distribution is mapped to the wavefront after the objective lens through the ray projection function $g(\theta)$:

$$r_p / f = g(\theta) \ . \tag{26}$$

With this mapping, the pupil apodization function is given by:[23]

$$P(\theta) = P(r_p) \sqrt{\frac{g(\theta)g'(\theta)}{\sin\theta}} = P(fg(\theta)) \sqrt{\frac{g(\theta)g'(\theta)}{\sin\theta}} \ . \tag{27}$$

The ray projection function for most of the commonly used sine condition objective lens is given by:

$$g(\theta) = \sin\theta \ , \tag{28}$$

$$r_p = f \sin\theta \ . \tag{29}$$

Thus the pupil apodization function can be found to be:

$$P(\theta) = P(fg(\theta)) \sqrt{\cos\theta} \ . \tag{30}$$

There are also objective lenses with other types of ray projection functions. For example, for objective lens that obeys Herschel condition,

$$g(\theta) = 2\sin(\theta/2) \tag{31}$$

$$P(\theta) = P(fg(\theta)) \ . \tag{32}$$

For objective lens that obeys Lagrange condition (also known as uniform projection condition),

$$g(\theta) = \theta \ , \tag{33}$$

$$P(\theta) = P(fg(\theta))\sqrt{\frac{\theta}{\sin\theta}} \ . \tag{34}$$

For Helmholtz condition objective lens,

$$g(\theta) = \tan\theta \ , \tag{35}$$

$$P(\theta) = P(fg(\theta))\left(\sqrt{\frac{1}{\cos\theta}}\right)^3 \ . \tag{36}$$

These types of objective lens are not as common as those objective lenses that obey sine conditions. Most of the studies use the pupil apodization function given in Eq. (30) for the sine condition objective lens.

With these equations, one can numerically compute the total intensity as well as the amplitude and phase distributions of different constituting components in the vicinity of focus. All components given in Eqs. (22)–(24) are independent of φ, maintaining the cylindrical symmetry of the focal field. From Eq. (23), a key feature of the focal field distribution can be clearly identified, which is the existence of strong longitudinal component (z-component). It is this strong z-component that is responsible for many of the unique properties of highly focused CV beams, particularly radially polarized beams. Another feature of highly focused CV beam is the spatial separation of the z-component (with J_0 in the integral) and the transverse components (with J_1 in the integral). This spatial separation of these orthogonally polarized components has been exploited in flat top focal field generation.[24]
For comparison purpose, the focal field of an x-polarized incident can be expressed as:[23]

$$\vec{E}(r,\varphi,z) = -iA\{[I_0 + \cos(2\varphi)I_2]\vec{e}_x + \sin(2\varphi)I_2\vec{e}_y - 2i\cos\varphi I_1\vec{e}_z\}, \tag{37}$$

where

$$I_0 = \int_0^{\theta_{max}} P(\theta)\sin\theta(1+\cos\theta)J_0(kr\sin\theta)e^{ikz\cos\theta}d\theta , \qquad (38)$$

$$I_1 = \int_0^{\theta_{max}} P(\theta)\sin^2\theta J_1(kr\sin\theta)e^{ikz\cos\theta}d\theta , \qquad (39)$$

$$I_2 = \int_0^{\theta_{max}} P(\theta)\sin\theta(1-\cos\theta)J_2(kr\sin\theta)e^{ikz\cos\theta}d\theta . \qquad (40)$$

The focal field is dominated by the transverse x-polarization component as one would expect.

One example of the field calculation of a highly focused CV beam is given in Fig. 12, which corresponding to radial polarization focused by an objective with NA = 0.95. From Eq. (23), we can see that for radial polarization incident the z-component experiences an apodization function of $\sin\theta$ given in Eq. (23) compared with the apodization function $(1+\cos\theta)/2$ for the dominant contribution I_0 of the transverse field components given in Eq. (38). This apodization function for radial polarization focusing places more weight towards the high spatial frequency components, consequently leading to smaller spot size. It also indicates that in order to have prominent effects of this longitudinal component, high NA is required and annular illumination is preferred.

It should be noted that there is an imaginary unit "i" in front of the expression for the z-component of the focal field in Eq. (23). This means that the longitudinal and transversal components are $\pi/2$ out of phase, which indicates that the z-component does not carry average power, even though there is a strong z-component of the field at the focus. However, this strong z-component of the field still stores electrical energy. This observation could be very useful for three-dimensional stable trapping of nanoparticles, especially metallic Rayleigh particles, as it offers the potential of eliminating the destabilizing optical pressure force that pushes the particles out of the focal volume.[25]

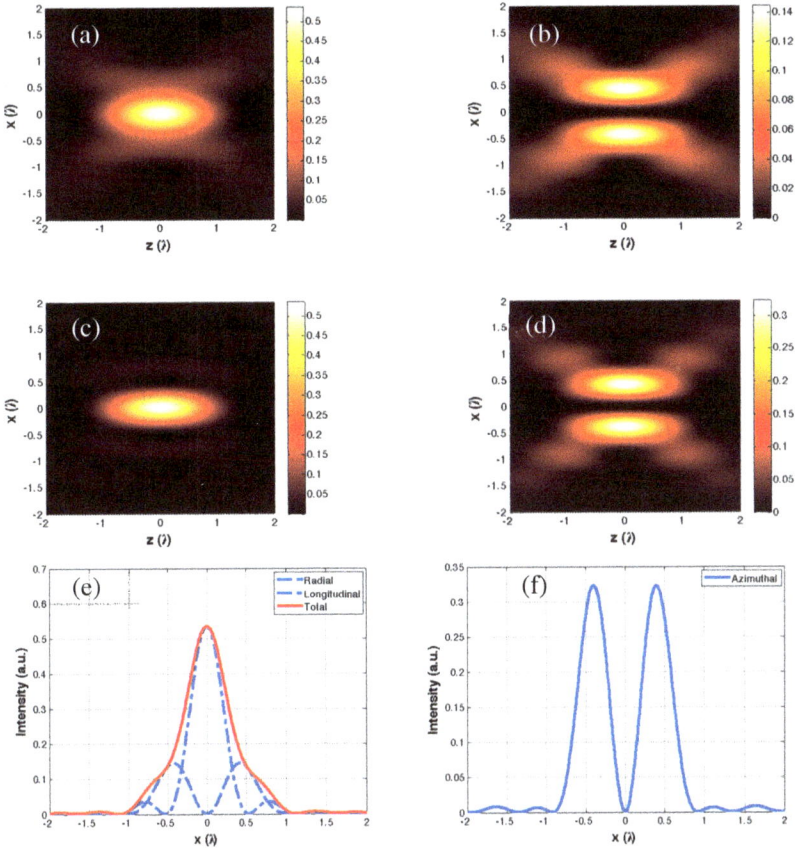

Fig. 12. Calculated focal field distributions for CV beams focused by objective lens with NA = 0.95. (a) Total intensity distribution for focused radial polarization; (b) Radial components of (a); (c) Longitudinal components of (a); (d) Total intensity distribution of focused azimuthal polarization; (e) Line scan of the distribution in the focal plane for (a); (f) Line scan of the distribution in the focal plane for (d), the focal field is solely made of azimuthal component. For this example, it can be seen that the longitudinal component is spatially separated from the transversal (radial and azimuthal) components.

Compared with highly focused linear polarization, the phase behavior in the vicinity of focus for a highly focused CV beam is also distinctively different. It is well known that a focused optical beam experiences a phase anomaly called the Gouy phase. The difference in the apodization functions for the integrals of focused CV beam in Eqs. (22)–(24)

compared with the apodization functions for the integrals of focused linear polarization in Eqs. (38)–(40) indicate that the Gouy phase behaves differently for a highly focused CV beam, particularly highly focused radial polarization. Adopting a second-order approximation of the tilted plane wave interpretation of the Gouy phase calculation, analytical expression of the Gouy phase has been derive.[26] It has also been shown that the Gouy shift of radially polarization illumination can be adjusted with different illumination conditions.[26] This finding has found important applications in nonlinear optical imaging such as the third harmonic generation (THG) or coherent anti-Stokes Raman Spectroscopic (CARS) microscopy that are strongly influenced by the existence of Gouy phase. For example, by adjusting the annular illumination geometry, it has been shown that THG from bulk materials may be detectable with better achievable spatial resolution using radially polarized illumination.[27]

The capability of producing smaller focus with radial polarization has been experimentally confirmed and lead to significant interests in applying CV beams in laser micro-fabrication and imaging. The supremacy of radial polarization can be understood from the point of view of electric dipole radiation. Consider a vertical electric dipole located at the focal point of a high NA aplanatic objective lens oscillating along the optical axis, the well-known angular radiation pattern of electric dipole is illustrated with the local polarization indicated (Fig. 13). The high NA objective lens collects the dipole radiation into the top half space and collimates the radiation. Clearly the polarization pattern at the lens pupil plane is aligned along the radial directions. If the optical path is reversed with the radial polarization pattern at the pupil plane as illumination, then the corresponding focal field should recover the propagating components in the upper half plane. If an identical objective lenses is also used in the lower half plane (4-π setup), the field from the electric dipole can be recovered up to all the propagating components. Electric dipole can be regarded as the smallest available point source. From this point of view, qualitatively speaking radial polarization should provide optimal focusing compared with other polarization distributions. This intuitive observation also leads to further interesting ways of utilizing radiation patterns from different emitters and emitter arrays. Generalized methods have been developed to engineer the focal field with various desired characteristics by taking advantage of the extensive body of literatures in the microwave and radio frequency antenna theory, which will be discussed in Chapter 4.

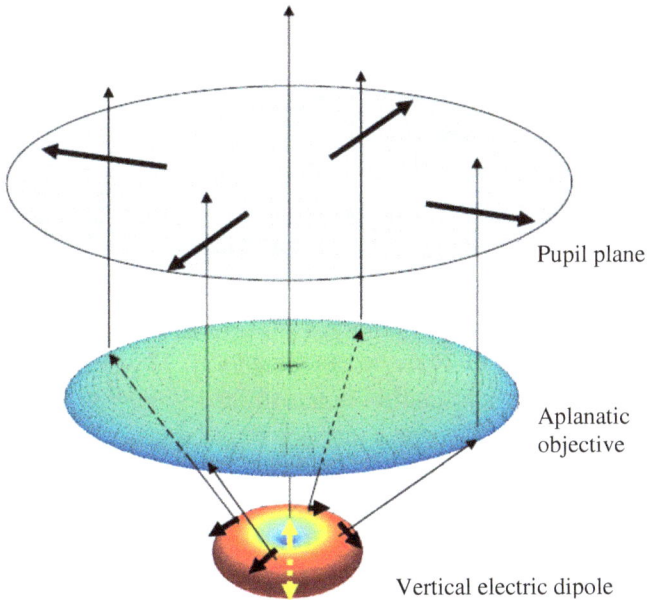

Fig. 13. Illustration of the polarization pattern at the pupil plane for the radiation from a vertical electric dipole collected by a high NA objective.[2]

6. Summary

In this chapter, fundamentals and recent developments in the field of cylindrical vector polarization are reviewed. Mathematical and graphical descriptions of CV beams by solving the full vector wave equations are presented. Various generation methods of CV beams have been developed and several products are available commercially. Beyond CV beams, optical vector beams with more complicated spatial polarization distributions exhibit many novel properties and physics that attracted rapidly increasing interests. The latest progress in this field will be discussed by Wang *et al.* in Chapter 2. Development of vectorial fiber lasers that can produce CV beam outputs have attracted increasing interest owing to their relatively high gain, compactness and flexibility. Due to the rotational polarization symmetry, CV beams were discovered to have many interesting properties when focused with high NA objective lens. Expressions that can be used to numerically calculate the focal field of highly focused CV beams are derived to illustrate some of the characteristics. Applications of cylindrical beams in diverse areas such as laser micro-machining, lithography, imaging, optical trapping and manipulation of nanoparticles, optical measurements and metrology have been pursued. Meixner *et al.* will review their latest work in the use of CV beams for imaging and spectroscopy of single nanoparticles and single quantum systems in Chapter 3. The use of vector beams to engineer optical focal field with specific desired characteristics, such as optical needle, optical bubble, flattop focusing, and 3D optical polarization etc., will be reviewed in Chapter 4. The unique polarization symmetry and focal field distribution also lead to interesting phenomena when highly focused CV beams interact with nanostructures, particularly plasmonic nanostructures. In Chapter 5, recent progresses on plasmonics with vectorial optical fields will be summarized and presented. Although the focusing properties and their corresponding applications have been the main focus of the field, the interests in CV beams and more generalized optical vector fields are much broader. The polarization diversity within the beam cross section of CV beams itself offers interesting opportunities in optical measurement. Chapter 6 summarizes examples of optical measurement techniques including ellipsometry, polarimetry and interferometry that explore such polarization symmetry and diversity. Second order statistics of CV beams propagating through turbulent atmosphere has been investigated and indicated the possibility of using CV polarization to mitigate adverse atmospheric effects with potential impacts on remote sensing and optical free space communications. In Chapter 7, Cai *et al.* provides a comprehensive review of partially coherent vector beams and their propagation. In view of these recent developments, it is safe to conclude that with simple and efficient generation methods becoming more and more accessible, the research into optical vector fields will further grow and the list of applications of optical vector beams will continue to expand rapidly.

References

1. E. Hasman, G. Biener, A. Niv, and V. Kleiner, *Progress in Optics* (E. Wolf (ed.), Elsevier), **47**, 215, (2005).
2. Q. Zhan, *Adv. Opt. Photon.* **1**, 1 (2009).
3. T. G. Brown, *Progress in Optics* (E. Wolf (ed.), Elsevier), **56**, 81, (2011).
4. D. Pohl, *Appl. Phys. Lett.* **20**, 266 (1972).
5. S. C. Tidwell, D. H. Ford and W. D. Kimura, *Appl. Opt.* **29**, 2234 (1990).
6. R. Dorn, S Qubis and G. Leuchs, *Phys. Rev. Lett.* **91**, 233901 (2003).
7. A. K. Spilman and T. G. Brown, *Appl. Opt.* **46**, 61 (2007).
8. Q. Zhan and J. R. Leger, *Opt. Commun.* **213**, 241 (2002).
9. "s-waveplate" from Altechna http://www.altechna.com/
10. R. Yamaguchi, T. Nose and S. Sato, *Japanese J. of Appl. Phys.* **28**, 1730 (1989).
11. M. Stalder and M. Schadt, *Opt. Lett.* **21**, 1948 (1996).
12. S. C. McEldowney, D. M. Shemo and R. A. Chipman, *Opt. Express.* **16**, 7295 (2008).
13. "radial polarizer" from ARCoptix http://www.arcoptix.com/
14. M. R. Beversluis, L. Novotny and S. J. Stranick, *Opt. Express* **14**, 2650 (2006).
15. W. Chen and Q. Zhan, *J. Opt.* **12**, 045707, (2010).
16. W. Chen, W. Han, D. C Abeysinghe, R. L Nelson and Q. Zhan, *J. Opt.* **13**, 015003, (2011).
17. V. G. Niziev, R. S. Chang and A. V. Nesterov, *Appl. Opt.* **45**, 8393 (2006).
18. L. Cai, J. Zhang, W. Bai, Q. Wang, X. Wei, and G. Song, *Appl. Phys. Lett.* **97**, 201101 (2010).
19. O. Weiss and J. Scheuer, *Appl. Phys. Lett.* **97**, 251108 (2010).
20. H. Yoshida, K. Tagashira, T. Kumagai, A. Fujii, and M. Ozaki, *Opt. Express* **18**, 12562 (2010).
21. R. Zhou, B. Ibarra-Escamilla, P. E. Powers, J. W. Haus and Q. Zhan, *Appl. Phys. Lett.*, **95**, 191111 (2009).
22. K. S. Youngworth and T. G. Brown, *Opt. Express* **7**, 77 (2000).
23. M. Gu, *Advanced optical imaging theory*, **75**, Springer-Verlag, New York (1999).
24. Q. Zhan and J. R. Leger, *Opt. Express*, **10**, 324, (2002).
25. Q. Zhan, *Opt. Express* **12**, 3377 (2004).
26. H. Chen, Q. Zhan, Y. L. Zhang and Y.P. Li, *Phys. Lett. A* **371**, 259 (2007).
27. S. Yang and Q. Zhan, *J. Opt. A: Pure and Appl. Opt.*, **10**, 125103 (2008).

CHAPTER 2

VECTOR OPTICAL FIELDS AND THEIR NOVEL EFFECTS

Hui-Tian Wang

MOE Key Laboratory of Weak Light Nonlinear Photonics and School of Physics
Nankai University, Tianjin 300071, China
National Laboratory of Solid State Microstructures
Nanjing University, Nanjing 210093, China
htwang@nankai.edu.cn
htwang@nju.edu.cn

In this chapter, the concept of vector optical fields is briefly introduced at first. Then we secondly introduce our method for generating vector optical fields based on the Poincaré sphere and wavefront reconstruction is presented. Novel effects arising from these vector optical fields, including the optical cage, the splitting caused by the axial-symmetry breaking, the Young's two-slit interference, and the optical orbital angular momentum will be reported followed by a brief summary of our findings.

1. Introduction

Since the invention of laser, it has led to extensive, tremendous and profound influences on scientific development, technological innovation, social progress, and human civilization. Laser plays an indispensable and paramount role in various realms from fundamentals to applications, such as physics, chemistry, information technology, materials processing, precision measurement, equipment manufacturing, and controlled fusion. Nevertheless, the potential of laser is still not fully excavated and exploited, due to the fact that the propagation behavior of light (motion of photons) is not effectively controlled. As is well known, photonic crystal is a kind of artificial microstructure that affects the propagation of light in the same way as the periodic potential in a semiconductor affects the motion of electrons by defining allowed and forbidden electronic energy bands. However, research on optical fields focuses almost on **scalar** optical fields, which are also referred to as homogeneously polarized optical fields, including the linearly, elliptically and circularly polarized optical fields. The so-called **scalar** optical field is a kind of optical field, i.e., anywhere on its wavefront has the same state of polarization (SOP). Naturally, a question might be raised that light has the intrinsic vector nature (its polarization) in and of itself, why the homogeneously polarized optical field is referred to as the **scalar** optical field. The reasons are based on the considerations as follows: (i) In contrast to the **vector** optical field as mentioned below; (ii) Physically, when treating its propagation behavior and the interaction of it with matter, its polarization feature can be omitted or the optical field can be considered as a whole; (iii) Scalar optical field is more simple and clear than homogeneously polarized optical field.

As the deepening of research on laser and the pushing of various demands, the **scalar** fields have emerged the limitation. If we are able to generate the designable structured optical field with novel properties by engineering the spatial SOP distribution in the wavefront, a window will be opened for manipulating the propagation behavior of light and the interaction of light with matter. Thereby, the manipulation of optical field and the relevant topics will certainly result in new effects and phenomena, introduce new concepts and principles, and produce new techniques. In particular, it is of great significance to deeply understand "light" and to exploit the potential of laser.

In this chapter, we focus on the generation of vector optical fields and their novel effects. The so-called **vector** optical field is a kind of structured optical field with the inhomogeneous SOP distribution.[1,2] The vector optical field is thereby also referred to as the inhomogeneously polarized optical field, i.e., the different locations across the wavefront have the different SOPs. The control of optical field in space domain focuses mostly on its phase or/and amplitude. As Allen *et al.* pioneered,[3] an optical vortex with a spiral phase, as a typical space-structured optical field, could carry the optical orbital angular momentum (OAM) caused by the azimuthal gradient of phase. When the optical vortex is focused into a ring focus, the optically isotropic and electrically neutral particles trapped are driven to move around the ring focus. The observed result suggests the fact that photons could indeed carry the optical OAM.[4-6] Since the optical spin angular momentum (SAM) was firstly pioneered by Beth in 1936,[7] the optical OAM is a profound understanding for the optical angular momentum again. The optical vortex, which can be easily generated by a simple control of phase in space domain, has given rise to many novel effects and various new applications, such as astronomical observation,[8] phase-contrast microscopy,[9] entanglement of the OAM states of photons in an infinitely dimensional discrete Hilbert space,[10] and quantum correlations in optical angle–OAM variables.[11]

If polarization being a fundamental and intrinsic nature of light, as a degree of controlling freedom, is able to be fully utilized, it is certainly of great importance for manipulating the optical field and controlling the interaction of light with matter. In recent years, the vector fields have been demonstrated to possess many unique features with respect to the scalar fields. For instance, the radially polarized field, as a typical vector field, can be focused into a far-field focal spot of $0.17\lambda^2$ beyond the diffraction limit of $0.26\lambda^2$.[12, 13] In particular, the azimuthally polarized vector field (another typical vector field) carrying a vortex phase can realize a sharper far-field focal spot of $0.15\lambda^2$.[14] Based on the design of the SOP distribution, the engineered vector fields can achieve the focus shaping,[15] the light needle of longitudinally polarized field,[16] the optical cage,[17, 18] the optical chain,[19] and so on. The vector fields have many important applications such as particle acceleration,[20] single molecule imaging,[21] near-field optics,[22] nonlinear optics,[23] and optical trapping and manipulation of particles.[24, 25] In Section 2, we introduce the method for generating the vector fields in detail. Section 3 focuses on the novel effects of the vector fields. Section 4 gives a summary.

2. Generation of vector optical fields

The methods for generating the vector fields can be divided into two kinds of direct (active) and indirect (passive) schemes.[1] The direct (active) method is from the output of novel laser with specially designed or modified laser resonator.[26, 27] The indirect one is based on the wavefront reconstruction of the output field from the traditional laser, with the aid of specially designed optical elements.[28-31] A spatial light modulator (SLM) allows flexibly designing the arbitrary spatial (phase or amplitude or both) modulation pattern to generate the desired vector fields.[32, 33] We present a convenient method for generating the vector fields, based on the Poincaré sphere and the wavefront reconstruction.[2, 25, 34]

In order to characterize the SOP of a polarized optical field, three independent parameters are needed, for example, (i) amplitudes of two orthogonally polarized components and phase difference between them or (ii) major and minor axes and the angle which specifies the orientation of polarization ellipse. Nevertheless, the Stokes parameters normalized by the light intensity are more convenient.[34, 35] More importantly, the Poincaré sphere is an efficient graphic tool for describing all possible SOP of polarized fields.[35] As shown in Fig. 1(a), a point $\mathbf{S}$ on the Poincaré sphere Σ of a unit radius can be determined by three independent Stokes parameters (s_1, s_2 and s_3) in a Cartesian coordinate system (satisfying $s_1^2 + s_2^2 + s_3^2 = 1$) and can also be defined by two independent angles (2ϕ, 2α) in a spherical coordinate system, and they hold the following relations

$$s_1 = \cos(2\alpha)\cos(2\varphi),$$
$$s_2 = \cos(2\alpha)\sin(2\varphi), \tag{1}$$
$$s_3 = \sin(2\alpha).$$

For a polarized field of a normalized intensity, any possible SOP corresponds to one point $\mathbf{S}$ on Σ and vice versa, as shown in Fig. 1(b). It should be emphasized that the factors 2 in front ϕ and α are introduced to ensure that one point on Σ corresponds to a unique SOP and vice versa. 2ϕ and 2α specify in fact the orientation and the ellipticity of the polarization ellipse, respectively. For instance, a linear polarization is represented by a point in the equator, while the right- and left-handed circular polarizations are represented by the north and south poles; the right-handed elliptical polarizations are represented by points on Σ which lie above the equator of Σ and the left-handed ones by points on Σ which lie below this equator, respectively. For a given polarized field, its SOP can also be described by the combination of a pair of orthogonal polarization base vectors. A representative pair of polarization base vectors are $\{\mathbf{e}_x, \mathbf{e}_y\}$ (with $\langle \mathbf{e}_x | \mathbf{e}_y \rangle = 0$), where $\mathbf{e}_x$ and $\mathbf{e}_y$ represent the linear

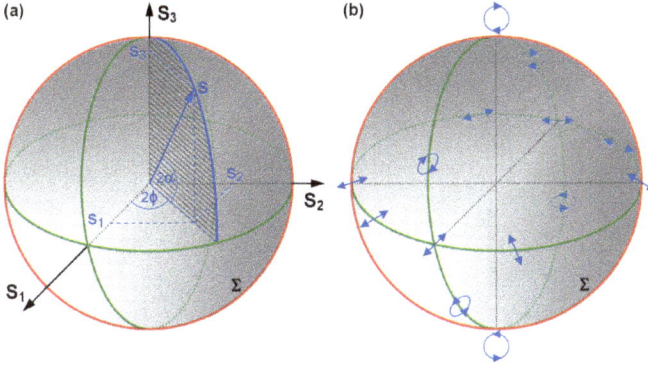

Fig. 1. (a) Poincaré sphere Σ and (b) different SOPs on Poincaré sphere Σ.

polarizations along the x and y axes, respectively. Another representative pair of polarization base vectors are right- and left-handed unit vectors $\{\mathbf{e}_R, \mathbf{e}_L\}$ (with $\langle \mathbf{e}_R | \mathbf{e}_L \rangle = 0$), where $\mathbf{e}_R$ and $\mathbf{e}_L$ represent the right- and left-handed circular polarizations, respectively. The two pairs of base vectors are related by $\mathbf{e}_x = 2^{-1/2}(\mathbf{e}_R + \mathbf{e}_L)$ and $\mathbf{e}_y = -2^{-1/2}j(\mathbf{e}_R - \mathbf{e}_L)$ as well as $\mathbf{e}_R = 2^{-1/2}(\mathbf{e}_x + j\mathbf{e}_y)$ and $\mathbf{e}_L = 2^{-1/2}(\mathbf{e}_x - j\mathbf{e}_y)$. Thus the SOP of a given point $\mathbf{S}(2\phi, 2\alpha)$ on Σ can be described by the unit vector $\hat{\mathbf{S}}(2\phi, 2\alpha)$, based on the pair of base vectors, $\{\mathbf{e}_x, \mathbf{e}_y\}$ or $\{\mathbf{e}_R, \mathbf{e}_L\}$, as follows

$$\hat{\mathbf{S}}(2\phi, 2\alpha) = \sin(\alpha + \pi/4)\exp(-j\phi)\mathbf{e}_R + \cos(\alpha + \pi/4)\exp(j\phi)\mathbf{e}_L$$

$$= \frac{1}{\sqrt{2}}[\sin(\alpha + \pi/4)\exp(-j\phi) + \cos(\alpha + \pi/4)\exp(j\phi)]\mathbf{e}_x \quad (2)$$

$$+ \frac{1}{\sqrt{2}}[\sin(\alpha + \pi/4)\exp(-j\phi) - \cos(\alpha + \pi/4)\exp(j\phi)]\mathbf{e}_y.$$

It should be pointed out that any two SOPs represented by any pair of antipodal points on Σ are a pair of orthogonal base vectors, due to $\langle \hat{\mathbf{S}}(2\phi, 2\alpha) | \hat{\mathbf{S}}(2\phi + \pi, -2\alpha) \rangle = 0$ from Eq. (2).

We present a passive method for generating the vector fields from the traditional linearly polarized laser field, based on the Poincaré sphere and the wavefront reconstruction. As shown in Fig. 2, the principle scheme includes a 4f system composed of a pair of identical lenses. A SLM is placed in the input plane of the 4f system. The designed sine/cosine holographic grating carrying a spatial phase distribution δ displays in the

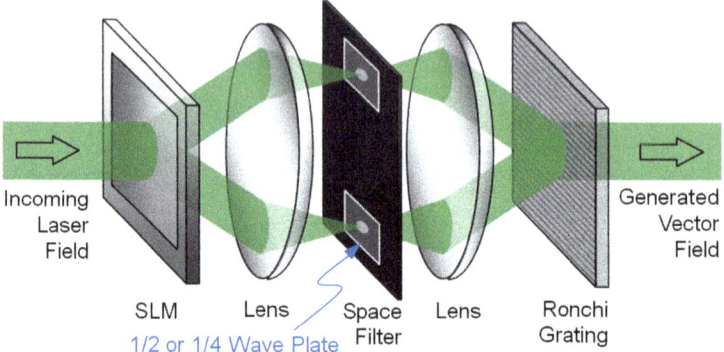

Fig. 2. Principle scheme generating the vector field.

SLM. The incoming laser field projected at the SLM is diffracted into the ±1th orders. Make the ±1th orders pass through a spatial filter (with two separate apertures) placed in the Fourier plane of the 4f system and then are converted by two 1/4 (or 1/2) wave plates into a pair of orthogonal linear polarizations (or of left- and right-handed circular polarizations), respectively. Finally, the ±1th orders are recombined by a Ronchi phase grating placed in the output plane of the 4f system and then achieve the vector filed. The period of the holographic grating is required to match to that of the Ronchi grating. The holographic grating has a transmission function of $t(x, y) = [1 + \cos(2\pi f_0 x + \delta)]/2$, where f_0 is the spatial frequency of grating.

In our method for generating the vector field, there are four key steps: (i) Dividing the incoming linearly polarized laser field into two parts, (ii) Making the two fields carry the space-variant phase δ, (iii) Letting the two fields pass through the different paths and then making them be orthogonal polarized, and (iv) Recombining the two fields to produce the vector field. In fact, the steps (i) and (ii) can be performed by the SLM simultaneously. Thus the SOP distribution of the generated vector field can be expressed as follows

$$\hat{\mathbf{P}} = \frac{1}{\sqrt{2}}[\exp(-j\delta)\mathbf{e}_1 + \exp(j\delta)\mathbf{e}_2], \tag{3}$$

where $\mathbf{e}_1$ and $\mathbf{e}_2$ stand for the base vectors of the SOPs of a pair of orthogonal polarized fields.

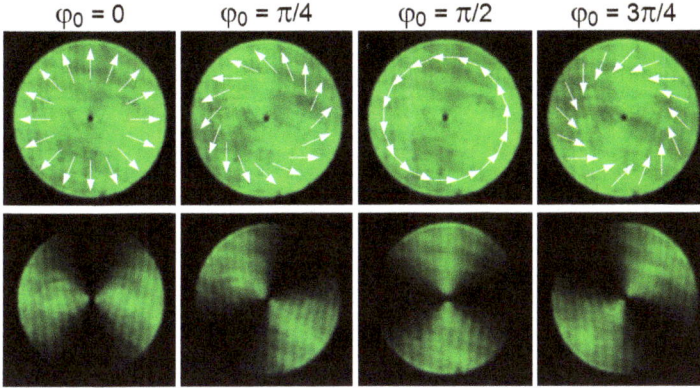

Fig. 3. Generated vector fields with the topological charge $m = 1$, for different values of $\varphi_0 = 0$, $\pi/4$, $\pi/2$, and $3\pi/4$. The top row shows the intensity patterns and the SOP distributions. The bottom row shows the intensity patterns behind a horizontal polarizer.

2.1. *Local linearly polarized vector fields*[2]

To generate the local linearly polarized vector fields, the two 1/4 wave plates should be used in the two different paths, making the SOPs of the two fields be the orthogonal right- and left-handed circular polarizations. This case implies that $\mathbf{e}_1$ and $\mathbf{e}_2$ in Eq. (3) should be $\mathbf{e}_1 = \mathbf{e}_R$ and $\mathbf{e}_2 = \mathbf{e}_L$. With Eq. (3), the SOP distribution of the generated vector field should be

$$\hat{\mathbf{P}} = \exp(-j\delta)\mathbf{e}_R + \exp(j\delta)\mathbf{e}_L = \cos\delta\,\mathbf{e}_x + \sin\delta\,\mathbf{e}_y. \tag{4}$$

We find from Eq. (4) that the generated vector field is indeed local linearly polarized because the x and y polarized components are always in phase. The distribution of linear polarization of the generated vector field depends only on the additional phase δ.

Even though δ is allowed to have arbitrary spatial distribution, we are first focus on the case when δ is only a function of the azimuthal angle φ, with a helical phase distribution of $\delta = m\varphi + \varphi_0$ (where m and φ_0 are the topological charge and the initial phase, respectively).

As shown in Fig. 2, a 532 nm collimated laser field with its linear polarization along the x direction illuminates a liquid crystal SLM with 1024×768 pixels (each pixel has a 14×14 μm^2 size). Figure 3 shows the generated single-mode vector fields with $m = 1$ for different values of

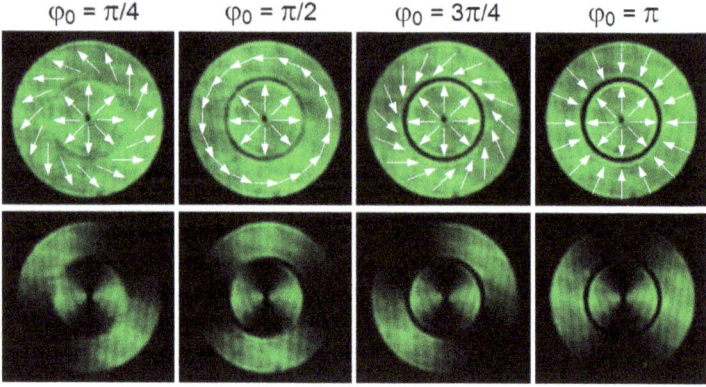

Fig. 4. Generated double-mode vector fields. All the inner modes correspond to $m = 1$ and $\varphi_0 = 0$, i.e., the radially polarized field. The outer modes correspond to $\varphi_0 = \pi/4$, $\pi/2$, $3\pi/4$, and π, with the same topological charge of $m = 1$. The top row shows the intensity patterns and the SOP distributions. The bottom row shows the intensity patterns behind a horizontal polarizer.

$\varphi_0 = 0$, $\pi/4$, $\pi/2$, and $3\pi/4$. The cases of $\varphi_0 = 0$ and $\pi/2$ correspond to the radially- and azimuthally-polarized fields, respectively. The arrows in the top row show the SOP distributions of the corresponding vector fields. The intensity distributions of the four vector fields have no difference. The central dark spot originates from the singularity of polarization. The intensity distribution behind a horizontal polarizer exhibits the extinction pattern. The radial extinction directions depend on the initial phase φ_0.

The generated double-mode vector fields are shown in Fig. 4. Two concentric modes are isolated by a dark annulus, which is formed by the SOP jump. The inner modes are the radially polarized fields with $m = 1$ and $\varphi_0 = 0$. In contrast, the outer modes correspond to $\varphi_0 = \pi/4$, $\pi/2$, $3\pi/4$, and π, respectively. The arrows in the top row show the SOP distributions. For the outer modes of the former three, the SOPs are the same as the latter three in Fig. 3, while the polarization direction of the outer mode with $\varphi_0 = \pi$ is opposite to that of the inner mode with $\varphi_0 = 0$. It can be found that the dark annulus becomes gradually clear as 0 increases from $\pi/4$ to π.

We explore the generation of the vector fields with larger topological charges. Figure 5 shows the generated single-mode vector fields with the

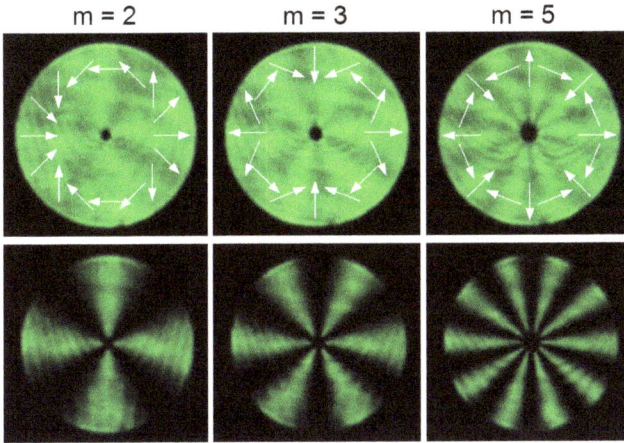

Fig. 5. Generated single-mode vector fields with larger topological charges of $m = 2, 3$, and 5, with the same initial phase of $\varphi_0 = 0$. The top row shows the intensity patterns and the SOP distributions of the generated vector fields. The bottom row shows the intensity patterns passing through a horizontal polarizer.

larger topological charges of $m = 2, 3$, and 5 when $\varphi_0 = 0$. Evidently, when no polarizer is used, the size of the central dark spot in the intensity pattern augments as m increases. After passing through a polarizer, the number of the extinction radial directions is equal to $2m$.

As shown in Fig. 6, the intensity pattern of any one of the generated double-mode vector fields is composed of two concentric inner and outer modes. The two modes are also separated by a dark annulus. Any inner mode is the radially polarized field (with $m = 1$ and $\varphi_0 = 0$) and the outer modes correspond to $m = 2, 4$, and 7 when $\varphi_0 = 0$. Unlike Fig. 4, however, the dark annulus is disconnected by one or several bright regions. For the bright regions, the number is equal to $m - 1$ and their azimuthal positions coincide with those positions where the inner and outer modes have the same SOP.

All the above investigations focus on the cases that the topological charge m is an integer. It should be interesting that the topological charge m is a noninteger. For example, Fig. 7 shows the cases of $m = 0.5$ and 1.5. The noninteger m results in the axial-symmetry breaking of the SOP and the intensity distribution. There exists a dark ray in the $\varphi_0 = 0$ direction, instead of a central dark spot. The presence of the dark ray originates

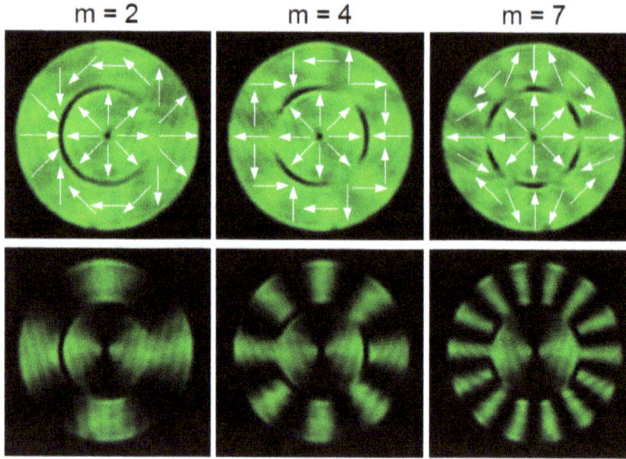

Fig. 6. Generated double-mode vector fields. All the inner modes correspond to $m = 1$ and $\varphi_0 = 0$, i.e., the radially polarized field. The outer modes correspond to larger topological charges of $m = 2$, 4, and 7, with the same initial phase of $\varphi_0 = 0$. The top row shows the intensity patterns and the SOP distributions of the generated vector fields. The bottom row shows the intensity patterns behind a horizontal polarizer.

from the uncertainty of SOP in the $\varphi = 0$ direction. After passing through a horizontal polarizer, only one extinction direction appears for $m = 0.5$ while three extinction directions appear for $m = 1.5$. It is clear that in the case of $\delta = m\varphi + \varphi_0$, the SOP of the generated vector field is solely azimuthal-variant.

We now explore the case when the additional phase δ is only a function of the radial radius ρ and has the form of $\delta = 2n\pi\rho / \rho_0 + \delta_0$ (n is referred to as the radial index and ρ_0 is the radius of the vector field). The SOP distribution of the generated vector field is still described by Eq. (4), with $\delta = 2n\pi\rho / \rho_0 + \delta_0$.

Figures 8(a)-(d) show the generated four vector fields with $\varphi_0 = 0$, corresponding to four radial indices of $n = 0.5$, 1.0, 1.5, and 2.0. The intensity distributions of the four vector fields have no difference and exhibit the uniform distribution. In particular, the radial-variant vector fields have no singularity for any value of n, which makes them intrinsically different from azimuthal-variant vector fields. The intensity patterns of the radial-variant vector fields exhibit always axial symmetry for any radial index n, which is different from the azimuthal-variant

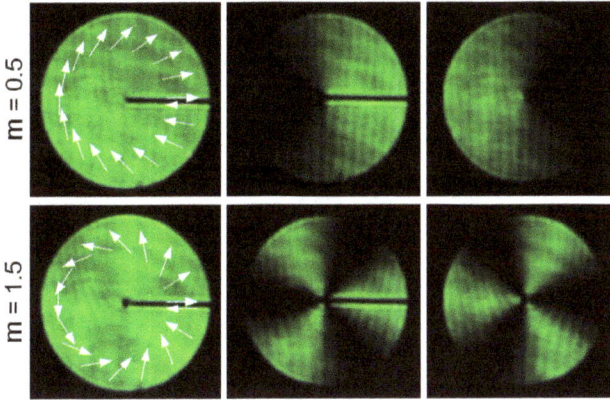

Fig. 7. Generated single-mode vector fields with the noninteger topological charges of m = 0.5 and 1.5, with $\varphi_0 = 0$. The top row shows the intensity patterns and the SOP distributions of the generated vector fields. The bottom row shows the intensity patterns behind a horizontal polarizer.

vector fields. As shown in Figs. 8(e)-(h), the intensity patterns behind a horizontal polarizer exhibit extinction rings, implying that the SOP of the generated vector field is indeed radial-variant. The number of extinction rings is precisely equal to $2n$ when n is an integer or a half integer. The radius of the jth extinction ring (where $j = 1, \ldots, 2n$) is given by $\rho_j = (2j - 1)(\rho_0/4n)$. Figures 8(i)-(l) show the corresponding schematics of SOPs.

We explore the role of φ_0 on the generated vector fields. For example, the four values of φ_0 are chosen to be 0, $\pi/4$, $\pi/2$, and $3\pi/4$ when the radial index n is fixed at 0.5. As shown in Figs. 9(a)-(d), the intensity patterns have no difference for the four radial-variant vector fields with the different values of φ_0. After passing through a horizontal polarizer, however, the intensity patterns become recognizable, and exhibit the homocentric extinction rings, as shown in Figs. 9(e)-(h). The radii of the extinction rings are $\rho_0/2$, $\rho_0/4$, 0, and $3\rho_0/4$ for $\varphi_0 = 0$, $\pi/4$, $\pi/2$, and $3\pi/4$, respectively. The intensity patterns of the four radial-variant vector fields have no singularity at the center $\rho = 0$, in which the local polarizations are determined by φ_0, as shown in Figs. 9(i)-(l). The local polarization at the center $\rho = 0$ is along the ray direction of $\varphi = \varphi_0$ and the local polarization at any location $\rho = \rho_0$ is in the ray direction of $\varphi = \varphi_0 + \pi$.

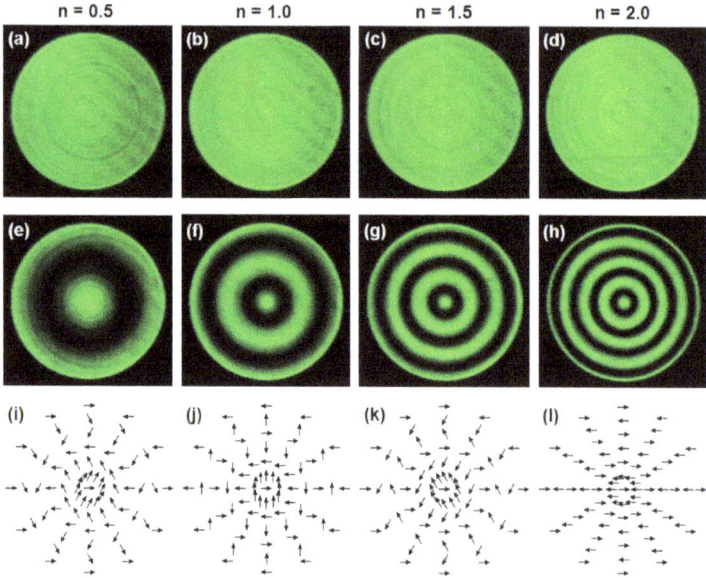

Fig. 8. Four radial-variant vector fields with n = 0.5, 1.0, 1.5, and 2.0 when φ_0 = 0. (a)-(d) Respective intensity patterns. (e)-(h) Respective intensity patterns behind a horizontal polarizer. (i)-(l) Respective schematics of the SOPs.

It might be helpful to generate azimuthal- and radial-variant vector fields, which are generated provided that δ is a function of both ρ and φ. The SOP distribution of the generated vector field can be still described by Eq. (4), with $\delta = 2n\pi\rho/\rho_0 + m\varphi + \varphi_0$. Figure 10 shows the generated vector fields when n = 1.0 and φ_0 = 0, with four different values of m = − 1.0, 1.0, 2.0, and 3.0. Figure 11 shows the generated vector fields when m = 1.0 and φ_0 = 0, with three different values of n = −1.0, 2.0, and 3.0. As shown in Figs. 10(a)-(d) and Figs. 11(a)-(c), the intensity patterns have a central singularity caused by the polarization uncertainty when m ≠ 0, like the azimuthal-variant vector fields and unlike the radial-variant vector fields. The intensity patterns behind a horizontal polarizer exhibit the Archimedean spiral structure. The number of arms is equal to $2|m|$ independent of n, as shown in Figs. 10(e)-(h) and Figs. 11(d)-(f). The chirality of the spiral is determined by the sign of the product of m and n. The chirality is a right-handed screw if $mn > 0$ whereas it is a left-handed screw if $mn < 0$. As shown in Figs. 10(i)-(l) and Figs. 11(g)-(i), the vector fields have the more complex and richer SOPs.

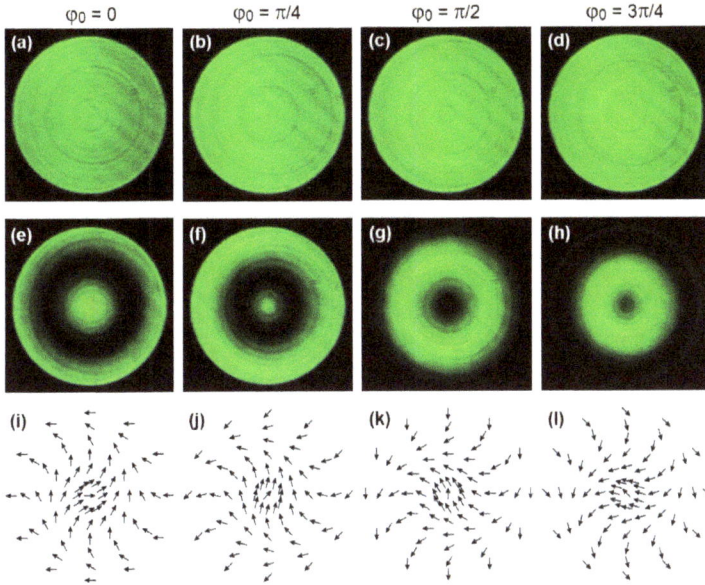

Fig. 9. Four radial-variant vector fields with $\varphi_0 = 0$, $\pi/4$, $\pi/2$, and $3\pi/4$ when $n = 0.5$. (a)-(d) Respective intensity patterns. (e)-(h) Respective intensity patterns behind a horizontal polarizer. (i)-(l) Respective schematics of the SOPs.

As any radial index n does not influence the axial symmetry, we only need to explore the effect of the fractional azimuthal index m on the azimuthal- and radial-variant vector fields. Figure 12 shows the cases of $m = 0.5$ (upper panel) and 1.5 (bottom panel), with $n = 1.0$ and $\varphi_0 = 0$. As Figs. 12(a) and (d), the intensity patterns have no difference and contain a dark ray with its initial point at the center along the radial direction $\varphi = 0$. The dark ray caused by the fractional m breaks the axial symmetry for the azimuthal- and radial-variant vector fields, as the azimuthal-variant vector fields. After passing through a horizontal polarizer, however, the intensity pattern exhibits the Archimedean spiral structure. The number of the Archimedean spiral is equal to $2|m|$ (even when m is a half integer), as shown in Figs. 12(c) and (f). All the arms are smoothly connected at the center. As the schematics shown in Figs. 12(c) and (g), the SOPs are markedly different from and more complex than that when m is an integer.

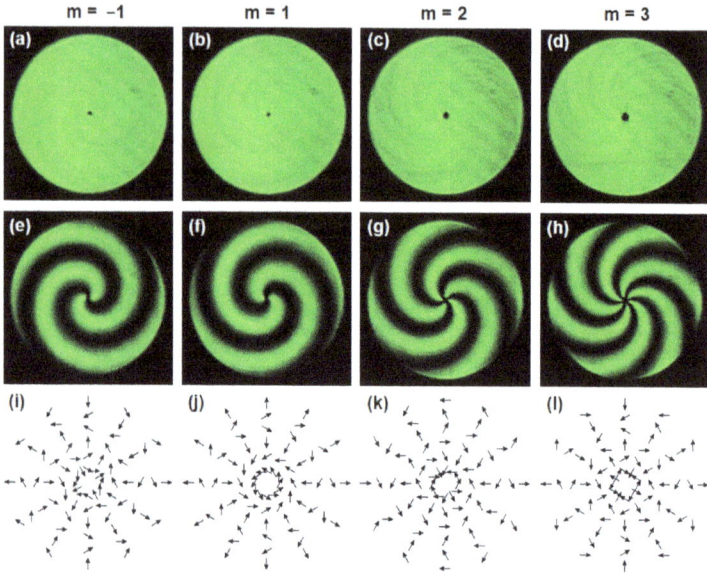

Fig. 10. Four hybrid azimuthal- and radial-variant vector fields with $m = -1$, 1, 2, and 3 when $n = 1.0$. (a)-(d) Respective schematics of SOPs. (e)-(h) Respective intensity patterns. (i)-(l) Respective intensity patterns behind a horizontal polarizer.

The spiral can be described by $\rho = A\varphi + B$, where $A = - m\rho_0 / 2n\pi$ and $B = (\rho_0 / 2n\pi)[(2k - 1)\pi / 2 + \theta - \varphi_0]$. Here, θ is an angle formed by the polarization direction of the polarizer and the ray direction $\varphi = 0$, and k denotes the order of the arms of the spiral with $k = 1, 2, ..., 2|m|$. The parameter B determines the spiral turning, while the magnitude of A controls the distance between successive turnings. The sign of A (the sign of the product of m and n) determines the chirality of the spiral. The extinction nodes along the radial direction are determined by $2|n|$.

2.2. Hybridly polarized vector fields[34]

As discussed above, based on the orthogonal right- and left-handed circular polarizations (which correspond to the north and south poles on Σ) as two base vectors, the generated vector field belongs to a kind of local linearly polarized vector field (with its SOPs corresponding to points in the equator on Σ). Therefore, if choosing the orthogonal linear

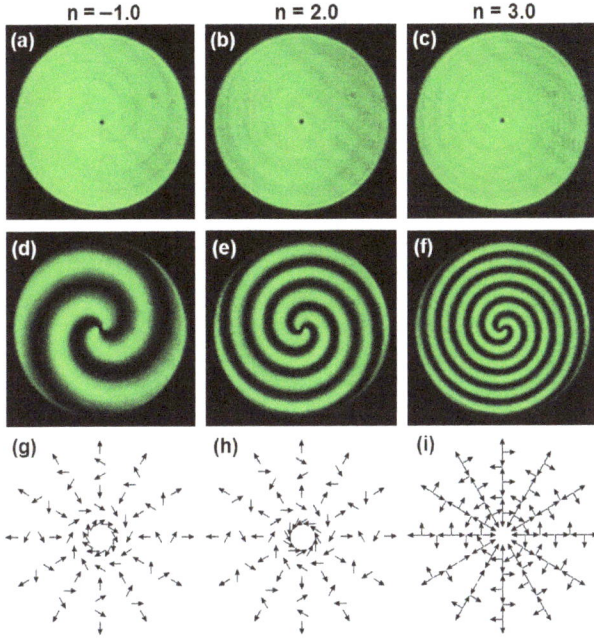

Fig. 11. Three hybrid azimuthal- and radial-variant vector fields with $n = -1$, 2, and 3 when $m = 1.0$. (a)-(c) Respective intensity patterns. (d)-(f) Respective intensity patterns behind a horizontal polarizer. (g)-(i) Respective schematics of SOPs.

polarizations corresponding to the antipodal points in the equator on Σ as a pair of base vectors, the SOPs of the generated vector field should correspond to the points in the meridian circle on Σ, which is perpendicular to the line linking the antipodal points. As a result, the SOP distribution of the generated vector field can be described as

$$\hat{\mathbf{P}}(\rho, \varphi) = \cos(\delta + \pi/4)\exp[-j(\phi + \pi/4)]\mathbf{e}_R$$
$$+ \sin(\delta + \pi/4)\exp[j(\phi + \pi/4)]\mathbf{e}_L$$
$$= \frac{1}{\sqrt{2}}[\cos\delta\cos(\phi + \pi/4) + j\sin\delta\sin(\phi + \pi/4)]\mathbf{e}_x \quad (5)$$
$$+ \frac{1}{\sqrt{2}}[\cos\delta\sin(\phi + \pi/4) - j\sin\delta\cos(\phi + \pi/4)]\mathbf{e}_y$$

We first consider the case of $\delta = m\varphi + \varphi_0$. In Eq. (5), ϕ determines which pair of orthogonal linear polarization as the base vectors. We can

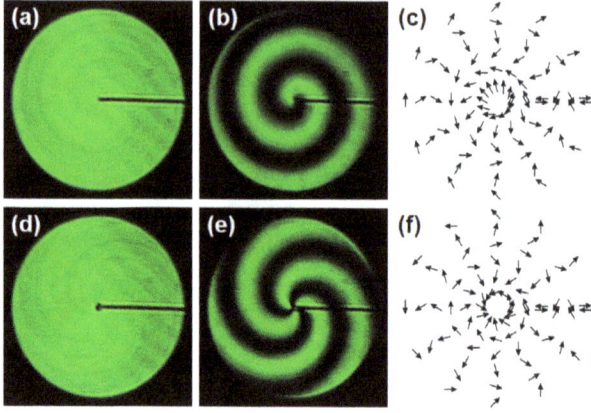

Fig. 12. Two hybrid azimuthal- and radial-variant vector fields with $m = 0.5$ and 1.5 when $n = 1.0$. (a) and (d) Respective intensity patterns. (b) and (e) Respective intensity patterns behind a horizontal polarizer. (c) and (f) Respective schematics of the SOPs.

find from Eq. (5) that the pair of base vectors $\{\cos\phi\,\mathbf{e}_x + \sin\phi\,\mathbf{e}_y, -\sin\phi\,\mathbf{e}_x + \cos\phi\,\mathbf{e}_y\}$ correspond to the pair of points $\{(2\phi, 0), (2\phi + \pi, 0)\}$ in the equator on Σ. The SOPs of the generated vector field described by Eq. (5) are all in the meridian circle of $2\phi + \pi/2$, which is perpendicular to the connecting line between the two points $\{(2\phi, 0), (2\phi + \pi, 0)\}$, due to the presence of the factor $\phi + \pi/4$ in Eq. (5).

To generate a kind of hybridly polarized vector fields (including linear, circular and elliptical polarizations), two $1/2$ wave plates must be used, as the experimental arrangement shown in Fig. 2. Since the linearly polarized field has the distinguishable direction, in principle, infinite pairs of orthogonal linearly polarized fields can be found in the equator on Σ shown in Fig. 1. In experiment, different pair of orthogonal linearly polarized base vectors can be realized by changing the orientations of two $1/2$ wave plates.

It should be emphasized that: (i) ϕ in Eq. (5) is the azimuthal angle of the polar coordinate system in the spatial frequency plane of the 4f system and specifies the polarization directions of the orthogonal linearly polarized fields generated by the $1/2$ wave plates in the ± 1st order paths, in particular, 2ϕ can characterize the meridian circle on Σ. (ii) φ in δ of Eq. (5) can indicate the azimuthal angle of the polar coordinate systems

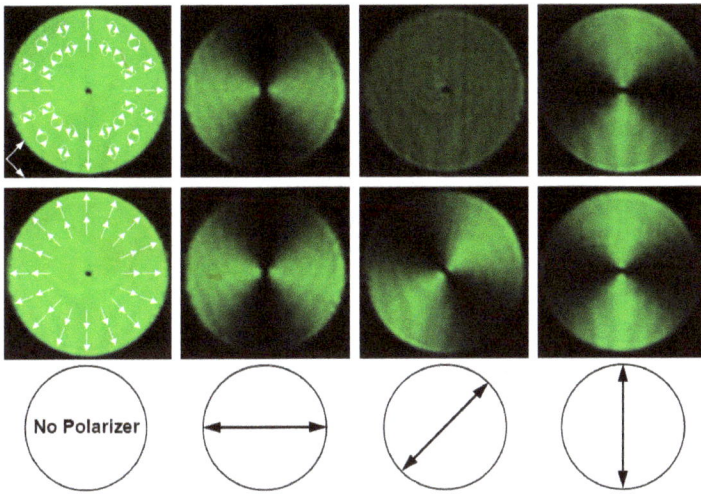

Fig. 13. Intensity patterns for the vector fields. First row shows the hybridly polarized vector field with $\phi = -\pi/4$ and $\delta = \varphi$. Second shows the radially polarized field. Third row shows the directions of the polarizer.

attached in both the input plane and the output plane of the 4f system. In addition, for generating the hybridly polarized vector fields, the pair of orthogonal arrows in the corner in each figure shows the directions of two orthogonal linearly polarized base vectors.

The intensity patterns of the generated hybridly polarized vector field based on the pair of base vectors $\{2^{-1/2}(\mathbf{e}_x-\mathbf{e}_y), 2^{-1/2}(\mathbf{e}_x + \mathbf{e}_y)\}$ when $\delta = \varphi$ (i.e., $m = 1$ and $\varphi_0 = 0$) are shown in the first row of Fig. 13. As a comparison, the radially polarized vector field is also shown in the second row of Fig. 13. Both vector fields have no difference in intensity pattern. After passing through a polarizer, however, the situation is quite different. For the hybridly polarized vector field, the intensity patterns behind the horizontal and vertical polarizers are recognizable, and the extinction directions orthogonal to the direction of the polarizer are the same as that for the radially polarized vector field. In contrast, the intensity pattern behind the $\pi/4$ polarizer has no extinction direction and its intensity becomes a half. For the radially polarized vector field, the intensity pattern behind the polarizer has always the extinction direction orthogonal to the direction of the polarizer.

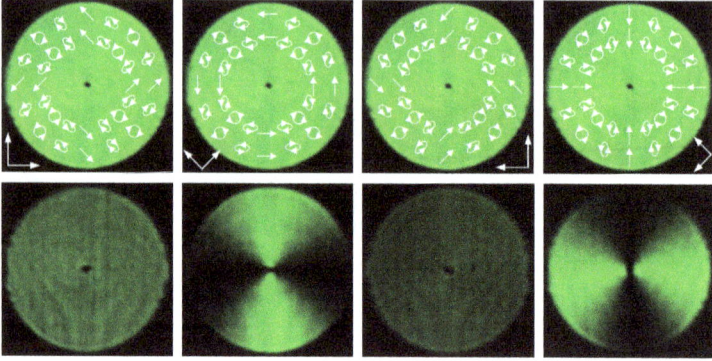

Fig. 14. Generated four different vector fields with the directions of the 1/2 wave plate in the +1st order path along 0, $\pi/4$, $\pi/2$ and $3\pi/4$, the SOPs (the upper row) and the intensity patterns behind a horizontal polarizer (the bottom row).

We explore the generation of the hybridly polarized vector field for four pairs of orthogonal linearly polarized base vectors, with $\delta = \varphi$, as shown in Fig. 14. Four pairs of base vectors are $\{\mathbf{e}_x, \mathbf{e}_y\}$, $2^{-1/2}\{(\mathbf{e}_x + \mathbf{e}_y), (-\mathbf{e}_x + \mathbf{e}_y)\}$, $\{\mathbf{e}_y, -\mathbf{e}_x\}$, and $2^{-1/2}\{(-\mathbf{e}_x + \mathbf{e}_y), -(\mathbf{e}_x + \mathbf{e}_y)\}$, which correspond to four pairs of points, $\{(0, 0), (\pi, 0)\}$, $\{(\pi/2, 0), (3\pi/2, 0)\}$, $\{(\pi, 0), (2\pi, 0)\}$ and $\{(3\pi/2, 0), (5\pi/2, 0)\}$, in the equator on Σ in Fig. 1, respectively. All the generated four intensity patterns have no difference. For the vector fields generated from the first and third pairs of base vectors, the two intensity patterns behind a horizontal polarizer are unrecognizable and have no extinction direction, and both intensities become a half. For the vector fields generated from the second and fourth pairs of base vectors, however, the intensity patterns behind a horizontal polarizer exhibit the extinction directions parallel and orthogonal to the direction of the polarizer, respectively. All the phenomena as mentioned above are easily understood by the schematic SOP distributions in the first row of Fig. 14.

For the vector field generated by the first [or second] pair of base vectors, its SOP at any location in the radial direction of $\varphi = 0$ is linearly polarized with the polarization direction of $2^{-1/2}(\mathbf{e}_x + \mathbf{e}_y)$ [or $\mathbf{e}_y$], and corresponds to the point $(\pi/2, 0)$ [or $(\pi, 0)$] in the equator on Σ. Within the range of $\varphi \in (0, \pi/4)$, the SOP is the left-handed elliptical polarization with its major axis of polarization ellipse in the direction of $2^{-1/2}(\mathbf{e}_x + \mathbf{e}_y)$ [or $\mathbf{e}_y$]. In the radial direction of $\varphi = \pi/4$, the SOP is left-handed circularly polarized. Within the range of $\varphi \in (\pi/4, \pi/2)$, the SOP is still the left-handed elliptical polarization with its major axis of the polarization ellipse in the direction of $2^{-1/2}(-\mathbf{e}_x + \mathbf{e}_y)$ [or $-\mathbf{e}_x$]. At the radial direction of $\varphi = \pi/2$, the SOP becomes linearly polarized again, while its polarization direction is in the direction of $2^{-1/2}(-\mathbf{e}_x + \mathbf{e}_y)$ [or $-\mathbf{e}_x$]. Within the range of $\varphi \in (\pi/2, 3\pi/4)$, the SOP becomes right-handed elliptically polarized, and the major axis of its polarization ellipse is the same as the linearly polarized direction in the radial direction of $\varphi = \pi/2$. In the radial direction of $\varphi = 3\pi/4$, the SOP is right-handed circularly polarized. Within the range of $\varphi \in (3\pi/4, \pi)$, the SOP is right-handed elliptically polarized and the major axis of its polarization ellipse is in the direction of $-2^{-1/2}(\mathbf{e}_x + \mathbf{e}_y)$ [or $-\mathbf{e}_y$]. At the radial direction of $\varphi = \pi$, the SOP becomes linearly polarized again and its polarization direction is in the direction of $-2^{-1/2}(\mathbf{e}_x + \mathbf{e}_y)$ [or $-\mathbf{e}_y$]. The variation of SOPs with φ ranging from π to 2π is very similar to the situation when φ varies from 0 to π.

For the vector field generated by the first [or second] pair of base vectors, the evolution process of SOPs with φ ranging from 0 to π corresponds to the point move in the meridian circle of $2\phi = \pi/2$ [$2\phi = 0$] on Σ, which starts from the equator point of $(\pi/2, 0)$ [$(\pi, 0)$], then pass through orderly the south pole, the equator point of $(3\pi/2, 0)$ [$(0, 0)$] and the north pole, and finally backs to the starting equator point of $(\pi/2, 0)$ [$(\pi, 0)$]. For the variation of SOPs with φ from π to 2π, the corresponding point move in the meridian circle of $2\phi = \pi/2$ on Σ will experience the same evolution process as mentioned above.

3. Novel effects

3.1. *Optical cages*[18]

To satisfy the requirements of various applications, the shape of optical focus is expected to be controllable and designable. As is well known, the shape can be modified by engineering the spatial structures of amplitude and/or phase. As reported in Refs. 12-18, the polarization is able to control the shape of optical focus, making the design of the focal field become more flexible. In particular, three-dimensional (3D) hollow focus surrounded by "optical barrier" of high intensity, as a 3D optical cage, has many potential applications due to its peculiar property, as the dark optical trap for capturing particles[17, 36-39] and as the erasing field for realizing superresolution fluorescence microscopy.[40, 41] Kozawa and Sato generated a 3D optical cage from a double-ring radially polarized field.[17] Arlt and Padgett produced a dark spot by superposing two Laguerre-Gaussian modes.[42] The 3D optical cages produced by the above two methods have the nonuniform optical barrier. Bokor and Davidson chose the counter-propagating radially polarized Laguerre-Gaussian fields to realize a 3D optical cage enclosed by the near-uniform optical barrier.[43] However, those methods require the sophisticated focusing system and elaborate the optical alignment. The radially polarized fields with a circular π phase plate[44] have also been used to produce a 3D optical cage surrounded by an optical barrier with a uniformity of ~85%.[45] We proposed an altercative method for producing a controllable 3D optical cage by using a double-mode local linearly polarized vector field.

As shown in Fig. 15, we consider a double-mode vector field, which is composed of an inner mode and an outer mode (both are local linearly polarized vector fields with the same topological charge $m = 1$ but the different initial phases of φ_{10} and φ_{20}).[18] With Eq. (4), the double-mode local linearly polarized vector field can be expressed as follows

$$
\mathbf{E}(\rho,\varphi) = \begin{cases} \cos(\varphi+\varphi_{10})\mathbf{e}_x + \sin(\varphi+\varphi_{10})\mathbf{e}_y, & 0 < \rho \le t\rho_0 \\ \cos(\varphi+\varphi_{20})\mathbf{e}_x + \sin(\varphi+\varphi_{20})\mathbf{e}_y, & t\rho_0 < \rho \le \rho_0 \end{cases}
$$
$$
= \begin{cases} \cos\varphi_{10}\mathbf{e}_\rho + \sin\varphi_{10}\mathbf{e}_\varphi, & 0 < \rho \le t\rho_0 \\ \cos\varphi_{20}\mathbf{e}_\rho + \sin\varphi_{20}\mathbf{e}_\varphi, & t\rho_0 < \rho \le \rho_0 \end{cases}
$$

$$(6)$$

where $\mathbf{e}_\rho$ and $\mathbf{e}_\varphi$ are the unit vectors in the polar and azimuthal directions of cylindrical coordinate system in the input plane, respectively. ρ_0 is the radius of the field and is also the entrance pupil of the objective lens used here. t is the ratio of the inner mode to the outer mode in radius. The angles φ_{10} and φ_{20} represent the polarization directions with respect to the radial direction for the inner and outer modes, respectively. The desired double-mode vector fields can be generated by the method presented in Ref. 2.

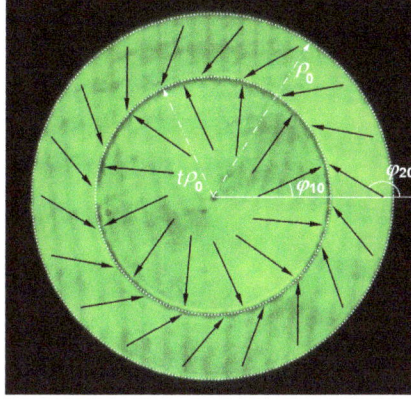

Fig. 15. Sketches of the intensity and SOP distributions for a double-mode vector field.

On the basis of the Richards-Wolf vector diffraction theory, when the double-mode vector field is focused by a high NA (numerical aperture) objective lens, the electric field in the vicinity of focus should be

$$\mathbf{E}(r,\psi,z) = \Big[\cos\varphi_{10} I_1(\alpha_1,\alpha_2) + \cos\varphi_{20} I_1(\alpha_2,\alpha_3) \Big]\mathbf{e}_r +$$
$$\Big[\sin\varphi_{10} I_2(\alpha_1,\alpha_2) + \sin\varphi_{20} I_2(\alpha_2,\alpha_3) \Big]\mathbf{e}_\psi + \qquad (7)$$
$$\Big[\cos\varphi_{10} I_3(\alpha_1,\alpha_2) + \cos\varphi_{20} I_3(\alpha_2,\alpha_3) \Big]\mathbf{e}_z ,$$

where $\mathbf{e}_r, \mathbf{e}_\psi$, and $\mathbf{e}_z$ are the three unit vectors in the polar, azimuthal, and axial directions of a cylindrical coordinate system with its $z = 0$ plane located at the focal plane. α_1 takes a very small value of 0.05 to avoid the center singularity, and α_3 and α_2 are $\alpha_3 = \sin^{-1}(\mathrm{NA})$ and $\alpha_2 = \sin^{-1}(t \cdot \mathrm{NA})$.

The three variables, I_1, I_2 and I_3, are defined as follows

$$I_1(\xi,\zeta) = A\int_\xi^\zeta \cos^{1/2}(\theta)\sin(2\theta) J_1(kr\sin\theta)\exp(jkz\cos\theta)d\theta ,$$
$$I_2(\xi,\zeta) = 2A\int_\xi^\zeta \cos^{1/2}(\theta)\sin(\theta) J_1(kr\sin\theta)\exp(jkz\cos\theta)d\theta , \quad (8)$$
$$I_3(\xi,\zeta) = 2iA\int_\xi^\zeta \cos^{1/2}(\theta)\sin^2(\theta) J_0(kr\sin\theta)\exp(jkz\cos\theta)d\theta ,$$

where A is a constant related to the focal length and the wavelength, J_0 and J_1 are the Bessel functions of the first kind of orders 0 and 1. Equation (7) gives the total focal field generated by the incident field

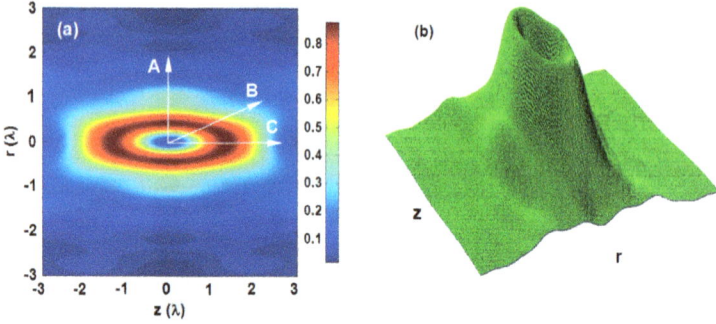

Fig. 16. The generated optical cage when $\varphi_{10} = 0.089\pi$ and $\varphi_{20} = 0.711\pi$, shown by (a) the contour figure and (b) the surface figure.

described by Eq. (6), and determines the field structure in the vicinity of focus. A 3D optical cage can be achieved by choosing the different combination of φ_{10} and φ_{20}.

Numerical simulation is performed to explore the focusing property of the double-mode vector fields, in which the parameters are chosen as follows: NA = 0.9 ($\alpha_3 = 0.356\pi$), $t = 0.707$ ($\alpha_2 = 0.220\pi$) ensuring the inner and outer modes to have the same area, and $\varphi_{10} = 0.089\pi$ and $\varphi_{20} = 0.711\pi$. The simulated total field in Fig. 16 shows indeed a 3D optical cage in the vicinity of the focus. Figures 16(a) and (b) give the contour figure and the surface figure of the 3D optical cage in the *r*-*z* plane near the focus. The optical cage is surrounded by a near uniform optical barrier, which exhibits a spheroid surface shape with an ellipticity of 3.13 and has a size of 1.36λ in the longitudinal direction and of 0.43λ in transverse direction. To further confirm the uniformity of optical barrier, we also calculate the intensity distributions in three typical directions, **A**, **B** and **C** as shown in Fig. 16(a), where **A** and **C** are along the radial and axial directions while **B** is a direction with an angle of 0.146π with respect to the axial direction (this direction has the lowest optical barrier). The optical barriers are the same in the **A** and **C** directions while the optical barrier in the **B** direction is 98% of that in the **A** and **C** directions.

The optical tweezer based on the focused field has proven to be a very useful tool for the manipulation of microscopic particles. As is well known, a bright focus cannot trap particles with refraction index lower than that of the surrounding medium, because the low-refractive-index

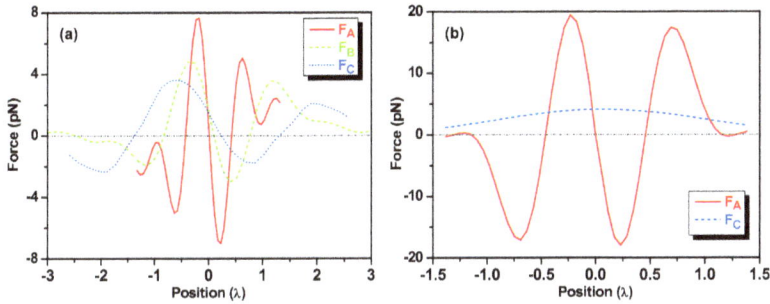

Fig. 17. The Calculated forces induced by the focused field acting on the low-refractive-index particle. (a) In three directions (**A**, **B** and **C**) for the 3D optical cage. (b) In radial (**A**) and axial (**C**) directions for the optical cage generated by the azimuthally polarized field.

particles will be repelled from the high-intensity region. In general, an optical cage with lower intensity in center is a promising technique for trapping the low-refractive-index particles. To confirm the trapping property, the improved finite-difference time-domain (FDTD) method[46] is used to calculate the optical force acting on the low-refractive-index particle. For simulations, the low-refractive-index particle used is the water droplets (with a radius of 60 nm) diffused in the acetophenone solution, as used in experiment.[47] The specimen solution is confined between glass coverslip and slide. The incident light has a wavelength of 532 nm and a power of 1 W. The refractive indices of the water and the acetophenone are 1.33 and 1.53 at 532 nm, respectively. Despite the fact that the theoretical treatment should take into account the influence of the interface effect on the focal intensity distribution,[48] we do not carefully consider the situation when the refractive index mismatches. The influence of interface of sample cell on focusing can be neglected due to the following two facts: (i) an oil-immersion objective is generally used for tight focusing, and consequently, the refractive indices of oil and coverslip are matched and (ii) the refractive index of coverslip is nearly close to that of acetophenone.

Figure 17(a) shows the components of the calculated optical forces on the droplet in the three directions (**A**, **B** and **C**, as shown in Fig. 16), $\mathbf{F_A}$, $\mathbf{F_B}$ and $\mathbf{F_C}$, as functions of position. It can be found that in the **A**

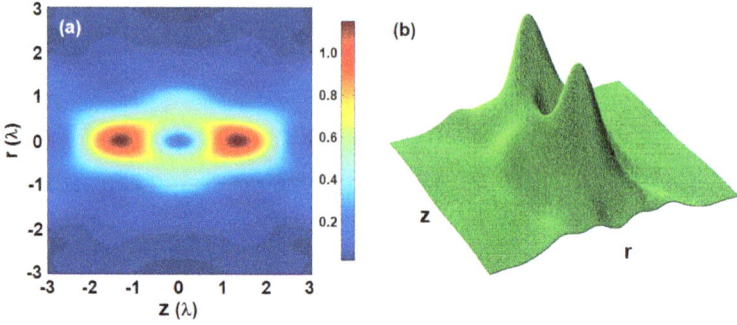

Fig. 18. Intensity distribution of the generated optical cage when $\varphi_{10} = 0.089\pi$ and $\varphi_{20} = 0.791\pi$, (a) in contour figure and (b) in surface figure.

direction, the droplet will be pulled back to the axis because the droplet always receives a restoring force pointing toward the axis (equilibrium position) once the droplet moves away from the axis along the radial direction. In the **C** direction, the equilibrium position is not exactly at the focal plane ($z = 0$) while is located behind the focal plane, since the scattering force always points toward the propagation direction of light between two kinds of optical forces (gradient and scattering forces). In addition, the water droplet inside the acetophenone undergoes a upward flotage, which can be in part counteracted by the scattering downward force when the light propagates downward, and thus the equilibrium position should be more close to the focal plane. Such a 3D optical cage could achieve the 3D trapping of the low-refractive-index particles. As a comparison, Fig. 17(b) shows the calculated optical force offered by the azimuthally polarized field under the same conditions. The trapping of the low-refractive-index particle is possible in the transversal (**A**) direction while impossible in the axial (**C**) direction, implying that the particle can be trapped in two dimensions only, differently from the optical cage.

In fact, the optical cage can be accurately controlled by changing the polarization angles, φ_{10} and φ_{20}. For instance, Fig. 21 (Fig. 22) shows the optical cage when $\varphi_{10} = 0.089\pi$ and $\varphi_{20} = 0.791\pi$ ($\varphi_{10} = 0.089\pi$ and $\varphi_{20} = 0.632\pi$). Clearly, the optical barrier in the transverse (axial) direction is a half of that in the axial (transverse) direction, as shown in Fig. 18

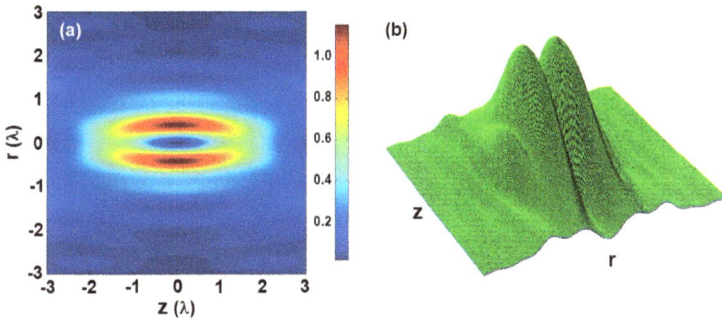

Fig. 19. Intensity distribution of the generated optical cage when $\varphi_{10} = 0.089\pi$ and $\varphi_{20} = 0.632\pi$, (a) in contour figure and (b) in surface figure.

(Fig. 19) radial direction. The shape of the optical cage can be indeed flexibly configured, implying that the trapping property is controllable, by engineering the SOP distribution of the double-mode vector field.

3.2. Axial-symmetry broken vector fields[49]

As is well known, the symmetry is a significant concept in modern science and a widely existing phenomenon in nature, and reflects beauty or perfection. The symmetry breaking is of great importance because it plays a crucial role in the origin of substance. Therefore, exploring the vector fields with broken axial symmetry should be a very interesting issue, although this issue is rarely invoked for vector fields. However, there have been some reports focusing on scalar fields with broken axial symmetry.[50-52] For instance, the diffraction behavior of a partially blocked scalar vortex field has been investigated.[50] The results reveal that a blocked input vortex field occupying half of the input plane results in an output field which is rotated by an angle of $\pi/2$ or $-\pi/2$ with respect to the input field, depending on the sign of the topological charge of the input vortex field. Padgett and collaborators[51, 52] have studied the peculiar effect of the traditional scalar field blocked by a sector aperture; in particular, Heisenberg's uncertainty principle for a pair of conjugate variables (angular position and orbital angular momentum).

As shown in Fig. 20(a), with Eq. (4), an axial-symmetry local linearly polarized vector field with a finite aperture can be written

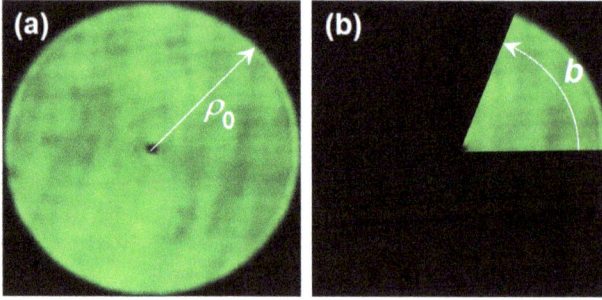

Fig. 20. (a) Axial-symmetry local linearly polarized vector field with a radius of ρ_0 and (b) its axial-symmetry-broken field by a sector aperture ranging from $\varphi = 0$ to b.

$$\mathbf{E}(\rho,\varphi) = A_0 \operatorname{circ}(\rho / \rho_0)[\exp(-j\delta)\mathbf{e}_R + \exp(j\delta)\mathbf{e}_L], \qquad (9)$$

with $\delta = m\varphi + \varphi_0$. Here A_0 is a constant, ρ and φ are the polar radius and azimuth angle, respectively. Without loss of generality, φ_0 is set to be $\varphi_0 = 0$. $\operatorname{circ}(\rho/\rho_0)$ is the circular function, and ρ_0 is the radius of the vector field. The axial-symmetry local linearly polarized vector field embodied by Eq. (9) can be easily generated by the same scheme as in Ref. 2. All the generated vector fields have the same radius of $\rho_0 = 2.5$ mm and the same wavelength of $\lambda = 532$ nm.

It is very interesting to explore the focusing behavior of the axial-symmetry-broken local linearly polarized vector field, by a lens with a focal length of $f = 500$ mm. The axial-symmetry vector field is broken by a sector aperture as follows

$$M(\varphi) = \begin{cases} 1 & \text{within } \varphi \in [0,b] \\ 0 & \text{otherwise} \end{cases}. \qquad (10)$$

The top and bottom rows in Fig. 21 show the input vector fields and their focal fields. As a comparison, the first column in Fig. 21 shows the case of the axial-symmetry local linearly polarized vector field with $m = 10$, and the intensity pattern of its focused field exhibits a focal ring. When the axial-symmetry vector field is partially blocked by a sector aperture whose apex being coincident with its field axis, the axial symmetry of the input vector field is broken, as shown in other three columns of Fig. 21. The second column in Fig. 21 shows the axial-symmetry-broken vector field occupying the first quadrant ($b = \pi/2$), its

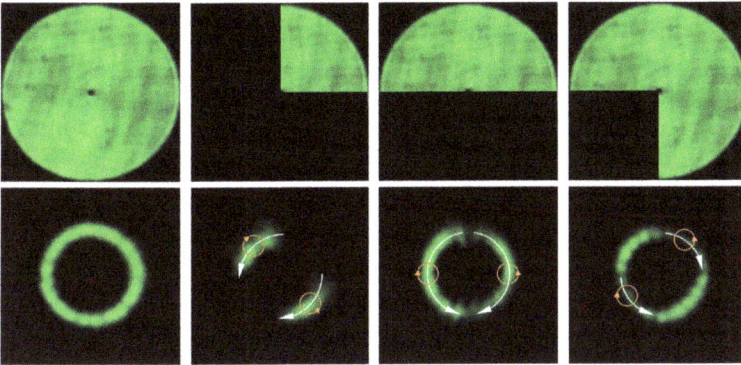

Fig. 21. Focal behavior of the locally linearly polarized vector fields with broken axial symmetry for $m = 10$ and $\varphi_0 = 0$. As a comparison, the first column shows the case of the locally linearly polarized vector fields with axial symmetry. The second, third, and fourth columns show the three cases of the locally linearly polarized vector fields with broken axial symmetry with $b = \pi/4$, $\pi/2$, and $3\pi/2$, respectively. The first row shows the input fields and the second row gives the measured results of the corresponding focal fields, respectively. The circles with arrows and the arcs with arrows indicate SAM and OAM, respectively.

focused field is azimuthally split into a pair of 1/4 rings occupying the second and fourth quadrants, which are rotated counterclockwisely and clockwisely by an angle of $\pi/2$ with respect to the input vector field. In addition, the SOPs of a pair of 1/4 rings are confirmed to be right- and left-handed circularly polarized, respectively, implying that both carry the opposite SAM, as shown by the circles with arrows.

The third and fourth columns of Fig. 21 show other two cases of $b = \pi$ and $3\pi/2$, respectively. In the case of $b = \pi$, the input vector field occupies the upper half of the input plane. The focused field is composed of a pair of near 1/2 rings occupying the left and right half-spaces in the focal plane. However, a pair of near 1/2 rings do not form a perfect ring. The SOPs of a pair of 1/2 rings are also right- and left-handed circularly polarized, respectively, implying that both carry the opposite SAM, as shown by the circles with arrows. For the case of $b = 3\pi/2$, the input field occupies 3/4 of the space (first, second, and third quadrants) of the input plane. The focused vector field seems to form a whole ring, but which exhibits a nonuniform intensity distribution along the ring (the second and fourth quadrants are stronger than the first and third quadrants in intensity). For the SOPs of its focused field, the two 1/4 rings in the first and third quadrants are right- and left-handed circularly polarized, respectively, implying that both carry the opposite SAMs, as shown by the circles with arrows. In contrast, the two 1/4 rings in the second and fourth quadrants are linearly polarized and then carry no SAM (their polarization states are not shown), because both are the superposition of right- and left-handed circularly polarized 1/4 rings, respectively.

It is of great importance to reveal the underlying physics behind the above observed phenomena. The sector-shaped function described by Eq. (10) can be decomposed into a series of Fourier components as[52-54]

$$M(\varphi) = (b/2\pi) \sum_{u=-\infty}^{+\infty} C_u \exp(ju\varphi), \qquad (11)$$

with

$$C_u = \mathrm{sinc}(ub/2)\exp(-jub/2). \qquad (12)$$

By combining with Eq. (10), the sector-shaped input vector field can be written as

$$\begin{aligned}
\mathbf{E}_{\mathrm{in}} &= A_0 M(\varphi)\mathrm{circ}(\rho/\rho_0)[\exp(-j\delta)\mathbf{e}_R + \exp(j\delta)\mathbf{e}_L] \\
&= A_0(b/2\pi)\mathrm{circ}(\rho/\rho_0)[E_R\mathbf{e}_R + E_L\mathbf{e}_L],
\end{aligned} \qquad (13)$$

with

$$\begin{aligned}
E_R &= \exp(-jm\varphi)\sum_{u=-\infty}^{+\infty} C_u \exp(ju\varphi) \\
&= \exp(-jm\varphi)\sum_{u=-\infty}^{+\infty} \mathrm{sinc}(ub/2)\exp[ju(\varphi-b/2)], \\
E_L &= \exp(jm\varphi)\sum_{u=-\infty}^{+\infty} C_u \exp(ju\varphi) \\
&= \exp(jm\varphi)\sum_{u=-\infty}^{+\infty} \mathrm{sinc}(ub/2)\exp[ju(\varphi-b/2)].
\end{aligned} \qquad (14)$$

Evidently, besides the common helical phases of $\exp(\pm jm\varphi)$, Eq. (14) suggests that the right- and left-handed circularly polarized components are both composed of an infinite series of vortices with helical phase $\exp(ju\varphi)$.[55] To investigate the focusing property of the axial-symmetry-broken local linearly polarized vector field, we should use the known focal-field distribution of a finite-aperture vortex field $\mathrm{circ}(\rho/\rho_0)\exp(jh\varphi)$, as follows[50, 56-58]

$$U_h(r,\psi) = (-j)^h \exp(jh\psi)\mathbf{H}_h\{\mathrm{circ}(\rho/\rho_0)\}, \tag{15}$$

where $\mathbf{H}$ is the operator of the Hankel transform

$$\mathbf{H}_h\{\mathrm{circ}(\rho/\rho_0)\} = \frac{2\pi\rho_0^2\, p^h}{(2+h)\Gamma(1+h)}\,_1F_2[1+h/2,(2+h/2,1+h);-p^2]. \tag{16}$$

Here $\Gamma(.)$ is the well-known Gamma function, $_1F_2[.]$ is the generalized hypergeometric function, and p is a dimensionless parameter $p = \pi\rho_0 r/(\lambda f)$ (λ is the wavelength and f is the focal length). The radius of the focus ring is determined by $S_h(r) \equiv \mathbf{H}_h\{\mathrm{circ}(\rho/\rho_0)\}$. By combining with Eq. (12), the focal field of the input field described by Eq. (13) can be written as

$$\mathbf{E}_f = A_0(b/2\pi)[\exp(jm\pi/2)V_R\mathbf{e}_R + \exp(-jm\pi/2)V_L\mathbf{e}_L], \tag{17}$$

with

$$V_R = \exp(-jm\psi)\sum_{u=-\infty}^{+\infty}\mathrm{sinc}(ub/2)\exp\{ju[\psi-(b/2+\pi/2)]\}S_{u-m}(r),$$

$$\tag{18}$$

$$V_L = \exp(jm\varphi)\sum_{u=-\infty}^{+\infty}\mathrm{sinc}(ub/2)\exp\{-ju[\psi-(b/2-\pi/2)]\}S_{u-m}(r).$$

It can be found that, due to the modulation effect of the factor $\mathrm{sinc}(ub/2)$, only $S_{-m}(r)$ (when $u = 0$) is dominant and represents a focus "ring." However, such a focus "ring" is truncated in the azimuthal direction by two sector apertures with the same azimuthal width of about b and their bisectors in the directions of $\psi = b/2 \pm \pi/2$ due to the presence of $\exp\{\pm ju[\psi - (b/2 \pm \pi/2)]\}$ in Eq. (18). The focus "ring" is separated into a pair of partial rings. Therefore, the axial-symmetry-broken local linearly polarized vector field with the azimuthal width of b and the bisector in the direction of $\psi = b/2$ is focused to form a pair of partial rings. The rings have the same azimuthal width of about b and are located at the azimuthal directions of $\psi = b/2 \pm \pi/2$. Each member of the pair of partial rings is rotated clockwisely and counterclockwisely by an angle of $\pi/2$ with respect to the input sector-shaped vector field. In particular, each member of the pair of partial rings carries not only the opposite SAMs but also the opposite OAMs of $\mp m\hbar$ (as shown by the

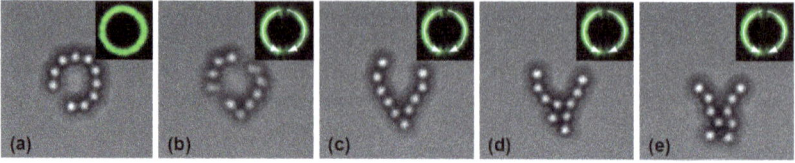

Fig. 22. Snapshots of the motion of particles trapped by the focused fields generated by the axial-symmetry-broken local linearly polarized vector field with $m = 10$ and $b = \pi$. The insets are the corresponding intensity distributions.

arcs with arrows in Fig. 21). It should be pointed out that the circular aperture has no dominant influence on the predicted effect, because it is related to the azimuthal topological structure, while has no relation to the radial distribution of the input field.

Finally, we would like to explore the trapping behavior of the focused field of such an axial-symmetry-broken local linearly polarized vector field and then to demonstrate the above analysis. To generate the optical trapping, a 63× dry lens with NA = 0.8 (corresponding to an effective NA = 0.6 in water) was used. First, we focus on an axial-symmetry local linearly polarized vector field to generate a perfect focus ring and then to expose the isotropic colloid particles with a diameter of 1.5 μm. We observe no motion of the trapped particles along the ring, as shown in Fig. 22(a), implying that the input axially symmetric local linearly polarized vector field indeed carries no OAM. However, the situation is quite different when the input axially symmetric vector field is switched to the axial-symmetry-broken one (as an example, considering the case of $b = \pi$). As shown by the time-lapse photographs shown in Figs. 22(b)–(e), the two 1/2 rings in the focal plane exert torques on the trapped particles to guide them through motion along the partial ring trajectories in opposite directions, suggesting that the two partial rings carry indeed the opposite OAMs. Ultimately, the particles trapped in the two 1/2 rings are transported to the ends of the two partial rings in the common area of the third and fourth quadrants. Consequently, the focused axial-symmetry-broken local linearly polarized vector field as a "transporter" can convey the particles. In particular, the experimental observation also demonstrates the above theoretical prediction.

Clearly, when focusing the axial-symmetry-broken local linearly polarized vector field carrying no angular momentum, there appears three kinds of separations in the azimuthal direction, as follows. (i) Separation in energy — the input energy is separated into a pair of partial rings; in particular, the input axial-symmetry-broken local linearly polarized vector field possesses only mirror symmetry about the angle bisector of the sector aperture, while its focused field exhibits the twofold rotation symmetry (central inversion symmetry) about the field axis and the mirror symmetry about the direction orthogonal to the bisector of the sector aperture. In addition, the original mirror symmetry of the input field is kept. (ii) Separation in SAM—the input axial-symmetry-broken local linearly polarized vector field carrying no SAM is divided into a pair of opposite SAMs carried by the right- and left-handed circularly polarized partial rings. (iii) Separation in OAM—the input axial-symmetry-broken local linearly polarized vector field carrying no OAM is divided into a pair of opposite OAMs of $\mp m\hbar$ (depending on the topological charge m), which are carried by the right- and left-handed circularly polarized components, respectively. However, conservation of angular momentum (whether SAM or OAM) is held. The input axial-symmetry-broken vector field lacks rotation symmetry (it goes without saying the axial symmetry), while its focused field exhibits rotation symmetry in addition to mirror symmetry. In fact, the rotation symmetry of the focused field suggests that the original field of the symmetry-broken input field should possess at least rotation symmetry. Finally, such an effect is closely related to the topological structure of the spatial distribution of polarization states in the input vector fields, because this effect will disappear when the topological charge is $m = 0$, which corresponds to a homogenously linearly polarized scalar field.

3.3. *Young's two-slit interference*[59]

One of the salient features of light is its coherence. The interference of the light fields carrying the spatial-variant phase has given rise to a lot of effects. For instance, the optical vortices result in the novel effects of the single-slit diffraction[60] and the two-slit interference,[61] due to the presence of OAM. A multi-pinhole interferometer was proposed to probe the OAM of optical vortex.[8, 62] In addition, the OAM can be used to unveil the lattice properties hidden in diffraction patterns using a simple triangular aperture.[63]

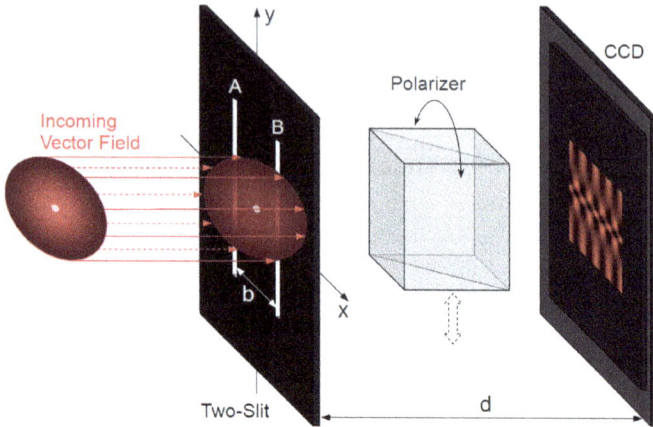

Fig. 23. Schematic of the Young's two-slit interference of the vector field.

It is very interesting to explore the two-slit interference of the vector optical field. Since the interference pattern carries the information of the SOP distribution and then unveils the vector characteristics of the optical field, the two-slit interference can be served as an alternative way for measuring the topological charge of the vector field.

Differently from the scalar field, the Young's two-slit interference of vector field should exhibit some novel effects, due to the inhomogeneous SOP distribution. As shown in Fig. 23, a monochromatic vector field described by Eq. (4) falls on the two slits A and B which are placed in the xy plane, equidistant from the origin and parallel to the y direction. The vector fields transmitted through two slits act as two secondary *line* sources and then are superposed on the observation screen (with a CCD camera) parallel to the xy plane. Let b be the separation of the two slits, a the slit width, d the distance of the observation screen from the xy plane. Suppose the slit width a is narrow enough, the slit length is infinite, and the separation b is small enough compared to d. Thus, inside the two slits the light polarization can be considered homogeneous in the x direction while is still spatial-variant in the y direction.

The local linearly polarized vector field can be decomposed into the orthogonal x- and y-polarized components, which can be independently treated. The x and y components transmitted through the two slits A and B depend only on y independent of x

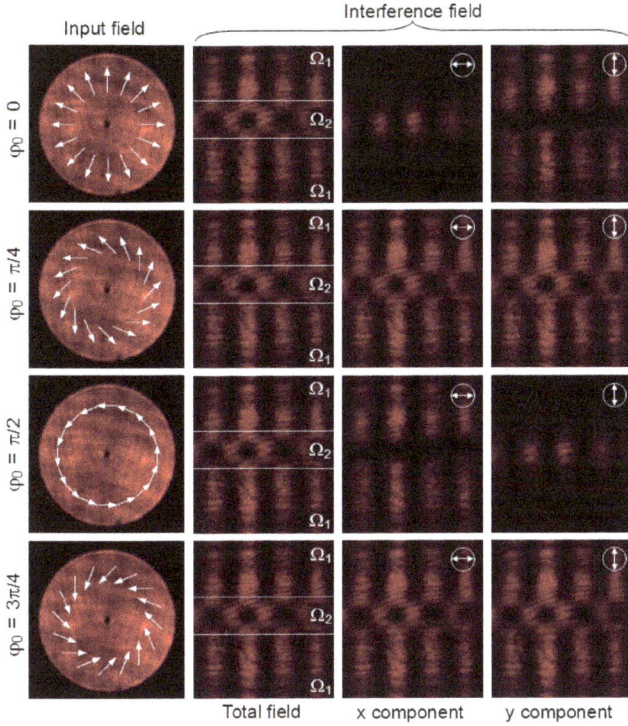

Fig. 24. The patterns of the Young's two-slit interference of the $m = 1$ vector fields with $\varphi_0 = 0$, $\pi/4$, $\pi/2$ and $3\pi/4$.

$$E_x^{A(B)}(y) = \cos[m\varphi_{A(B)} + \varphi_0],$$
$$E_y^{A(B)}(y) = \sin[m\varphi_{A(B)} + \varphi_0], \qquad (19)$$

where $\varphi_A = \arctan(-2y/b)$, $\varphi_B = \arctan(2y/b)$, and $\varphi_A = -\varphi_B$. The intensity distributions of the x and y components in the interference pattern are written as

$$I_x(x, y) \propto 1 + \cos(2m\varphi_B)\cos(2\varphi_0) + [\cos(2m\varphi_B) + \cos(2\varphi_0)]\cos\phi,$$
$$I_y(x, y) \propto 1 - \cos(2m\varphi_B)\cos(2\varphi_0) + [\cos(2m\varphi_B) - \cos(2\varphi_0)]\cos\phi, \qquad (20)$$

where $\phi = 2\pi bx/\lambda d$ and the terms with $\cos\phi$ represent the interference effect. The distribution of the total intensity is

$$I(x, y) = I_x(x, y) + I_y(x, y) \propto 1 + \cos(2m\varphi_B)\cos\phi. \qquad (21)$$

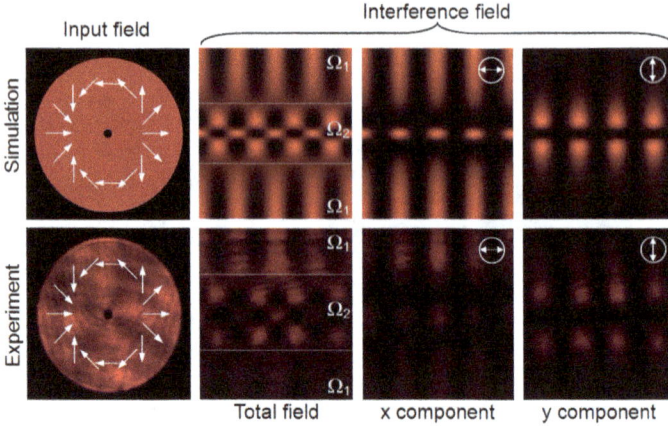

Fig. 25. The patterns of the Young's two-slit interference of the vector field with ($m = 2$, $\varphi_0 = 0$).

The factor $\cos(2m\varphi_B)$ in Eqs. (19) and (20) and the factor $\cos(2\varphi_0)$ in Eq. (20) are from the contribution of the vector field, which embody the intrinsic feature of the two-slit interference of the vector field.

We now explore the Young's two-slit interference of the vector field. A He-Ne laser at $\lambda = 632.8$ nm as a light source is used to generate the vector fields described by Eq. (4), by using the method in Ref. 2. The generated vector field falls normally on the two slits with $a = 0.2$ mm and $b = 0.8$ mm. A CCD with a distance $d = 500$ mm from the slits is used to acquire the interference pattern. A polarizer can be inserted between the slits and CCD, and then to acquire the interference patterns of the x and y components.

Figure 24 shows the measured interference patterns of four $m = 1$ vector fields with $\varphi_0 = 0$, $\pi/4$, $\pi/2$ and $3\pi/4$. Figures 25 and 26 show the situations of two vector fields with ($m = 2$, $\varphi_0 = 0$) and ($m = 3$, $\varphi_0 = 0$), respectively.

As shown in Figs. 24-26, different from the well-known Young's two-slit interference of the scalar light field, the interference patterns of the vector fields exhibit the spatial structure in the y (slits) direction besides in the x direction. The interference pattern of the total field depends only on m independent of $\varphi_0 = 0$. In contrast, the interference patterns of the x and y components depend on both m and φ_0. In

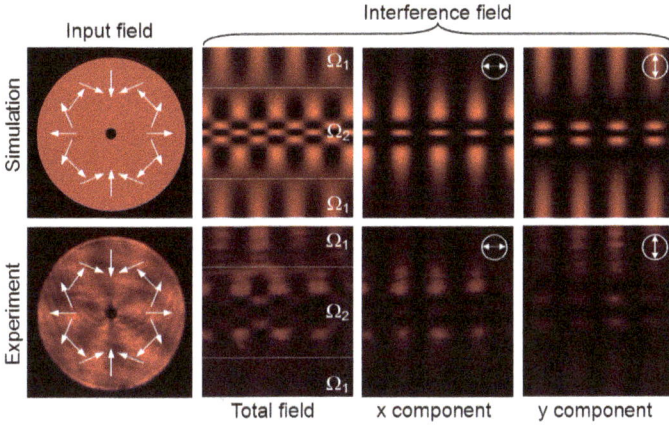

Fig. 26. The patterns of the Young's two-slit interference of the vector field with ($m = 3$, $\varphi_0 = 0$).

particular, the shift between the interference fringes of the x and y components depends on φ_0. The novel behaviors of the Young's two-slit interference of the vector fields can be understood as follows.

We find from Eqs. (20) and (21) that the interference fringes described by $I_x(x, y)$, $I_y(x, y)$, and $I(x, y)$ have the same period of $\Lambda = \lambda d/b$ in the x direction. The interference fringe of $I(x, y)$ is independent of the initial phase φ_0. With Eq. (20), we can find $I_x(x,y)|_{\varphi_0 = \pm\pi/2} = I_y(x,y)|_{\varphi_0 = 0}$, $_{\pm\pi}$ and $I_y(x,y)|_{\varphi_0 = \pm\pi/2} = I_x(x,y)|_{\varphi_0 = 0, \pm\pi}$, implying that the interference fringes of the x and y components have a shift of a half of fringe period ($\Lambda/2$) when $\varphi_0 = 0, \pm\pi/2$ and $\pm\pi$. When $\varphi_0 = \pm\pi/4$ or $\pm3\pi/4$, however, we have $I_x(x,y)|_{\varphi_0 = \pm\pi/4, \pm3\pi/4} = I_y(x,y)|_{\varphi_0 = \pm\pi/4, \pm3\pi/4}$, suggesting that the interference fringes describing by $I_x(x, y)$ and $I_y(x, y)$ are completely identical, so that the interference fringe of $I(x, y)$ has the same pattern as that of the x or y component. Except for $\varphi_0 = 0$ (or $\pm\pi$), $\pm\pi/4$ (or $\pm3\pi/4$) and $\pm\pi/2$, the shift between the interference fringes of the x and y components is between 0 and $\Lambda/2$.

The interference pattern can be separated to two kinds of regions in the y direction, by the boundaries of $y = \pm(b/2)\tan(\pi/4m)$, where the total intensity is independent of x and is a half of the maximum intensity. The first region is a range of $|y| > (b/2)\tan(\pi/4m)$, called as the region Ω_1 as shown in Figs. 24-26, away from the centerline $y = 0$. The second one belongs to the region of $-(b/2)\tan(\pi/4m) < y < (b/2)\tan(\pi/4m)$, called the region Ω_2, in the immediate vicinity of the centerline $y = 0$.

In region Ω_1, we can consider $y \rightarrow \pm\infty$ ($\varphi_B \rightarrow \pm\pi/2$). So Eqs. (20) and (21) can be degenerates to $I_x(x) \propto 1 + \cos(m\pi)\cos(2\varphi_0) + [\cos(m\pi) + \cos(2\varphi_0)]\cos\phi$, $I_y(x) \propto 1 - \cos(m\pi)\cos(2\varphi_0) + [\cos(m\pi) - \cos(2\varphi_0)]\cos\phi$, and $I(x) \propto 1 + \cos(m\pi)\cos\phi$, respectively. When $2\varphi_0$ and $m\pi$ are out of phase, $I_x(x) \equiv 0$. In contrast, the interference fringe of the y component is similar to the traditional two-slit interference, and the interference fringe for an odd m has a shift of a half of fringe period with respect to that for an even m. When $2\varphi_0$ and $m\pi$ are in phase, the interference behaviors are opposite to the above situation. If $\varphi_0 = \pm\pi/4$ and $\pm3\pi/4$, the interference fringes of the x and y components have no difference for any integer m. The interference fringe of the total field is independent of φ_0, but the interference fringes between the odd m and the even m have a relative shift of $\Lambda/2$.

More interestingly, the interference behavior is in the region Ω_2. The intensity is modulated by both x and y, unlike in the region Ω_1. For the x or y component, the field intensity (bright or dark) at the origin of $(x, y) = (0, 0)$ is determined by φ_0 sorely independent of m. For instance, if $\varphi_0 = \pm\pi/2$ ($\varphi_0 = 0$ or π) the x and y intensities at the origin have the minimum and the maximum (the maximum and the minimum), respectively. In contrast, if $\varphi_0 = \pm\pi/4$ or $\pm3\pi/4$, the x and y intensities at the origin are both the maximum. Based on this, we can determine the initial phase φ_0 of the input vector field. The total intensity is always the maximum at the origin, that is to say, the origin is always bright for any m and φ_0. When $\cos\phi = 1$ ($2\pi bx/\lambda d = 2p\pi$, where p is an integer), the total intensity exhibits the maximum for $2m\varphi_B = 2p\pi$ while the minimum for $2m\varphi_B = (2p + 1)\pi$. If $\cos\phi = -1$ [$2\pi bx/\lambda d = (2p \pm 1)\pi$], the total intensity is the minimum for $2m\varphi_B = 2p\pi$ while the maximum for $2m\varphi_B = (2p + 1)\pi$. In the region Ω_2, therefore, the interference pattern exhibits a chessboard structure, in particular, the total number of the bright and dark regions is equal to $2m - 1$ in the $x = p\Lambda$ or $x = (p \pm 1/2)\Lambda$ lines. Based on this, we can determine the topological charge m of the input vector field.

The above results reveal that the interference pattern of the total intensity is allowed to determine the topological charge of the input vector field, while the interference patterns of the x and y components can be used to determine the initial phase of the input vector field. This method should be also effective for characterizing the topological properties of the complex vector fields, for example, the vector fields with hybrid SOPs[34] and the vector vortex fields with the higher-order SOPs.[64]

3.4. *Optical orbital angular momentum (OAM)*[25]

A light field can carry the optical SAM or/and OAM. As an intrinsic part of the nature of a light field, SAM is associated with circular polarization and has two possible quantized values of $\pm\hbar$.[51, 55] Since Allen *et al.* pioneered the original concept of optical OAM,[3] as a fundamentally new optical degree of freedom,[65] OAM has attracted extensive attention and academic interest due to its practical and potential applications in various realms.[4, 8, 10, 11, 66-75] The concept of the optical OAM has now been extended to other natural waves such as a radio wave,[76, 77] sonic wave,[78, 79] x ray,[80] electron beam,[81] and a matter wave,[82, 83] so that the optical OAM is undoubtedly an extensively interesting issue.

A scalar vortex field with a helical phase of $\exp(-jm\varphi)$ could carry an optical OAM of $m\hbar$.[3, 5, 51, 55, 57, 65, 84-88] Evidently, for scalar fields with homogeneous SOP distribution, the space-variant phase is a prerequisite for possibility of producing optical OAM caused by the phase. The question is whether the polarization as a fundamental nature of light can also be used to produce optical OAM. It is imaginable that the light fields with the space-variant SOP distribution have the possibility of carrying OAM. Consequently, vector fields with space-variant SOP distribution[1, 2, 33, 34] offer an opportunity of producing optical OAM associated with the polarization nature.

We predict in theory and validate in experiment a new class of optical OAM associated with the curl of polarization independent of phase. Differently from the well-known OAM associated with the phase gradient independent of polarization, the results reveal that this novel OAM can be carried by a radial-variant vector field with hybrid SOPs.

A light field at an angular frequency ω has a vector potential **A**, as $\mathbf{A}(x, y) = A(x, y)[\alpha(x, y)\mathbf{e}_x + \beta(x, y)\mathbf{e}_y]\exp(jkz - j\omega t)$ with $|\alpha|^2 + |\beta|^2 = 1$, where the complex amplitude A can be described by real-valued module

u and phase ψ. α and β stand for the SOP distribution of a light field. Under the paraxial limit and the Lorenz gauge, the cycle-average momentum flux $\mathbf{P}$ is written as $\mathbf{P} \propto \langle \mathbf{E} \times \mathbf{H} \rangle$ in terms of the electric field $\mathbf{E} = j\omega\mathbf{A} + j(\omega/k^2)\nabla(\nabla \cdot \mathbf{A})$ and the magnetic field $\mathbf{H} = (\nabla \times \mathbf{A})/\mu_0$. The transversal component of $\mathbf{P}$ is divided into

$$\mathbf{P}_\perp^{(1)} \propto 2u^2\nabla\psi,$$
$$\mathbf{P}_\perp^{(2)} \propto ju^2(\alpha\nabla\alpha^* - \alpha^*\nabla\alpha + \beta\nabla\beta^* - \beta^*\nabla\beta), \tag{22}$$
$$\mathbf{P}_\perp^{(3)} \propto j\nabla \times [u^2(\alpha\beta^* - \alpha^*\beta)\mathbf{e}_z].$$

The cross product of $\mathbf{P}$ with $\mathbf{r}$ (radius vector) gives the angular momentum flux $\mathbf{J} \propto \mathbf{r} \times \mathbf{P}$. Accordingly, the z component of $\mathbf{J}$, J_z, is composed of three parts

$$J_z^{(1)} \propto 2u^2\partial\psi/\partial\varphi,$$
$$J_z^{(2)} \propto -ju^2(\alpha^*\partial\alpha/\partial\varphi + \beta^*\partial\beta/\partial\varphi - \text{c.c.}), \tag{23}$$
$$J_z^{(3)} \propto jr\partial[u^2(\alpha\beta^* - \alpha^*\beta)]/\partial r,$$

where $J_z^{(1)}$ is the well-known OAM associated with the azimuthal phase gradient, $J_z^{(2)}$ and $J_z^{(3)}$ are associated with the SOP distribution. If α and β are real valued, implying that SOP at any location across the field section is local linearly polarized, both $J_z^{(2)}$ and $J_z^{(3)}$ are zero. If either α or β is at least a complex-valued function, $J_z^{(2)}$ and $J_z^{(3)}$ are possibly nonzero values.

$J_z^{(2)}$ arising from the azimuthal variation of SOPs is the simple superposition of contributions from two orthogonal field components. We are interested in $J_z^{(3)}$, a new class of optical OAM. We define a parameter σ, describing SOP, as $\sigma = j(\alpha\beta^* - \alpha^*\beta)$. For instance, $\sigma = +1$ or -1 represents the right- or left-handed circular polarization, whereas $\sigma = 0$ is the linear polarization. Surprisingly, $J_z^{(3)}$ originates from the *curl* of the vector $\sigma u^2\mathbf{e}_z$. The optical OAM can be indeed associated with space-variant SOPs independent of phase. To achieve nonzero-valued $J_z^{(3)}$, two prerequisites should be satisfied: (i) $\sigma \neq 0$, i.e., local SOPs at locations across the field section should not be all linearly polarized, and (ii) σ or SOPs should be radial-variant rather than azimuthal-variant. A nonzero-valued $J_z^{(3)}$ requires the light field to be a vector field with the radial-variant hybrid SOPs (including linear, elliptical, and circular

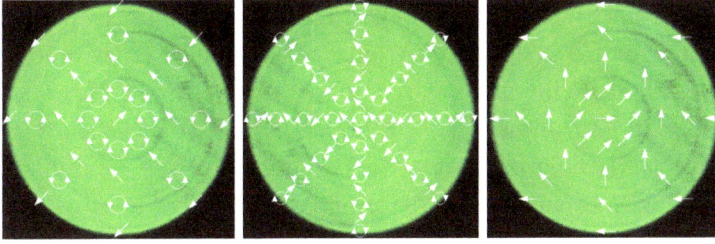

Fig. 27. Intensity patterns of two generated radial-variant vector fields with hybrid SOPs and schematics of SOPs. The first and second columns correspond to ($n = 0.5$, $\varphi_0 = 0$) and ($n = 1.0$, $\varphi_0 = \pi/4$), respectively. For comparison, a local linearly polarized vector field with ($n = 0.5$, $\varphi_0 = 0$) is also depicted in the third column.

polarizations). The crucial issue is how to generate the desired vector field satisfying the above requirements.

We should emphasize that as a common requirement for the light field carrying optical OAM, a certain physical quantity (phase or SOP) of light field must be space-variant. The helically phased scalar fields carry the known OAM caused by the azimuthal phase gradient independent of polarization (see Eq. (23)). In contrast, the radial-variant vector fields with hybrid SOPs, which have no polarization singularity, carry the optical OAM associated with the curl of polarization independent of phase (see Eq. (23)).

Based on our method mentioned in Section 2,[3, 35] to generate the vector field with the radial-variant hybrid SOPs, the additional phase δ should be a function of ρ only as $\delta = 2n\pi\rho/\rho_0 + \varphi_0$. The two 1/2 wave plates should be used to produce a pair of orthogonal linearly polarized fields, as shown in the generating unit of vector field surrounded by the dashed-line box in Fig. 28 below.

We now generate the radial-variant vector fields with the hybrid SOPs. As shown in the first and second columns of Fig. 27, the two generated radial-variant vector fields exhibit the uniform distribution in intensity, although SOPs have the hybrid distribution. For instance, in the first column, as ρ increases from $\rho = 0$ to ρ_0, the SOPs change from $\pi/4$ linear polarization at $\rho = 0$ to right-handed elliptical polarization within a range of $\rho \in (0, \rho_0/4)$, right-handed circular polarization at $\rho = \rho_0/4$, right-handed elliptical polarization within a range of $\rho \in (\rho_0/4, \rho_0/2)$,

$3\pi/4$ linear polarization at $\rho = \rho_0/2$, left-handed elliptical polarization within a range of $\rho \in (\rho_0/2, 3\rho_0/4)$, left-handed circular polarization at $\rho = 3\rho_0/4$, left-handed elliptical polarization within a range of $\rho \in (3\rho_0/4, \rho_0)$, and finally, to $-3\pi/4$ linear polarization at $\rho = \rho_0$. To make a comparison, the local linearly polarized vector field is also shown in the third column of Fig. 27. The generated radial-variant vector fields with hybrid SOPs indeed satisfy the two prerequisites previously mentioned for generating OAM associated with the curl of polarization.

To confirm the feasibility of OAM associated with the curl of polarization, the focused vector field as an optical tweezer is a useful tool. Figure 28 shows the trapping experimental scheme, wherein the laser source at 532 nm has a power of 20 mW. All the generated vector fields have the same radius of $\rho_0 = 2.5$ mm, a 60× objective (with NA = 0.7) is used to focus the vector field, and the neutral colloidal microspheres with a diameter of 3.2 μm are dispersed in a layer of sodium dodecyl sulfate solution between a glass coverslip and a microscope slide.

We first implement simulations for the focusing property of the radial-variant vector fields. The parameters used in simulations are the same as those used in experiment. The simulation results indicate that the radial-variant vector fields (not only local linear polarization but also hybrid SOPs) are tightly focused into a ring focus using a high NA objective. As an example, the simulated intensity pattern of the focus ring where $n = 10$ is shown in the inset (a) of Fig. 28. The inset (b) of Fig. 28 shows the simulated radial dependences of intensity (solid line) and of OAM $J_z^{(3)}$ associated with the curl of polarization (dashed line) in the focal plane for the radial-variant vector field with hybrid SOPs where $n = 10$. The maximum OAM $J_z^{(3)}$ and the strongest intensity locate at the same radial position, which is of such great importance that trapped particles in the ring focus can acquire the maximum torque when rotating around the ring focus.

The trapping experiments indicate that the ring traps generated by radial-variant vector fields (not only with the local linear polarization but also with hybrid SOPs) can trap an arbitrary number of particles. The reason originates from the fact that the ring focus has a continuously changeable radius because n can be an arbitrary real number, which is

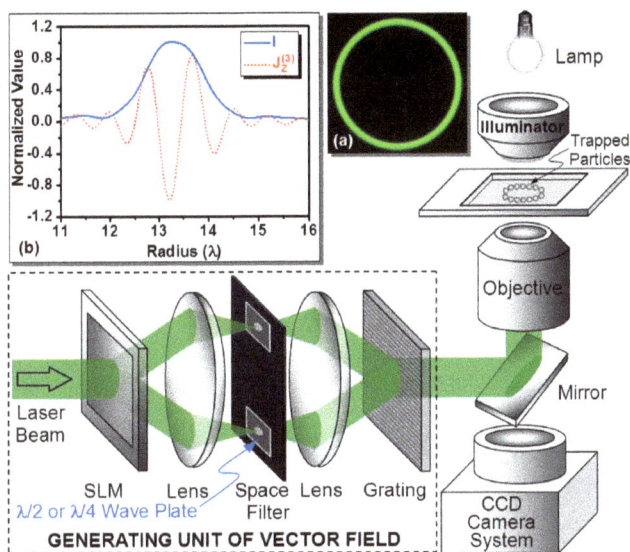

Fig. 28. Experimental configuration for conforming OAM associated with the curl of polarization. The dashed-line box shows the generating unit of radial-variant vector fields. Inset (a) shows an example of simulated ring focus. Inset (b) shows the properties of the ring focus for a radial-variant vector field with hybrid SOPs ($n = 10$), where the solid and dashed lines are the radial dependences of intensity and OAM, respectively.

quite different from the discrete radius of the ring focus generated by the azimuthal-variant vector field. For the particles trapped in the ring optical tweezer produced by radial-variant vector fields with local linear polarization, no motion is observed around the ring focus, implying that radial-variant vector field with local linear polarization carries no optical OAM.

We are very interested in the trapping property of the ring traps generated by the radial-variant vector field with hybrid SOPs. As the time-lapse photographs shown in the upper row of Fig. 29, the trapped particles when $n = 10$ move clockwise around the ring focus, with an orbital period of ~8.4 s. When n is switched from positive ($n = 10$) to negative ($n = -10$), the motion direction of the trapped particles is synchronously reversed, as shown in the bottom row of Fig. 29. The observed results imply that the ring traps have the capability to exert torque to the trapped isotropic particles. Consequently, the radial-variant

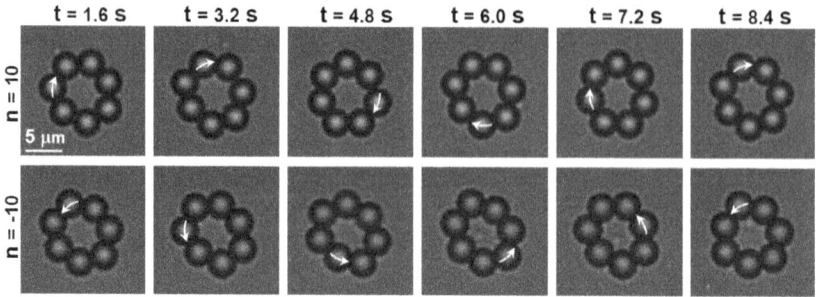

Fig. 29. Snapshots of the motion of trapped particles around the ring focus generated by radial-variant vector fields with hybrid SOPs, caused by polarization-curl-induced OAM.

vector fields with hybrid SOPs carry the optical OAM associated with the curl of polarization independent of phase.

For the interaction of light with matter, the polarization can influence anisotropic materials, but not isotropic materials, which is true for scalar fields. As we predicted, the induction of the vector fields breaks this limit to make the polarization nature of light field influence optically isotropic materials. Since SAM and OAM decouple in the paraxial limit[89] and SAM results in the rotation of an anisotropic particle around its own,[88] our observed rotation of the trapped isotropic particles should be dominated by OAM.

Our results create a link between two important issues on the optical OAM and the vector field. Our idea may spur further independent insight into the generation of natural waves carrying OAM and the expansion of functionality of optical systems, thereby facilitating the development of additional surprising applications.

4. Summary

This chapter introduced the vector optical fields and their novel effects. As a special family of optical fields, the vector optical fields are a kind of optical fields with the spatially inhomogeneous distribution of states of polarization across the field section. We gave a detailed description regarding to a passive method for generating the vector fields, based on the Poincaré sphere and the wavefront reconstruction. This method could in principle generate the vector field with arbitrary SOP distribution.

We also discussed novel effects of vector fields. By engineering the spatial SOP structure, the vector field could constructs the controllable 3D optical cages. The spatial SOP structure results in novel interference behaviors in the Young's two-slit configuration. The axial-symmetry breaking induces three kinds of azimuthal separations: energy, optical SAM and OAM. The radial-variant hybrid SOP distribution (or radial-variant SAM or curl of polarization) produces a new class of optical OAM independent of phase. The vector fields were discovered to have many interesting and unique properties. These properties result in a variety of effects and applications, such as adverse atmospheric effect, remote sensing and optical free-space communications, quantum optics and information, nonlinear optics, atom optics, optical communication, optical tweezers and micromechanics or microfluidics, biosciences, and astronomy. The concept of the vector optical field can be extended to other natural waves such as radio wave, sonic wave, x-ray, electron beam, and matter wave. The vector field is certainly the extensively interesting and greatly important topics. Nevertheless, the simple and efficient methods for generating the vector fields become more and more urgent, due to the requirements of various practical applications. In particular, the vector fields have the great scientific significance to deeply understand "light" and to exploit the potential of laser.

Acknowledgments

This work was supported by the 973 Program of China under Grant No. 2012CB921900 and the National Natural Science Foundation of China under Grants 10934003. Author would like to thank all collaborators who contributed to the related works.

References

1. Q. Zhan, *Adv. Opt. Photon.* **1**, 1 (2009).
2. X. L. Wang, J. Ding, W. J. Ni, C. S. Guo, and H. T. Wang, *Opt. Lett.* **32**, 3549 (2007).
3. L. Allen, M. W. Beijersbergen, R. J. Spreeuw, and J. P. Woerdman, *Phys. Rev. A* **45**, 8185 (1992).
4. D. G. Grier, *Nature* **424**, 810 (2003).

5. A. T. O'Neil, I. MacVicar, L. Allen, and M. J. Padgett, *Phys. Rev. Lett.* **88**, 053601 (2002).

6. V. Garcés-Chávez, D. McGloin, M. J. Padgett, W. Dultz, H. Schmitzer, and K. Dholakia, *Phys. Rev. Lett.* **91**, 093602 (2003).

7. R. A. Beth, *Phys. Rev.* **50**, 115 (1936).

8. G. Berkhout and M. Beijersbergen, *Phys. Rev. Lett.* **101**, 100801 (2008).

9. A. Jesacher, S. Fürhapter, S. Bernet, and M. Ritsch-Marte, *Phys. Rev. Lett.* **94**, 233902 (2005).

10. A. Mair, A. Vaziri, G. Weihs, and A. Zeilinger, *Nature* **412**, 313 (2001).

11. J. Leach, B. Jack, J. Romero, A. K. Jha, A. M. Yao, S. Franke-Arnold, D. G. Ireland, R. W. Boyd, S. M. Barnett, and M. J. Padgett, *Science* **329**, 662 (2010).

12. K. S. Youngworth and T. G. Brown, *Opt. Express* **7**, 77 (2000).

13. R. Dorn, S. Quabis, and G. Leuchs, *Phys. Rev. Lett.* **91**, 233901 (2003).

14. X. Hao, C. Kuang, T. Wang, and X. Liu, *Opt. Lett.* **35**, 3928 (2010).

15. Q. Zhan and J. R. Leger, *Opt. Express* **10**, 324 (2002).

16. H. Wang, L. Shi, B. Lukyanchuk, C. Sheppard, and C. T. Chong, *Nature Photon.* **2**, 501 (2008).

17. Y. Kozawa and S. Sato, *Opt. Lett.* **31**, 820 (2006).

18. X. L. Wang, J. Ding, J. Q. Qin, J. Chen, Y. X. Fan, and H. T. Wang, *Opt. Commun.* **282**, 3421 (2009).

19. Y. Q. Zhao, Q. Zhan, Y. L. Zhang, and Y.-P. Li, *Opt. Lett.* **30**, 848 (2005).

20. C. Varin and M. Piche, *Appl. Phys. B* **74**, S83 (2002).

21. L. Novotny, M. R. Beversluis, K. S. Youngworth, and T. G. Brown, *Phys. Rev. Lett.* **86**, 5251 (2001).

22. A. Ciattoni, B. Crosignani, P. Di Porto, and A. Yariv, *Phys. Rev. Lett.* **94**, 073902 (2005).

23. A. Bouhelier, M. Beversluis, A. Hartschuh, and L. Novotny, *Phys. Rev. Lett.* **90**, 013903 (2003).

24. Q. Zhan, *Opt. Express* **12**, 3377 (2004).

25. X. L. Wang, J. Chen, Y. N. Li, J. P. Ding, C. S. Guo, and H. T. Wang, *Phys. Rev. Lett.* **105**, 253602 (2010).

26. Y. Kozawa and S. Sato, *Opt. Lett.* **30**, 3063 (2005).

27. J. L. Li, K. Ueda, M. Musha, A. Shirakawa, and Z. M. Zhang, *Opt. Lett.* **32**, 1360 (2007).

28. K. C. Toussaint, S. Park, J. E. Jureller, and N. F. Scherer, *Opt. Lett.* **30**, 2846 (2005).

29. E. G. Churin, J. Hoßfeld, and T. Tschudi, *Opt. Commun.* **99**, 13 (1993).

30. Z. Bomzon, G. Biener, V. Kleiner, and E. Hasman, *Opt. Lett.* **27**, 285 (2002).

31. K. J. Moh, X.-C. Yuan, J. Bu, D. K. Y. Low, and R. E. Burge, *Appl. Phys. Lett.* **89**, 251114 (2006).

32. M. A. A. Neil, F. Massoumian, R. Juškaitis, and T. Wilson, *Opt. Lett.* **27**, 1929 (2002).

33. C. Maurer, A. Jesacher, S. Fürhapter, S. Bernet, and M. Ritsch-Marte, *New J. Phys.* **9**, 78 (2007).
34. X. L. Wang, Y. N. Li, J. Chen, C. S. Guo, J. P. Ding, and H. T. Wang, *Opt. Express* **18**, 10786 (2010).
35. M. Born and E. Wolf, *Principles of Optics*, 7th ed. (Cambridge U. Press, 1999).
36. A. Ashkin, J.M. Dziedzic, J.E. Bjorkholm, and S. Chu, *Opt. Lett.* **11**, 288 (1986).
37. K.T. Gahagan and G.A. Swartzlander, *Opt. Lett.* **21**, 827 (1996).
38. T. Kuga, Y. Torii, N. Shiokawa, T. Hirano, Y. Shimizu, and H. Sasada, *Phys. Rev. Lett.* **78**, 4713 (1997).
39. R. Ozeri, L. Khaykovich, N. Friedman, and N. Davidson, *J. Opt. Soc. Am. B* **17**, 1113 (2000).
40. S. W. Hell and J. Wichmann, *Opt. Lett.* **19**, 780 (1994).
41. P. Török and P. R. T. Munro, *Opt. Express* **12**, 3605 (2004).
42. J. Arlt and M. J. Padgett, *Opt. Lett.* **25**, 191 (2000).
43. N. Bokor and N. Davidson, *Opt. Lett.* **31**, 149 (2006).
44. W. Chen and Q. Zhan, *Opt. Commun.* **265**, 411 (2006).
45. N. Bokor and N. Davidson, *Opt. Commun.* **279**, 229 (2007).
46. J. Q. Qin, X. L. Wang, D. Jia, J. Chen, Y. X. Fan, J. P. Ding, and H. T. Wang, *Opt. Express* **17**, 8407 (2009).
47. S. K. Mohanty, R. S. Verma, and P. K. Gupta, *Appl. Phys. B* **87**, 211 (2007).
48. Y. Zhao, G. Milne, J. S. Edgar, G. D. M. Jeffries, D. McGloin, and D. T. Chiu, *Appl. Phys. Lett.* **92**, 161111 (2008).
49. X. L. Wang, K. Lou, J. Chen, B. Gu, Y. N. Li, and H. T. Wang, *Phys. Rev. A* **83**, 063813 (2011).
50. J. A. Davis and J. B. Bentley, *Opt. Lett.* **30**, 3024 (2005).
51. M. Padgett, J. Courtial, and L. Allen, *Phys. Today* **57**, 35 (2004).
52. S. Franke-Arnold, S. M. Barnett, E. Yao, J. Leach, J. Courtial, and M. Padgett, *New J. Phys.* **6**, 103 (2004).
53. E. Yao, S. Franke-Arnold, J. Courtial, S. Barnett, and M. Padgett, *Opt. Express* **14**, 9071 (2006).
54. B. Jack, M. J. Padgett, and S. Franke-Arnold, *New J. Phys.* **10**, 103013 (2008).
55. L. Allen, M. J. Padgett, and M. Babiker, *Prog. Opt.* **39**, 291 (1999).
56. C. S. Guo, X. Liu, J. L. He, and H. T. Wang, *Opt. Express* **12**, 4625 (2004).
57. J. E. Curtis and D. G. Grier, *Phys. Rev. Lett.* **90**, 133901 (2003).
58. Z. S. Sacks, D. Rozas, and G. A. Swartzlander Jr., *J. Opt. Soc. Am. B* **15**, 2226 (1998).
59. Y. N. Li, X. L. Wang, H. Zhao, L. J. Kong, K. Lou, B. Gu, C. H. Tu, and H. T. Wang, *Opt. Lett.* **37**, 1790 (2012).
60. D. P. Ghai, P. Senthilkumaran, and R. S. Sirohi, *Opt. Lasers Eng.* **47**, 123 (2009).
61. H. I. Sztul and R. R. Alfano, *Opt. Lett.* **31**, 999 (2006).
62. C. S. Guo, S. J. Yue, and G. X. Wei, *Appl. Phys. Lett.* **94**, 231104 (2009).

63. J. M. Hickmann, E. J. S. Fonseca, W. C. Soares, and S. Chávez-Cerda, *Phys. Rev. Lett.* **105**, 053904 (2010).
64. G. Milione, H. I. Sztul, D. A. Nolan, and R. R. Alfano, *Phys. Rev. Lett.* **107**, 053601 (2011).
65. G. Molina-Terriza, J. P. Torres, and L. Torner, *Nature Phys.* **3**, 305 (2007).
66. D. N. Neshev, T. J. Alexander, E. A. Ostrovskaya, Y. S. Kivshar, H. Martin, I. Makasyuk, and Z. G. Chen, *Phys. Rev. Lett.* **92**, 123903 (2004).
67. N. K. Efremidis, K. Hizanidis, B. A. Malomed, and P. D. Trapani, *Phys. Rev. Lett.* **98**, 113901 (2007).
68. S. Feng and P. Kumar, *Phys. Rev. Lett.* **101**, 163602 (2008).
69. S. Barreiro and J.W. R. Tabosa, *Phys. Rev. Lett.* **90**, 133001 (2003).
70. M. S. Bigelow, P. Zerom, and R.W. Boyd, *Phys. Rev. Lett.* **92**, 083902 (2004).
71. M. F. Andersen, C. Ryu, P. Cladé, V. Natarajan, A. Vaziri, K. Helmerson, and W. D. Phillips, *Phys. Rev. Lett.* **97**, 170406 (2006).
72. A. R. Altman, K. G. Köprülü, E. Corndorf, P. Kumar, and G. A. Barbosa, *Phys. Rev. Lett.* **94**, 123601 (2005).
73. E. Nagali, L. Sansoni, F. Sciarrino, F. D. Martini, L. Marrucci, B. Piccirillo, E. Karimi, and E. Santamato, *Nature Photon.* **3**, 720 (2009).
74. C. Paterson, *Phys. Rev. Lett.* **94**, 153901 (2005).
75. M. Harwit, *Astrophys. J.* **597**, 1266 (2003).
76. T. B. Leyser, L. Norin, M. McCarrick, T. R. Pedersen, and B. Gustavsson, *Phys. Rev. Lett.* **102**, 065004 (2009).
77. B. Thidé, *Phys. Rev. Lett.* **99**, 087701 (2007).
78. R. Marchiano and J.-L. Thomas, *Phys. Rev. Lett.* **101**, 064301 (2008).
79. P. Z. Dashti, F. Alhassen, and H. P. Lee, *Phys. Rev. Lett.* **96**, 043604 (2006).
80. S. Sasaki and I. McNulty, *Phys. Rev. Lett.* **100**, 124801 (2008).
81. M. Uchida and A. Tonomura, *Nature* **464**, 737 (2010).
82. S. Tung, V. Schweikhard, and E. A. Cornell, *Phys. Rev. Lett.* **97**, 240402 (2006).
83. K. C. Wright, L. S. Leslie, A. Hansen, and N. P. Bigelow, *Phys. Rev. Lett.* **102**, 030405 (2009).
84. Y. Roichman, B. Sun, Y. Roichman, J. Amato-Grill, and D. G. Grier, *Phys. Rev. Lett.* **100**, 013602 (2008).
85. A. Niv, Y. Gorodetski, V. Kleiner, and E. Hasman, *Opt. Lett.* **33**, 2910 (2008).
86. K. Miyakawa, H. Adachi, and Y. Inoue, *Appl. Phys. Lett.* **84**, 5440 (2004).
87. H. Adachi, S. Akahoshi, and K. Miyakawa, *Phys. Rev. A* **75**, 063409 (2007).
88. Y. Q. Zhao, J. S. Edgar, G. D. M. Jeffries, D. McGloin, and D. T. Chiu, *Phys. Rev. Lett.* **99**, 073901 (2007).
89. M. V. Berry, *J. Opt. A* **11**, 094001 (2009).

CHAPTER 3

CYLINDRICAL VECTOR BEAMS FOR SPECTROSCOPIC IMAGING OF SINGLE MOLECULES AND NANOPARTICLES

Regina Jäger, Anna M. Chizhik, Alexey I. Chizhik, Frank Wackenhut, Alfred J. Meixner*

Department of Physical and Theoretical Chemistry, University of Tübingen Auf der Morgenstelle 18, 72076 Tübingen, Germany
*E-mail: alfred.meixner@uni-tuebingen.de

Due to their unique polarization properties, cylindrical vector beams (CVBs) have gained an increasing interest for applications in optical microscopy. They have proven to be an efficient technique for probing the three-dimensional orientation and dimensionality of the excitation transition dipole moment of single quantum systems. Therefore they can additionally reveal interesting properties of the quantum systems such as dynamical effects and origins of the photoluminescence therein. In this chapter we present the principle and some examples of the experiments using CVBs for excitation.

1. Introduction

The research on single quantum systems, like atoms, molecules or quantum dots, and their interaction with light is not only of fundamental interest but also necessary for future applications using single photon sources or photons-on-demand sources. Since it is possible to detect single quantum emitters in solids,[1, 2] fluorescence microscopy of these structures has become a routine technique and is widely used, for example in molecular biology.[3] Especially for applications, the knowledge of all parameters of a single quantum system is desirable. Aside from properties like position or emission characteristics, the three-dimensional (3D) orientation of a single quantum absorber's excitation transition dipole moment (TDM) is of special interest as it influences, for example, the absorption and emission efficiency and the corresponding lifetime. The orientation of a single emitter can even be used as a probe to obtain information about the local environment or about the orientation of labeled biomolecules allowing precise distance measurements via Förster resonance energy transfer (FRET) efficiency.[4] In addition, molecular machines have been studied involving the orientation of chromophores.[5, 6]

To obtain the spatial orientation of the emitter's absorption/emission dipole moment, different techniques can be applied. One commonly used approach is the analysis of defocused single molecule images.[7, 8] This defocusing technique is based on the fact that depending on the focal plane, microscope objectives render the intensity distribution of single emitters differently, resulting in a different recorded emission pattern. In focus, the molecules orientated in-plane result in a spot-like pattern while molecules oriented along the optical axis result in a ring-shaped pattern. By slightly defocusing, the ring pattern remains unchanged while a pair of side-wings appears in spot-like patterns. The orientation of these side-lobes visualizes the in-plane angle while the asymmetry of the central feature allows determining the azimuthal angle with respect to the optical axis.[9] This technique is not only applicable to systems with a one-dimensional linear transition dipole moment, like molecules, but also to systems with a two-dimensional emission dipole moment, like CdSe/ZnS core-shell quantum dots.[10] Another technique to determine the 3D dipole moment orientation is far-field polarization microscopy. As the absorption and emission of a single dipole is highly polarized, the orientation of this dipole can be determined by corresponding polarization-dependent measurements, i. e., varying the excitation polarization[11, 12] or measuring the fluorescence intensity as a function of its polarization.[13] Further techniques are for example the use of a linearly polarized doughnut mode[14] or annular illumination with a linear Gaussian beam.[15]

As polarization properties play a key role in investigations of single quantum systems, cylindrical vector beams (CVBs), in particular radially and azimuthally polarized doughnut modes (RPDM and APDM, respectively), have attracted growing attention in recent years. Whereas the RPDM creates a dominant longitudinal electric field component in the focal region, the focused APDM consists of purely transversal electric field vectors.[16] Therefore, these modes have proven to be an efficient tool to probe the orientation and dimensionality of the excitation TDM of single quantum systems such as dye molecules,[17-21] silica nanoparticles,[22-25] silicon nanocrystals,[26] gold nanoparticles[27, 28] and CdSe/ZnS quantum dots.[29] Single molecules have been studied in non-confined media as well as in microresonators.[19, 30] Also dynamic processes as the excited state tautomerization in single molecules[31-34] or flipping of the TDM in single silica nanoparticles[25] can be studied applying CVBs for the excitation. Due to the strong longitudinal component of the RPDM in the center of the focal region, this mode has been proposed as an efficient optical trap for metallic nanoparticles[35-37] and to increase the transverse resolution.[38, 39] The strong longitudinal field is also of great interest in near- field microscopy[40, 41] or for electron acceleration.[42] Within the following sections the basics and some experiments will be described to point out the principle and advantages of using CVBs for single quantum system studies.

2. Theoretical background

First we will briefly describe the theoretical background of the mode formation. The fundamental transverse electromagnetic mode, TEM_{00} is the simplest type of mode, called a Gaussian beam. Solving the scalar Helmholtz equation with the boundary conditions of a laser cavity and applying the paraxial approximation, the solution of the eigenstates can be expressed in Cartesian coordinates as Hermite-Gaussian modes (HG_{mn}) or in cylindrical coordinates as Laguerre-Gaussian modes (LG_{pl}). The CVBs can be described by a linear combination of these modes. As a result, one can obtain linearly, azimuthally and radially polarized doughnut modes,[43] whereas, due to their peculiar polarizations, we concentrate on the APDM and RPDM. These CVBs can be obtained by combining the TEM_{10} and TEM_{01} modes in the following way:

$$APDM = -HG_{01}n_x + HG_{10}n_y \,, \tag{1}$$

$$RPDM = HG_{10}n_x + HG_{01}n_y \,. \tag{2}$$

The corresponding HG-modes are illustrated in Fig. 1 to demonstrate the mode formation. Here, n_x and n_y are unit vectors. A further approach describing the superposition of modes in the formation of a generalized CVB is described in detail in the papers of Zhan *et al.*[36, 44]

There are two ways to generate these modes: either inside a laser cavity or extra-cavity using a mode converter to transform the Gaussian laser beam. Intra-cavity formation of higher order laser modes can be achieved in a number of ways, e.g., axial intra-cavity birefringence[45, 46] or dichroism,[47] diffractive phase plates or polarization selective end mirrors,[48] and interferometric methods.[49] These methods require a more complex resonator design and/or special fabrication techniques. Thus, for most research areas, the passive creation of higher order laser modes from a Gaussian beam outside the laser resonator delivers an adequate beam quality and is easier to achieve. Different kinds of mode converters can be used to achieve an APDM or RPDM. A Liquid crystal mode converter[50] is convenient when different wavelengths are required as excitation source. By applying different voltages to the individual cells, the liquid crystals therein can be oriented in the desired direction. Thus, the polarization of the incident beam is varied spatially. As long as only one excitation wavelength is needed for the investigations, an easier way to generate CVBs is a home-made mode converter consisting of specially arranged $\lambda/2$ waveplates. Assembling four[38, 51] or eight[52] segments of $\lambda/2$ plates at discrete angles, the cylindrical vector beams APDM and RPDM can be obtained converting a linearly polarized Gaussian beam. However, due to the structural discontinuity at the edges of the segments, different higher order modes are generated, of which the undesired modes have to be removed. This mode cleaning can be done either by a pinhole[51] or by a near-confocal Fabry-Perot interferometer.[38] The mode creation via a four-segment mode converter is sketched in Fig. 1.

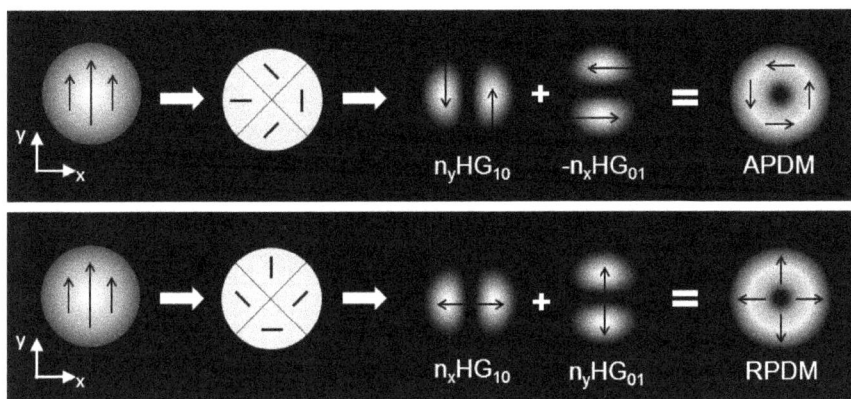

Fig. 1. APDM (top) and RPDM (bottom) generation by a mode converter consisting of four λ/2 plates transforms a linearly polarized Gaussian beam into the two orthogonal first-order Hermite-Gauss modes which, together, make up the ring-shaped higher order modes. Depending on the orientation of the mode converter in respect to the original beam polarization either an APDM or an RPDM is formed.

To calculate the field distribution of a focused Gaussian beam, Wolf and Richards established a fundamental geometrical method,[53, 54] which also takes polarization into account. Their calculations can therefore be adapted to CVBs as well.[16] Basically, an aplanatic lens forms a spherical wavefront with a radius corresponding to the focal length f. The definition of the parameters is illustrated in Fig. 2. The electric field $E_f(\theta, \phi)$ at a point Q on the reference sphere is defined by the spherical coordinates (f, θ, ϕ) with the origin of the coordinate system (x, y, z) in the focus. The electric field at a point P (ρ, φ, z) in the focal volume, defined by cylindrical coordinates is given by:[53-55]

$$E(\rho,\varphi,z) = -\frac{ikf}{2\pi} \int_0^{2\pi} \int_0^{\theta_{max}} E_f(\theta,\phi)e^{iksz} \sin\theta d\theta d\phi . \qquad (3)$$

Integrating point by point, we can calculate the electric field distribution in the focal volume, while (ρ, φ, z) are the cylindrical coordinates of point P in the focal volume, k represents the wavenumber, the solid angle θ_{max} is determined by the NA of the objective lens (NA = $n\sin\theta_{max}$; n is the refraction index of the surrounding medium), θ is the incident angle of a single ray, s represents a unit vector in the

direction of propagation, and the angle ϕ defines the integral space
$(0 \leq \phi < 2\pi)$.

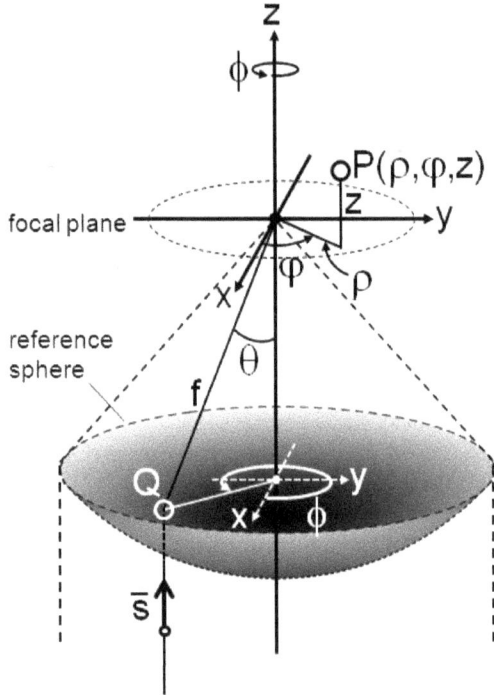

Fig. 2. Illustration of the coordinates for the basic calculations of the electric field
distribution according to Eq. (3). The electric field $E_f(\theta, \phi)$ at a point Q on the reference
sphere is defined by the radius of the reference sphere f, the incident angle θ and the
relative angle ϕ, which defines the integral space. Every point P in the focal volume is
defined in cylindrical coordinates by the angle φ and lengths ρ and z.

Figure 3 shows how the electric field vectors of three different points
is composed by the electric field components of the two marked rays
starting at the reference sphere and interfering in the focal plane. The
blue arrows represent the field vectors in the center of the focal spot,
where the longitudinal parts can interfere constructively as they all travel
the same optical path length while the transversal parts annihilate each
other. The red arrows show the field vectors for a spot shifted laterally
from the center. There, the beams must travel different path lengths and
the transversal components can interfere cumulatively, achieving

constructive interference for a phase shift of π. The electric field distributions in the focus of the APDM and RPDM reveal the difference between these modes. While the APDM shows only in-plane (x/y) polarization components of the electric field, the RPDM has a dominant field component E_l^2 in the longitudinal direction (z) (cf. Fig. 3). Note, that the ratio E_l^2/E_t^2 increases with increasing NA. This is of particular importance for all applications when a strong electric field component perpendicular to the sample surface is required e.g., to map TDMs of molecules perpendicular to the sample surface or to excite a tip plasmon in tip-enhanced near-field optical microscopy.

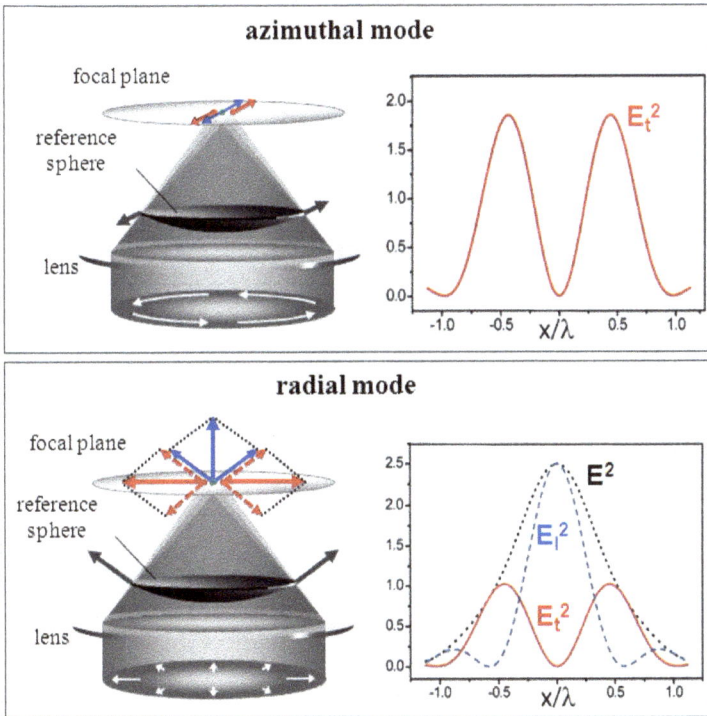

Fig. 3. Illustration of the formation of the electric field vectors in the focus of an objective lens illuminated by an APDM (top) and an RPDM (bottom), with the corresponding cross sections of the electric field distributions to the right (E_t: transversal and E_l: longitudinal electric field).

Results of calculations of the electric field intensity distributions in the focal volume of an immersion oil objective lens (NA = 1.25, λ = 488 nm) are presented in Fig. 4. The profiles in the x-z plane of the CVBs and the linear Gaussian mode show the differences when focusing on a surface changing the refractive index. In the cross sections it can be seen that the refractive index change has almost no influence on the transversal component, while the magnitude of the longitudinal component increases in the central part. In the profile of an RPDM the longitudinal and transversal parts can be clearly separated. This results in a tighter focus for an RPDM than a linear Gaussian mode.

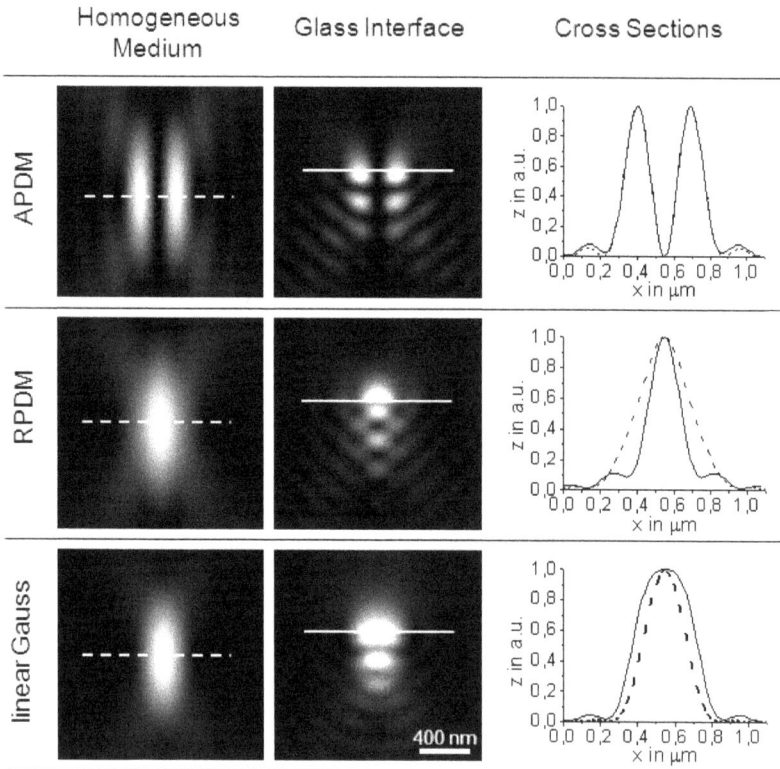

Fig. 4. Calculated xz-profiles of the electric field distributions of an APDM, RPDM and linear Gaussian mode in a homogeneous medium opposed to a glass interface. On the right: the cross sections through the focal plane in a homogeneous medium (dashed line) and on a glass interface (solid line) are presented normalized towards each other.

The focusing element plays an important role for investigations of single quantum systems (molecules, quantum dots etc.). Considering that the highest NA of an objective lens that can be obtained in air is 0.95, there could still be improvement. Geometrically the highest NA conceivable in air is 1 for an entire half-space. Replacing the focusing element by a parabolic mirror (PM), the maximum collection angle corresponding to an NA of 1 can be achieved inherently. PMs have been used as collectors for decades, but avoided for imaging techniques due to poor off-axis imaging properties.[55] On the other hand, while objective lenses suffer of chromatic aberrations, parabolic mirrors can be produced with highly reflective surfaces for a wide wavelength range. It has been shown by Dorn *et al.*[38] that the RPDM can be focused to a significantly smaller spot size $(0.16(1)\lambda^2)$ than a linear polarized beam $(0.26\lambda^2)$ by an objective lens. In comparison, using a PM as focusing element, the focus becomes even 1.2 times smaller $(0.134\lambda^2)$ for an RPDM.[56] As the longitudinal component of the electric field dominates the total field distribution in the focal plane for high NA, the PM is suitable for SNOM on non-transparent samples. In addition, for investigations at cryogenic temperatures PMs have an advantage over conventional objective lenses as immersion oil cannot be used to increase the NA. Further details on investigations of the focusing properties of parabolic mirrors compared to objective lenses can be found elsewhere.[55, 56]

Now we will discuss the interaction of a single quantum system with the oscillating electric field in the focus of such CVBs. For weak excitation of a molecule we can calculate the fluorescence rate according to the linear regime of Fermi's Golden Rule:

$$\Gamma(\vec{r},\nu) = \frac{2\pi}{\hbar}\left|\vec{\mu}\vec{E}(\vec{r})\right|^2 \rho(\vec{r},\nu). \tag{4}$$

With the absorption TDM $\vec{\mu}$, the electric field vector $\vec{E}(\vec{r})$ for the excitation of a molecule at the position $\vec{r}$ and the local density of states $\rho(\vec{r},\nu)$. The following calculations are based on the assumption that the absorption and emission TDMs are equal and the quantum efficiency is unity. To compute the excitation patterns, the fluorescence rate is multiplied by the collection efficiency function (CEF) of the focusing element. The CEF describes the ratio between the total intensity emitted by the dipole and the collected intensity depending on the position of the dipole.[55]

For the detection of the patterns either an avalanche photodiode (APD) or a CCD camera can be used (see also Sectio 3. Instrumentation). Depending on the arrangement of filters in the confocal setup different information can be obtained. In general the resulting patterns can be considered a convolution of the focal field distribution and the excitation TDM or the polarizability tensor in the case of metallic nanostructures. For the sake of simplicity, the polarizability tensor can be assumed to interact like a TDM. It is important to know the real polarizability in the metallic nanoparticle of interest to validate the results.

As the focal spot of the laser beam is much larger than the quantum system, the latter is excited by different parts of the doughnut laser beam during raster scanning. Thus, it can only be excited if the polarization in the excitation beam coincides with the absorber's TDM (or polarizability tensor) orientation. Therefore, one can efficiently map the excitation pattern of the quantum system.

To study the interaction of a single quantum absorber with CVBs, we would first like to consider the excitation patterns observed upon scanning quantum absorbers possessing different orientations and dimensionalities of their excitation TDMs through the focal area of an APDM and RPDM. Figure 3 shows the field intensity distributions as a function of the radial coordinate within the focal region of an APDM and RPDM, respectively. The distributions are calculated for an objective lens with a numerical aperture of 1.25. While an APDM has only in-plane electric field components with a doughnut shaped intensity distribution, an RPDM also comprises longitudinal electric field components, which form a distinct maximum in the center of the focal area of a high NA objective lens (Fig. 3).[38]

Mapping the excitation of a single quantum absorber, possessing a fixed linear TDM, by an APDM results in a so called double lobe pattern, consisting of two nearby bright spots of elliptical shape, provided the TDM is oriented horizontally or tilted (Fig. 5(b)). Due to the absence of the longitudinal component of the field, absorbers with vertically oriented TDM cannot be excited and are therefore not observable. Figures 5(b) shows three calculated patterns of an individual absorber possessing a fixed linear TDM excited with an RPDM. Absorbers having different TDM orientations cause different excitation patterns, allowing the precise determination of a TDM's orientation by comparing experimental and simulated patterns. Such images are typically observed upon excitation of single dye molecules[17, 18] or SiO_2 nanoparticles,[22, 23] possessing a stable linear TDM.

In more rare cases, an individual molecule can have a degenerate TDM distributed in a plane (for instance, due to high symmetry of the molecular structure)[57] or can exhibit fast flipping between two TDMs forming an angle close to 90° (in case of tautomerization).[31-34] As a result of the excitation of such a molecule with CVBs, the pattern exhibits a ring shape when the TDM-plane is oriented horizontally. Vertically oriented molecules are depicted as double lobe patterns upon excitation with an APDM or an elongated spot in the case of an RPDM. Figures 5(d) and (e) show characteristic examples of simulated patterns, modeled with two orthogonal linear dipoles.

Typical examples of quantum systems possessing degenerate TDMs distributed isotropically in space are fluorescence spheres. These nanometer-sized polymeric spheres contain a large number of fluorescent molecules (~200), with random orientation. Therefore the sphere possesses emission and excitation isotropy.[30, 58, 59] In contrast to the systems with fixed linear or degenerate TDM-planes, such an isotropic system exhibits the same shape of excitation pattern, independent of its orientation. A dye doped nanosphere excited with an APDM renders a ring-shaped pattern due to interaction of horizontal components of the molecules' TDM with the in-plane component of the field (Fig. 5(f)). The excitation of a sphere with an RPDM gives rise to an intense spot in the center of the image caused by the strong longitudinal component and a weaker ring due to the interaction with the in-plane component of the field (Fig. 5(g)). Note that this kind of pattern is also obtained in the case of single CdSe/ZnS quantum dots (see Section 6.3 for the details) as well as for metallic nanospheres when in resonance with the excitation wavelength, as the latter exhibit an isotropic polarizability during the acquisition time.

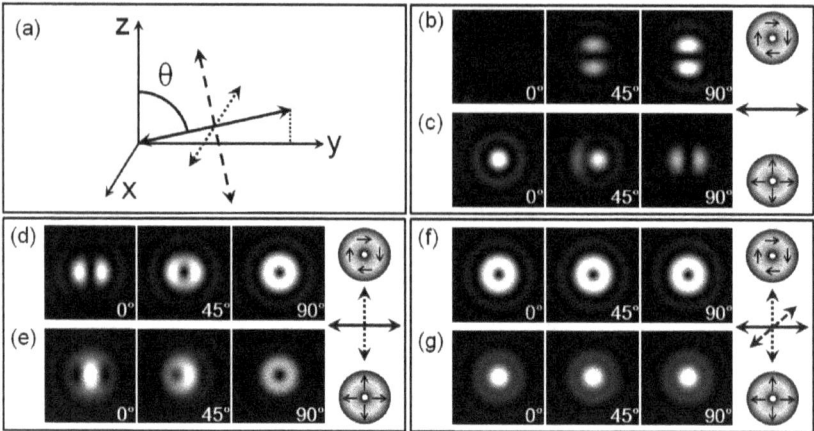

Fig. 5. Panel (a) exhibits the coordinate system which was used for theoretical simulations (b)-(g). Images (b) and (c) show the simulated excitation patterns of a single quantum system with a stable linear excitation TDM, excited with an APDM and RPDM, respectively. The arrow represents the orientation of the transition dipole. Images (d) and (e) demonstrate calculated excitation patterns of an absorber possessing a degenerate excitation TDM, oriented in a plane, which can be modeled as two perpendicular linear transition dipoles. Figures (f) and (g) show the simulated excitation patterns of an absorber, possessing excitation isotropy. Note, that the intensity of the calculated excitation patterns is normalized to one scale in each row.

We can conclude that the APDM and RPDM can be used to determine the orientation and dimensionality of the investigated single quantum systems. In contrast, when a linearly polarized Gaussian laser beam is used, it is not possible to determine these parameters directly. The field distribution of a linearly polarized Gaussian beam has an in-plane and a longitudinal component. The calculated electric field distributions focused by a microscope objective (NA of 1.25) are depicted in Fig. 6. The longitudinal part can be neglected as it is significantly weaker in intensity than the transversal component. In consequence, the total field distribution is similar to the transversal component alone, which is linearly polarized. Therefore, the fluorescence patterns of single quantum systems excited by a linearly polarized Gaussian beam have shapes like bright spots. To illustrate the difficulties using linearly polarized Gaussian beams, the excitation fluorescence patterns of single PI ((N-(2,6-diisopropylphenyl)-perylene-3,4-dicarboxymide) molecules are presented in Fig. 6. The intensity depends on the orientation of the TDM of the quantum system with respect to the polarization direction of the excitation laser beam, which can be simplified to consist of only the linearly polarized in-plane component. The smaller the angle between the polarization of the laser beam and the excitation TDM, the brighter the pattern is. It should be noted that the brightest spot in Fig. 6, which is also bigger in size than the other patterns, is assumed to be agglomeration of several molecules.

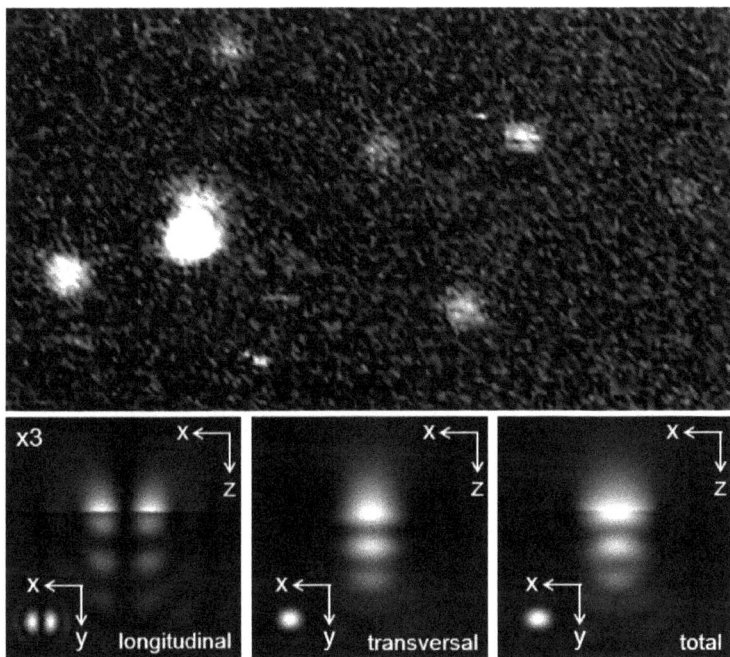

Fig. 6. Excitation fluorescence patterns of single PI molecules embedded in PMMA matrix excited with a linearly polarized Gaussian beam. Note that the brightest spot, which is also bigger in size than the other patterns, is assumed to be an agglomeration of several molecules. Bottom: calculated field intensity distributions of the longitudinal components, transversal components and the total intensity of a linearly x-polarized Gaussian beam focused by an objective lens onto a glass substrate. The total distribution in the focus is the sum of both field components. The insets show the corresponding xy-profiles on the surface. For comparison, the intensity of the longitudinal component is multiplied by a factor of 3.

Besides depending on the electric field distribution in the focus, the exact resulting patterns are also influenced by the used wavelength, the refractive index of the surrounding medium of the system (see also Section 5.1) and the refractive index of the used substrate. The effect of using different focusing elements should not be underestimated as well, as it renders the shape of the focus by guiding the vector beams, resulting in constructive or destructive interference. The electric field distribution in the focus has to be considered for each focusing element separately.

3. Instrumentation

To be able to detect the convolution patterns of single quantum systems with specially shaped excitation beams, the setup must be equipped with polarization maintaining components. For the results presented here, home-build inverted confocal microscopes have been used with high numerical aperture objective lenses (NA = 1.25 or 1.4) to detect the orientation of the TDM of single quantum absorbers. A scheme of the setup is depicted in Fig. 7. Different lasers can be coupled into the microscopes as excitation sources, e.g., an optically pumped semiconductor laser at 488 nm or a tunable argon-krypton laser. After the Gaussian beam has passed the mode-conversion part, the resulting doughnut laser mode is focused by the objective lens onto the sample. The fluorescence (or interference) pattern of the quantum systems is collected through the same objective lens and focused either onto an avalanche photodiode (APD) or a CCD camera for spectral analysis. Confocal images are acquired by raster scanning the sample through the focal spot of the microscope objective. The mode-conversion part consists of the mode-converter and a pinhole placed in the focal point between two lenses, thus cutting off other higher order modes. Both, a mode converter formed by four $\lambda/2$ plates[38, 51] oriented as shown in Fig. 1 or a Liquid Crystal mode converter[50] have been used. To investigate a sample with different wavelengths, it is more convenient to use a commercial liquid-crystal polarization converter, as otherwise a separate mode converter has to be constructed for every wavelength.

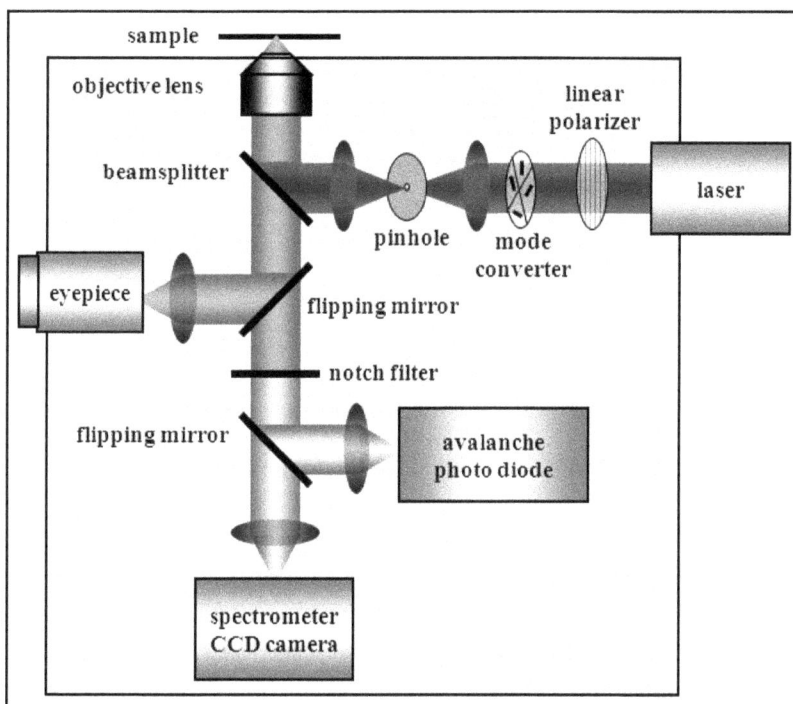

Fig. 7. Scheme of the experimental setup. The laser beam passes the mode conversion part and is guided into an inverted confocal microscope, where the sample is raster scanned through the focus and can be observed through an eyepiece. The optical response of the sample is detected by either an avalanche photo diode or a spectrometer with CCD camera.

The samples are prepared on standard glass cover slides (170 μm thickness), which must be cleaned carefully to avoid any kind of contamination fluorescence. To this end, the cover slides are stored at least 12 hours in chromosulfuric acid and subsequently cleaned with tri-distilled water and methanol uvasol®. To obtain single quantum absorber concentrations on the sample, solutions of 10^{-9} mol/l are added to non-fluorescent transparent polymer solutions with a concentration of $\leq 1\%$ and spin-coated with a rotational speed of 8000 rpm onto the cleaned cover slides. The polymer, e.g., polymethylmethacrylate (PMMA), polyvinyl alcohol (PVA) or polystyrene (PS) provides a matrix to suppress translation and rotation of the quantum absorber. If it is necessary to investigate systems without the influence of a surrounding polymer, the sample can be placed under nitrogen flow. Usually it is advisable to use a polymer film, otherwise movement of the particles or molecules cannot be neglected, especially for the determination of their orientation. According to atomic force microscopy measurements, the obtained polymer film has a thickness of approximately 50 nm.[33] In order to prove the samples exhibit single molecule characteristics, common techniques such as time traces and photon antibunching experiments[60] can be used.

4.   Fluorescence spheres to probe the quality of CVBs

The fastest way to examine the quality of the generated CVB is to record raster-scan images of fluorescence spheres. These nanometer sized polystyrene spheres contain a large number of fluorescent molecules (approximately 200), which are oriented randomly. Thus, all polarization directions of the excitation light can interact during the scan with an excitation TDM of a molecule. The scan image then directly shows the electric field intensity distribution of the focus. For the images shown in Fig. 8(b) and (c), fluorescent spheres purchased by Molecular Probes with a diameter of 20 nm were used. A wavelength of 488 nm was used as excitation source, therefore spheres loaded with Nile Red molecules (Fig. 8(a)) were chosen. The insets in the scan images show the corresponding electric field intensity distributions. In agreement with the predictions introduced in Section 2, the resulting patterns from the Nile Red spheres interacting with an APDM exhibit a ring shaped pattern (Fig. 8(b)), as the focus of this mode consists only of transversal components. The focus of an RPDM, on the other hand, has a strong longitudinal field in the center of the focus and a weak transversal field distribution in a circular pattern surrounding the central spot (Fig. 3). As a consequence, the scan of a Nile Red sphere shows mainly the interaction with the strong longitudinal electric field in the center (Fig. 8(c)).

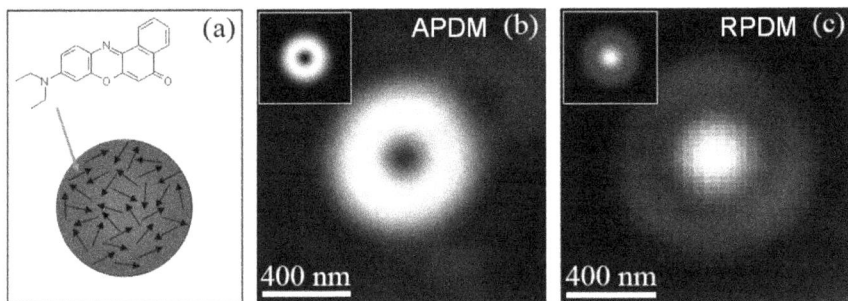

Fig. 8. (a) The structure of a fluorescent Nile Red sphere, the black arrows show the random orientations of the TDMs of the Nile Red molecules inside the sphere. (b) and (c) are fluorescence images of a single sphere excited by an APDM and RPDM, respectively, the insets show the corresponding calculated field intensity distributions.

5. Single molecules

The interaction of CVBs with the dipole moment of a single quantum system has been introduced in Section 2 theoretically. Here we compare experimental results with the theoretical predictions, determining the orientation of single molecules (see Fig. 9). The presented examples show polymer-embedded PI ((N-(2,6-diisopropylphenyl)-perylene-3,4-dicarboxymide) molecules, which were excited by an RPDM at a wavelength of 488 nm.

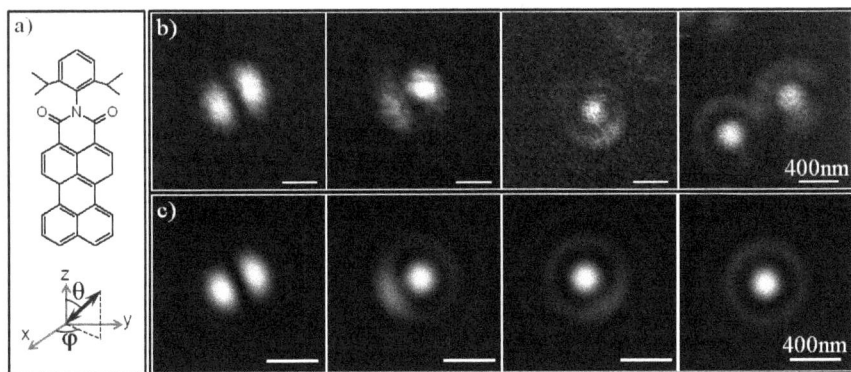

Fig. 9. (a) Structure of a PI molecule and the reference coordinate system. (b) Experimental and (c) the corresponding calculated excitation patterns of five PI molecules resulting from illumination with an RPDM. The corresponding θ/φ angles are: $85°/320°$, $60°/45°$, $40°/210°$, $15°/25°$, where $\varphi = 0°$ is in the horizontal direction.

It is clearly visible that the experimentally recorded images show patterns ranging from double lobe to spot-like shapes. The recorded patterns of randomly oriented single molecules excited by an APDM are more difficult to analyze, as for TDMs oriented out of the focal plane the intensity vanishes. However, different intensities do not automatically correspond to an out of plane orientation. Rather, one has to take into account that the patterns may originate from absorbers slightly out of focus or that the background intensity as well as the quantum efficiency of the molecules is not constant during the acquisition time of the image. Nevertheless, the use of an APDM to characterize the in-plane orientation is convenient for molecules which are oriented nearly parallel to the x-y plane, as the APDM has a strong in-plane component compared to the RPDM (see Fig. 3). However, from experience, when preparing single molecule samples as described in Section 3, the molecules often happen to be oriented perpendicular to the sample surface (depending on the exact parameters). Therefore, investigations using an APDM are more convenient in the case of molecules with low quantum efficiency.

However, so far no conclusion on the parameters of the emission TDM can be made. For this purpose, we combine the CVBs for excitation with the well-known method of polarization microscopy,[13, 17] in which a linear polarizer is placed in front of the detector. By recording a series of images of the emitters with different orientations of the polarizer, a direct relation between the absorption TDM orientation and the emission TDM orientation can be obtained. Corresponding to this method, Fig. 10 shows three images of single dye PI molecules fixed in a polymer matrix and excited with an APDM.

Without the linear polarizer (Fig. 10(a)) the intensities of the single molecule patterns correspond to the out-of-plane orientation of their excitation TDM. In particular, the weak intensity of molecule 2 gives evidence that its transition dipole moment is oriented nearly perpendicular to the sample surface, while molecule 1 is aligned almost in the sample plane.

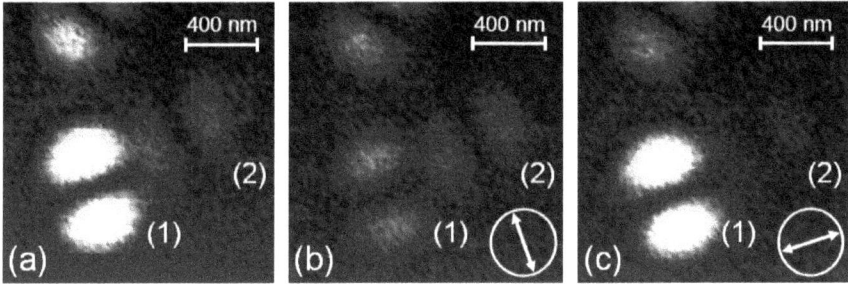

Fig. 10. Three single PI molecules excited with APDM. (a) without linear polarizer in front of the photodetector, (b) and (c) with linear polarizer in front of the photodetector oriented according to the arrow. The intensities of the images are normalized to one scale.

When the linear polarizer is placed in front of the APD according to the orientation of the arrows shown in Figs. 10(b) and (c), only the emission possessing the selected polarization direction will be detected. We see that the maximum emission from a single molecule is detected when the linear polarizer is aligned parallel to the double lobe pattern, i.e., to the projection of the excitation dipole on the sample surface. On the other hand, when the polarizer axis is nearly perpendicular to the dark axis of the pattern, the single molecule emission detected by the APD dramatically decreases (Fig. 10(b), molecule 1) or completely vanishes (Fig. 10(c), molecule 2). Thus, the combination of CVBs and classical polarization microscopy shows that the emission and excitation TDMs of this type of dye molecule are parallel to each other, as for the most molecules that have been investigated so far.

5.1. *Single molecules as nanoscale probes for interfaces*

In this section, we demonstrate that the modification of the optical environment around a single dye molecule with a fixed TDM changes the shape of its excitation pattern when the molecule is scanned through the focal region of an RPDM. Figures 11(a)-(g) and (h)-(n) show simulated excitation patterns of a single molecule located on a glass-air interface or in bulk glass, respectively. The patterns are calculated for different molecular orientations with respect to the sample surface ranging from 0° to 90°, according to the schemes shown in Figs. 11(o)-(u).

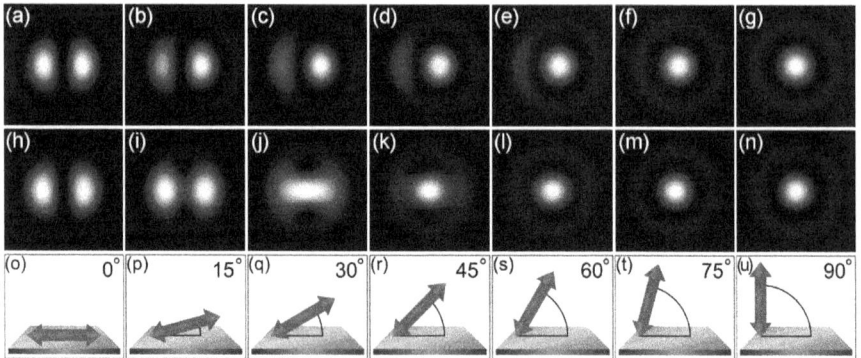

Fig. 11. Calculated excitation patterns of a single molecule scanned through the focal region of an RPDM incident on a glass-air interface (a)-(g) and in bulk glass (h)-(n). The patterns were calculated for different orientations of the molecule's transition dipole moment according to the scheme shown in (o)-(u). For better comparison, all the patterns are normalized to the same intensity scale.

The shape alone of the simulated patterns for the case of a vertically or horizontally oriented molecule does not allow determining whether the molecule is located on a glass-air interface or in bulk glass. However, if the TDM of the molecule is tilted with respect to the sample surface, the presence of the dielectric interface drastically changes the shape of the single molecule pattern. In this case the molecule can be excited with both in-plane and longitudinal components of the field, which results in a non-symmetric double lobe pattern if the molecule is located on a glass-air interface Figs. 11(d)-(f). Changing the molecule's orientation results in a redistribution of the intensity between two lobes and a modification of their size. When the molecule is placed in bulk glass, the pattern remains symmetric for any molecular orientation. Deflection of the molecule from a horizontal position results in the merging of the two lobes in a single elongated spot which becomes radially symmetrical when the molecule is oriented vertically with respect to the sample surface. Thus, our calculations show that the excitation of an individual molecule scanned through the focal region of an RPDM allows us to determine if the molecule is located on a glass-air interface or in bulk glass, provided its TDM is tilted with respect to the sample surface. It should be noted that the influence of the optical homogeneity of the medium on the shape of single molecule excitation patterns is especially pronounced when the tilt of the molecule's TDM lies between 10° and 60° according to the scheme shown in Figs. 11(o)-(u).

In order to experimentally verify the calculated field intensity distributions and excitation patterns, we measured the excitation patterns of single molecules with fixed linear TDMs. The molecules were embedded in a thin polymer film on the surface of a glass cover slide in order to achieve random orientation of the molecules' TDMs. The negligible thickness of the polymer layer (around 5 nm) with respect to the size of the focal spot and nearly identical values of the refractive indices of polymer and glass (1.491 and 1.518, respectively) allow us to neglect any modification of the field intensity distribution by the polymer film. After selecting a sample area with spatially isolated individual molecules, we successively recorded patterns of the same molecules with two different optical media on top of the sample.

In a first experiment, we investigated a sample with air over the polymer layer. Figures 12(d)-(f) show four prominent examples of single molecule patterns recorded under this condition. For a better comparison, the single molecule patterns are marked in the figure as M_1, M_2, M_3 and M_4. Figures 12(g)-(i) show the corresponding simulated single molecule excitation patterns giving best fits to the experimental data. The fits suggest that the molecules M_1, M_2 and M_3 are tilted with respect to the sample surface by angles of 16°, 37° and 58°, exhibiting non-symmetric double lobe patterns, while molecule M_4 is nearly perpendicular to the sample surface, which results in a radially symmetric spot in the center of the focal region. An excitation pattern of molecule M_3 is partly truncated because of fluorescence blinking during image acquisition.

In the second experiment, we recorded excitation patterns of the same molecules, which had been investigated in the previous experiment on a glass-air interface, but embedded in a homogeneous optical medium. In order to achieve refractive index matching conditions we covered the polymer layer with a droplet of optical adhesive. Since the refractive index of the liquid adhesive is nearly identical to that of glass (1.527 and 1.518, respectively), the molecules can be considered to be placed in a medium with a constant refractive index. Figures 12(j)-(l) display the patterns of the same molecules M_1-M_4. It is remarkable that all the patterns reveal symmetrical shapes, in contrast to those obtained in the first experiment. Figures 12(m)-(o) show simulated patterns of a molecule, placed in a homogeneous refractive index medium, modeled

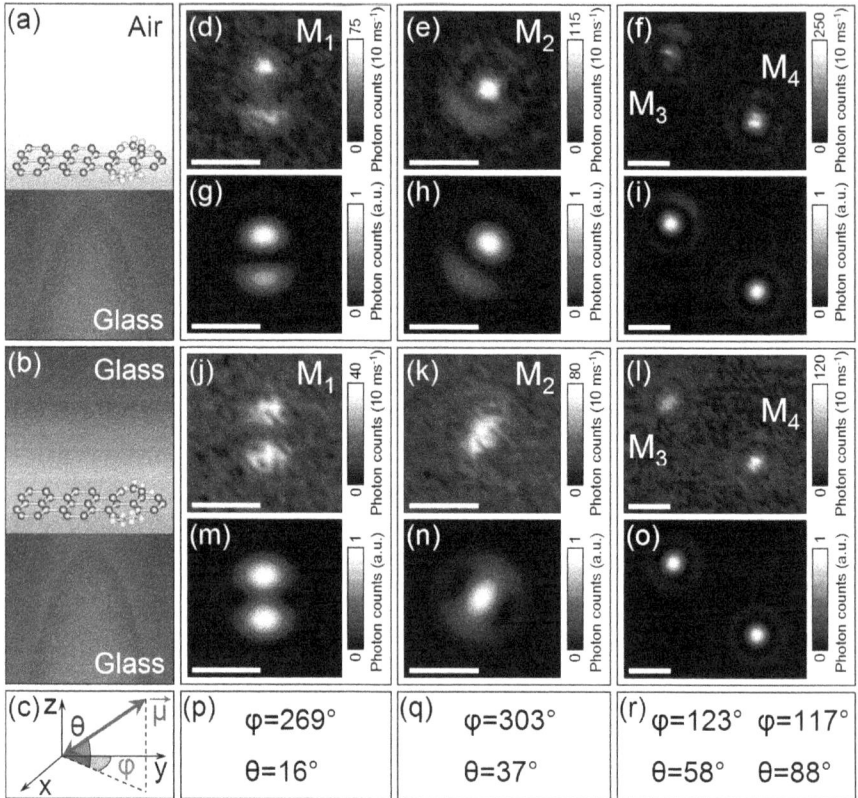

Fig. 12. (a) and (b) Schematic representation of a single molecule placed on a glass-air interface and in a homogeneous optical environment, respectively. Panel (c) shows the coordinate system which was used for theoretical simulations. Experimental single molecule excitation patterns obtained upon scanning the same individual molecules through the focal region of a radially polarized laser beam first placed on a glass-air interface (d)-(f) and subsequently embedded in a homogeneous optical environment by immersion in a droplet of optical adhesive with a matching refractive index (j)-(l). The molecules are marked M_1 - M_4. The simulated excitation patterns corresponding to the best fits to the experimental data are shown in (g)-(i) for the molecules on a glass-air interface and in (m)-(o) in bulk glass. The scale bars in the bottom left corners of the panels correspond to a length of 400 nm. Bottom panels (p)-(r) show the angular values used for fitting to the experimental data in each of the columns according to panel (c).

for the same orientation coordinates which gave the best fits in case of the glass-air confinement. As was expected from theoretical modeling (Figs. 12(i) and (o)), molecule M_4, which is oriented nearly vertically with respect to the sample surface, revealed almost identical patterns in both experiments. On the other hand, molecules M_1, M_2 and M_3 experienced a strong modification of their excitation pattern when moved from a glass-air interface into a medium with constant refractive index. In particular, the asymmetric double lobe patterns visible in Figs. 12(d), (g) and (e), (h) respectively, are transformed to symmetric patterns with either partially overlapping double lobes (Figs. 12(j), (m)) or a single elongated spot (Figs. 12(k) and (n)).

Thus, experimental and theoretical results demonstrate excellent agreement and show that excitation patterns obtained by scanning a single molecule with a fixed linear TDM through the focal region of an RPDM possess different shapes if the molecule is placed either on a glass-air interface or in bulk glass. This variation of the patterns is especially pronounced if the tilt of the molecule's TDM lies between $10°$ and $60°$. This finding allows us to show experimentally that focusing an RPDM near a planar dielectric interface strongly changes its field intensity distribution, resulting in the observed modification of the excitation pattern.

5.2. *CVBs for the investigation of the Host-Guest compounds*

In this section, we show investigations using CVBs on a first example of an all-organic perhydrotriphenylene (PHTP)-based host-guest compound (HGC), featuring channels which are closed by stopcock molecules (SC). The stopcock consists of a tetraphenylporphyrin (TPP) substituted with a long alkoxy chain (Fig. 13) which enters the channels of PHTP. We demonstrate that using an APDM for the excitation it is possible to determine the orientation of the single stopcock molecules in the channels.

HGCs have gained considerable interest in the past years since they allow the preparation of dye systems in well-defined geometries.[61-67] To avoid out-diffusion of guest molecules, the channels of the host are closed by so-called stopcock molecules, which consist of two covalently linked moieties: one of them entering a channel, the other one being too large to enter and thus closing it.[61] Stopcocks may be used to provide HGCs with additional photophysical features such as directed energy transfer to and from the channel ends, thus making HGCs suitable as photon harvesting antenna systems.[61] Among organic channel-forming host systems, PHTP has found special attention.[66-73] Upon co-crystallization with rod-shaped guest molecules, parallel stacks of PHTP molecules form nanochannels with a diameter of about 5 Å, in which the guests are co-linearly aligned. An inter-channel distance of 15 Å ensures electronic separation of the chromophores.[66-72,74] PHTP-based HGCs have been intensively investigated in the last few years in particular with respect to their potential for long-range energy-transfer.[75-77] In contrast to the inorganic HGCs, the channels of the organic host are only formed in the presence of the guest compound, thus they are completely filled with guests and no cavities are found. The description of the synthesis of stopcock molecules, preparation of PHTP-4,4'-dibromobiphenyl crystals with SC and theoretical calculations can be found in Ref. 78.

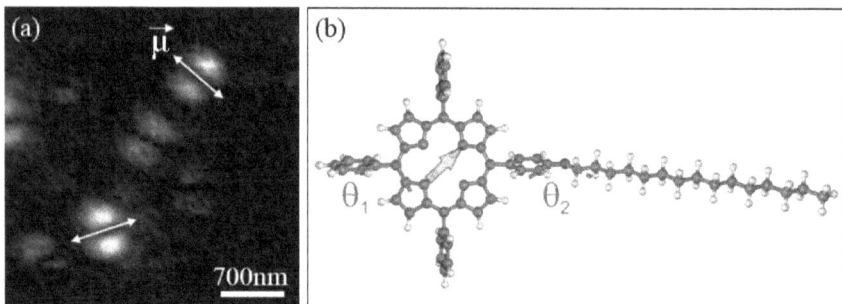

Fig. 13. (a) Fluorescence microscopy image of six isolated SC molecules in PMMA, excited by an APDM. The arrows indicate the orientation of the TDM projection on the sample surface. (b) DFT optimized geometry of the used SC. The arrow indicates the direction of the TDM. θ_1 indicates the torsion angle between the planes of the phenyl ring and the TPP core. θ_2 indicates the torsion angle between the plane of the phenyl ring and the O-C_{alkyl} bond.

Single stopcock molecules were characterized in a polymer matrix (PMMA) prior to inclusion, using confocal microscopy under APDM excitation.[17] The image in Fig. 13(a) reveals significant differences between the emission intensities of different molecules, which is mainly related to the different tilts of the molecular TDMs with respect to the sample surface. One of the molecules (at the bottom-left corner) shows the effect of sudden bleaching as to be expected for single molecules. Only clear double lobe patterns were observed indicating that no aggregations of the molecules were formed.

PHTP crystals with stopcock molecules (PHTP-SC) were investigated by recording white light wide field transmission (Fig. 14 (a), (c)) and fluorescence (Fig. 14 (b), (d)) images. Strong fluorescence is emitted from the bases of the crystals, whereas the prismatic faces exhibit significantly lower intensity. Light guiding effects can be excluded as a reason for this observation, since additional test experiments with inclusion compounds without stopcock molecules, but with chromophores in the channels, did not show enhanced emission from the crystal end faces.[79]

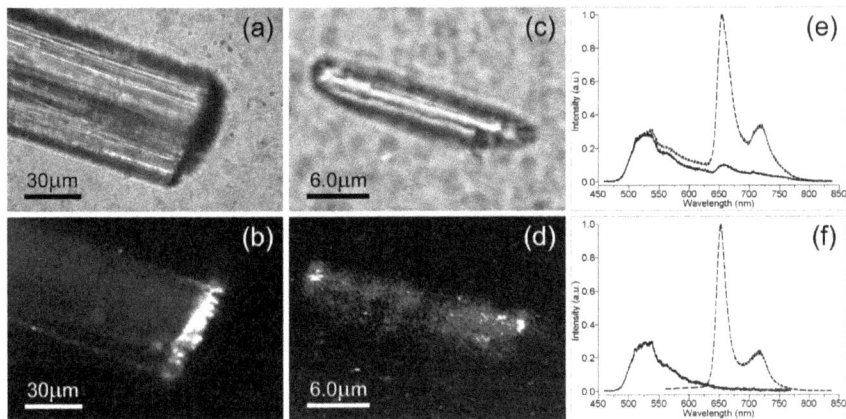

Fig. 14. White light wide field transmission (a), (c) and fluorescence images (b), (d) of a large (a), (b) and a small (c), (d) PHTP crystal with included stopcock (SC) molecules. (e) Normalized fluorescence spectra acquired at the end (dashed curve) and at the middle part (solid curve) of a crystal. (f) Emission spectra of SC molecules in PMMA (dashed curve) and of a stopcock-free PHTP crystal (solid curve).

More detailed information on the insertion of the stopcock molecules into the channels of PHTP is provided by the APDM technique described above, which yields the orientation of the porphyrin TDM (S_1) with respect to the axis of the channel into which the alkoxy tail of the SC is inserted. For SC molecules which are not included in a channel, a random distribution of TDM orientations is expected. Evaluation of the double lobe patterns (Fig. 15) with respect to their orientation against the crystal axis yields a ratio of parallel to perpendicular orientations of the TDM vs. the crystal axis of about 2:1, whereas intermediate orientations, i.e., with both parallel and perpendicular components, are hardly observed. This observation might be explained by the fact that the crystal basis is not a smooth planar surface but that some channels stick out more than others. TPP cores whose alkyl chains are inserted in protruding channels will attach face-on to the crystal basis. This leads to projections of the TDMs which are perpendicular to the crystal axis. TPP cores whose alkyl chains are inserted in channels surrounded by protruding ones will adsorb to the side walls of adjacent protruding channels. The resulting projections of the TDMs will be preferentially parallel to the crystal axis. The statistical analysis of the excitation patterns can be found in Ref. 78.

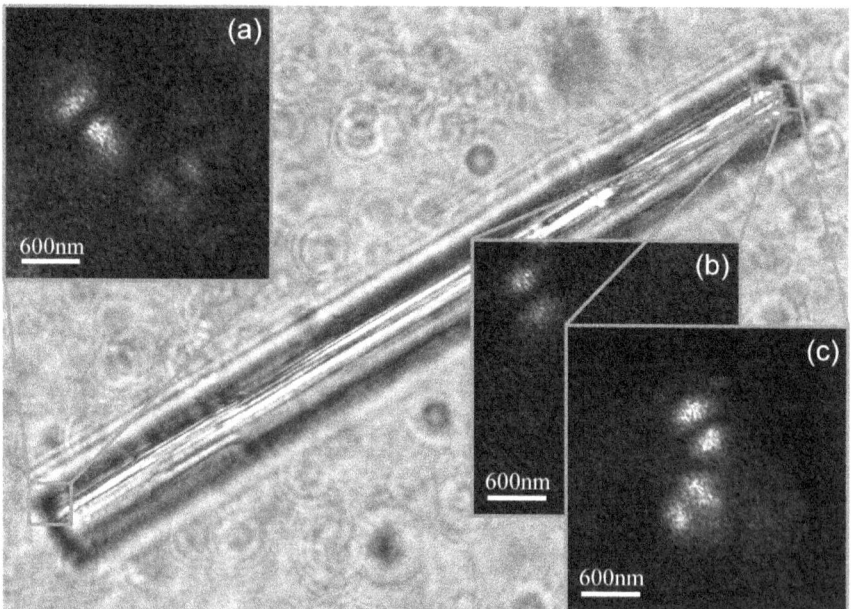

Fig. 15. Fluorescence patterns of single stopcock molecules closing the channel entrances of PHTP crystal excited with an APDM. The background is an optical image of the PHTP-SC-TPP crystal.

In conclusion, we have prepared the first example of an all-organic channel-forming host-guest-compound in which the channel entrances are closed by porphyrin based molecular stopcocks. The localization of the stopcocks in the channel entrances was confirmed by confocal microscopy in combination with an APDM. We found two possible configurations of the stopcock molecules closing the channel entrances of the crystals: TPP cores of the stopcock molecules adsorbed to the side walls of adjacent protruding channels and TPP cores whose alkyl chains are inserted in protruding channels attached face-on to the crystal basis.

5.3. *Imaging of excited-state tautomerization*

CVBs are also a powerful tool to visualize dynamical effects when the TDM of the molecule changes its orientation during the scanning process. As the molecules are fixed in a polymer matrix, TDM changes caused by rotation or motion of the molecules can be excluded (Section 3). The benefit of using single molecule confocal microscopy in combination with CVBs is the possibility to discriminate, in one and the same molecule, between its chemically and magnetically identical conformers. However, it is important to consider the image recording speed in relation to the investigated dynamical process.

Tautomerism is an outstanding example of the aforementioned multiconformity. It is a basic intramolecular process that has been studied extensively for years, especially for porphyrin type molecules by different techniques like NMR, X-ray and other spectroscopic methods.[80-92] Porphyrin-type molecules occur in nature, e.g., in hemoglobin for oxygen transport or chlorophyll in photosynthesis. Therefore, the energy transfer in porphyrin dimers has also been studied to model photosynthesis processes. Porphyrins are used as pigments, in photodynamic tumor therapy,[93-97] they serve as biomimetic enzymes, biosensors[98] and in solar cells.[99, 100] They are also proposed to be used for information storage holography[101] or as a molecular current router.[102] Porphyrins are also of special interest as they serve as a basic matrix for tautomerism, which in this case proceeds isolated within the porphyrin ring. The most stable form of such a structure is the trans conformation when the hydrogen atoms are oriented opposite. Two types of mechanism can occur in a porphyrin: the hydrogen atoms move either concertedly or step wise, as depicted in (Fig. 16). When the hydrogen atoms move synchronously, the transition between two trans forms can be measured. In the case of a stepwise mechanism an additional cis tautomer is detectable, which can form both possible chemically identical trans tautomers. So far, only averaged information over ensembles of such molecules could be obtained. Single molecule confocal microscopy utilizing CVBs allows precise investigation of conformal changes within one single molecule.

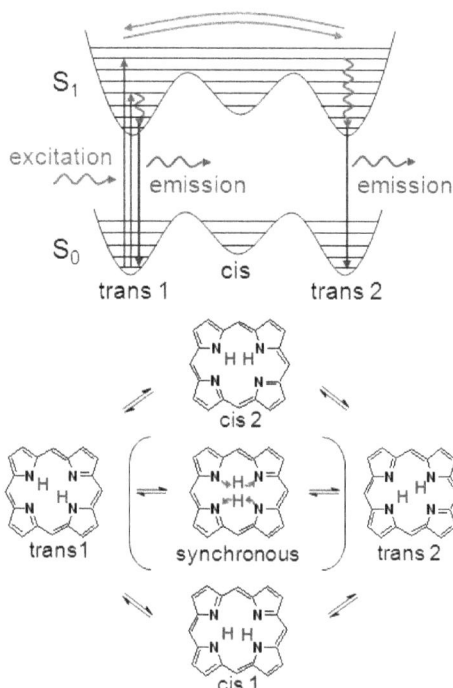

Fig. 16. Sketch of possible paths of the photoinduced tautomerism process. Below, the scheme of possible tautomeric mechanisms within a porphyrin ring. Either the hydrogen atoms move concertedly or step-wise.

For the excited state (photoinduced) tautomerism process we have to consider that the molecule has to overcome an energy barrier to undergo the tautomerism process. Figure 16 shows a sketch of the two possibilities: either the molecule can be excited in a state higher than the energy barrier followed by the tautomerism process (conversion from trans 1 tautomer into trans 2), or the energy is not sufficient and after excitation the molecule relaxes back in the same tautomeric form (trans 1). Moreover the surrounding of a single molecule leads to distortions of the energy potentials. A single molecule might therefore undergo the photoinduced tautomerism process even if, theoretically, the excitation energy is too low. In the same way, due to the local environment, the photoinduced tautomerism can also be suppressed.

Here we demonstrate investigations on two kinds of porphyrin-type molecules: a symmetrically substituted phthalocyanine (octa-(-2-ethylhexyloxy)phthalocyanine, H_2Pc) and octaethylporphyrin (OEP) (depicted in Figs. 17 and 18 respectively). The probability to excite a cis tautomer is very low, we therefore assume only to detect the two possible trans tautomers. The orientation of the TDMs of these forms are perpendicular to each other[33] and as the tautomerism proceeds faster than the acquisition time of a typical confocal image (minimum 3 ms per pixel), the detected patterns should show an overlapped pattern of the two trans forms with perpendicular TDMs (according to a degenerate excitation TDM forming a plane as introduced in Section 2).

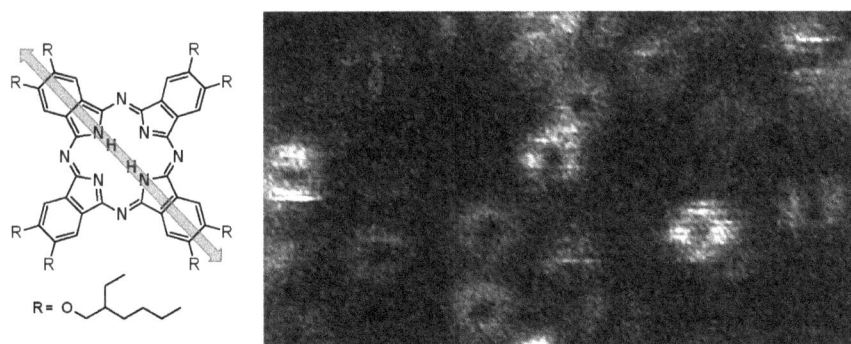

Fig. 17. Left: structural formula of the investigated H_2Pc (octa-2(-ethylhexyloxy) phthalocyanine)) molecules, the arrow indicating the orientation of the excitation TDM. Right: Excitation patterns of single H_2Pc molecules obtained by raster-scanning through an APDM. Due to the high speed of the tautomerism, the molecules exhibit ring shaped patterns.

Figure 17 shows a scanned image of H_2Pc molecules excited by an APDM showing ring-like, double lobe or intermediate patterns, in

accordance with the predictions for a degenerated TDM. A comparison of
the images obtained by APDM excitation with the ones where RPDM has
been used to confirm the observations.[34]

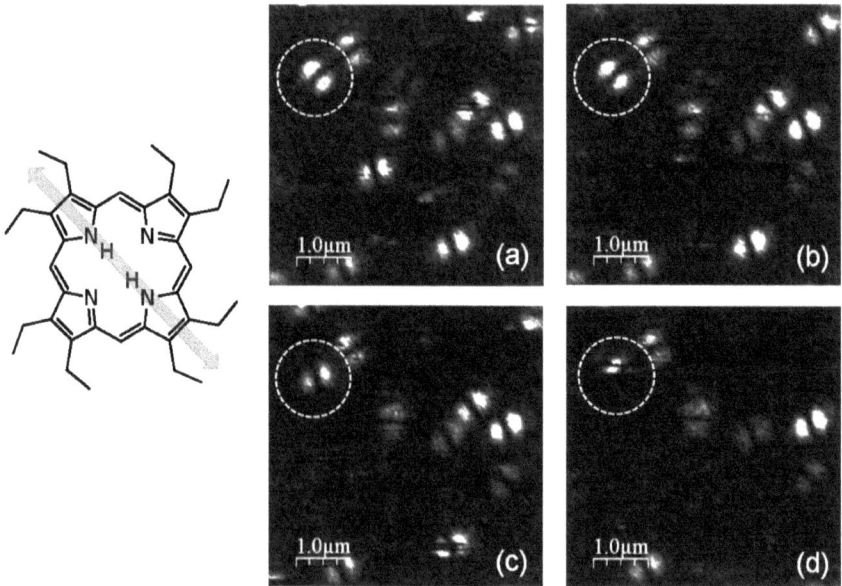

Fig. 18. Left: structural formula of OEP (octaethylporphyrin), while the arrow indicates
the orientation of the excitation TDM. Right: Series of excitation patterns of single OEP
molecules obtained by raster scanning through an APDM. The images were recorded one
after another. One molecule shows flipping of its excitation TDM (in dashed white circle)
from one trans tautomer to the second. All other molecules exhibit a stable linear
excitation TDM.

OEP molecules are expected to show the same behavior, as the TDMs
of the trans tautomers are also oriented perpendicular.[33] Interestingly, only
7 % of the investigated molecules show a flipping of their TDM
orientation, while most of the OEP molecules exhibit only one fixed TDM
oriented in a single direction (Fig. 18). According to calculations for a
degenerated TDM, it is possible to obtain double lobe patterns if one of
the TDMs is oriented vertically to the sample plane. This possibility can
be probed by exciting the same molecules with RPDM, as this mode
provides information about the three dimensional orientation with its
strong longitudinal field. The OEP molecules, however, show stable

linear TDM patterns even when excited by an RPDM. Figure 18 shows a molecule that exhibits sudden flipping of the excitation TDM during scanning (400 s acquisition time for one image) from one trans tautomer to the other and flipping back. These images show the two separated trans tautomers.

Comparing the results from OEP molecules and H₂Pc molecules, we have to consider the energy needed to undergo the tautomerism process. Calculations relating the tautomerization energy to the incident photon energy in the experiments show a clear difference: H_2Pc requires approximately 228 kJ/mol and OEP 250 kJ/mol for a conformation change.[33] As the excitation laser wavelength corresponds to an energy density of 245 kJ/mol, it is obvious that H_2Pc shows a fast tautomerism process, while only some OEP molecules are able to change their conformation. Due to local effects, some OEP molecules are able to tautomerise even if the excitation energy is theoretically not sufficient to undergo the excited state tautomerism process.

6. Single nanoparticles

In this section we will present the results of investigations on the excitation TDMs of SiO_2 NPs, Si/SiO_2 and CdSe/ZnS core-shell nanocrystals. We will show that, depending on the origin of the PL (quantum confined exciton or radiative defect), the TDM of the particle may possess different dimensionality in space. Additionally, dynamical effects such as the fluctuation of single particle TDM orientations will be discussed.

6.1. *Single SiO₂ nanoparticles*

Nanostructured silicon dioxide (SiO_2) has attracted a great deal of attention for its application in microelectronics,[103, 104] synthesis of silica-encapsulated metal nanoshells,[105] and fabrication of fluorescent silica-coated quantum dots as biological labels with enhanced surface conjugation and reduced toxicity.[106] In the past decades, significant efforts have been devoted to understand the fundamental mechanisms of the photoluminescence (PL) in SiO_2.[22, 107-113] It has been shown that defects in a silica bulk structure induce localized states in the band gap of SiO_2, which are possible sources of radiative recombination.[107] Although details are still unclear, a consensus has emerged that the PL from SiO_2 structures in the visible spectral range arises from non-bridging oxygen centers,[108, 109] neutral oxygen vacancies[110] and hydrogen-related species.[111] Recently it has been shown that PL in single SiO_2 nanoparticles (NPs), which were obtained by full oxidation of silicon nanocrystals,[112] originates from localized states in the silica structure.[22] Furthermore, single particle PL measurements revealed a strong coupling between the electronic transition and a collective vibration in the SiO_2 network due to a sudden relaxation mechanism involving charges.[22]

In order to obtain detailed information on the origin and the properties of the defect luminescence of SiO_2 NPs and to reveal possible similarities with the recently published single particle spectroscopy of Si nanocrystals surrounded by a SiO_2 shell,[113] we have investigated the PL of single SiO_2 NPs. The excitation of a single SiO_2 NP scanned through the focal region of an APDM yields an image consisting of two closely spaced bright spots of elliptical shape, resembling the double lobe pattern as in the case of single molecules. In the same way as for single molecules, the RPDM can be used to determine the three dimensional orientation of the TDM of the NP.[22] Figure 19 displays an example of SiO_2 NPs embedded in a PMMA matrix.

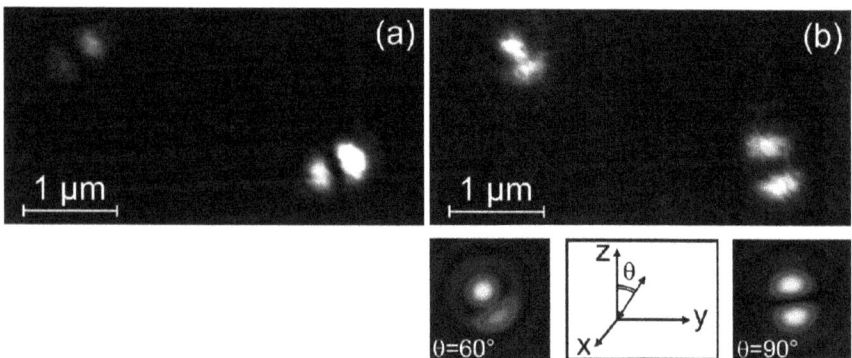

Fig. 19. SiO_2 NPs in PMMA observed with an excitation wavelength of 488 nm. The left panel (a) shows two NPs excited with an APDM (maximum intensity: 614 counts per pixel in 5 ms). Image (b) shows the same NPs excited with an RPDM (maximum intensity: 176 counts per pixel in 5 ms). The bottom panel shows examples of simulated emission patterns for two orientations of the TDM after excitation with an RPDM.

We now turn to polystyrene (PS) as a matrix material. First, we would like to state that, dynamical effects seem more likely to occur in PS than in PMMA. An example is given in Fig. 20, showing a series of images, taken one after the other of the small section of $4.3 \times 4.3 \ \mu m^2$, containing essentially only two SiO_2 NPs. The sample was excited with an APDM, and the time between two consecutive images was 100 s.

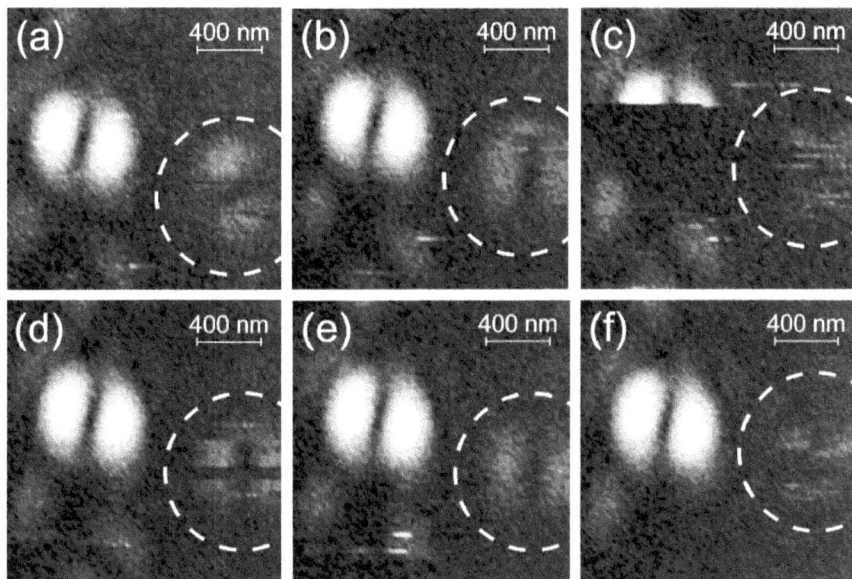

Fig. 20. Images of single SiO_2 NPs embedded in a PS matrix and excited in an APDM. Images (a)-(f) display the same sample area scanned sequentially with an acquisition time of 100 s. Each image essentially shows two NPs, one represented by the bright double lobe and a fainter one marked by the dashed circle. The following dynamical processes are observed: image (c) reveals fluorescence intermittency of the brighter NP (also called blinking). Blinking is also observed for the fainter NP in image (d). In addition, we observe sudden flipping of the TDM of the fainter NP as seen by comparing images (a) and (b) as well as (e) and (f).

Regarding Figs. 20(a)-(f), we observe a bright double lobe in the left half of each image, corresponding to a single SiO_2 NP. In image (c), the particle experiences fluorescence intermittency or blinking as manifested by the sudden quenching of PL lasting for at least 24 s. In image (d), the same NP appears bright again. Another interesting dynamical effect is demonstrated by the fainter double lobe pattern marked by a dashed circle, representing second, fainter NP. Comparing images (a) and (b), a sudden flipping of the orientation of this second particle's TDM can be observed. A similar change in dipole moment orientation has occurred between images (e) and (f). In picture (c), the image of this NP is somewhat smeared, indicating that the orientation of the TDM was not stable during the scan. It should be mentioned that a flipping of the TDM was observed only rather rarely.

In summary, by recording the single SiO_2 NPs excitation patterns, we have shown that individual SiO_2 NPs possess a linear TDM, which is stable in time for most of the particles. At the same time, while the particles are fixed in a polymer matrix, we observed a rare reorientation of the particles' TDM, which is caused by energy fluctuation of the defects in the NP. These results suggest that there is only one optically active defect in the SiO_2 NPs.

6.2. *Single silicon nanocrystals*

In order to study the origin of the PL in silicon nanocrystals, we have also investigated the dimensionality (according to the theory described in Section 2) of the excitation TDM of single silicon nanocrystals surrounded by SiO_2 shell (Si/SiO_2 NCs).[26] Pure silicon nanocrystals were deposited on the surface of a quartz cover slide and naturally oxidized to form an SiO_2 shell, passivating surface defects. The investigations of the dimensionality (see Section 2) of the Si/SiO_2 NCs' excitation TDM gives insight into determine the origin of the fluorescence. While the fluorescence originating from the surface defects shows a linear excitation TDM, the quantum confined (QC) excitons in spherical Si/SiO_2 NCs possess higher dimensionality of the TDM. We were able to observe QC PL from individual Si NCs, illustrated in Fig. 21. The fluorescence images (a) and (b), obtained under excitation with an APDM, each show a different sample area of $3.2 \times 3.2 \ \mu m^2$. Several excitation patterns clearly resembling a double lobe or a ring structure can be observed. Besides the two prominent ring patterns in the center of image (a), three closely spaced Si NCs are found in the lower left of image (b), which clearly exhibit a more-dimensional TDM. The visualization of the TDM using CVBs for excitation allows us to identify the origin of the observed PL in a single Si nanocrystal. For APDM excitation, defect-related PL is associated with a double lobe pattern (pointing to a linear TDM)[22] while quantum-confined PL gives rise to a ring-like structure (indicating a 3D TDM).[13, 29, 114]

For the following discussion, it should be kept in mind that the present free-standing Si NCs consist of a crystalline Si core surrounded by a native layer of SiO_2. All nanocrystals experienced the same history, i.e., after synthesis, they were all subjected to the same gentle oxidation in ambient air. Furthermore, the information given by each fluorescence image was obtained in one single scan. Thus, we can clearly state the coexistence of individual Si NCs exhibiting QC PL and others showing defect luminescence, both within the same sample. Regarding the abundance of double lobes and rings, defect-based PL is observed 1.5 times more frequently (despite the careful preparation and a rather large mean particle size of 4.0 nm). It is interesting to note that, in the present study, a switching from a ring-like structure to a double lobe pattern could be observed only once during the course of two consecutive scans. This shows that both DC PL and QC PL may occur in one and the same particle and that both mechanisms could indeed be competing processes. Comparing the stability of the fluorescence patterns of both recombination processes, it is clearly observed that fluorescence intermittency (blinking) occurs more frequently in the case of quantum confinement. The interruption by many dark periods results in a poor rendition of the ring-like structure. In order to visualize the frequent blinking with higher resolution, the number of pixels per fluorescence image was increased by a factor of four. As can be seen in Fig. 21 the intensity of the double lobe and ring-like fluorescence patterns is more or less equal even though the defect-based PL is characterized by a short lifetime so that we would expect stronger fluorescence intensity than for QC PL. This can be explained as follows: first, in contrast to polymer-embedded nanocrystals, a preferred orientation of the TDM normal to the substrate can be assumed due to the breaking of symmetry. The free-standing NCs are attached directly to the substrate surface (SiO_2) and out-of-plane (almost vertical) orientations of the TDM may prevail resulting in a weakening of the double lobe images under APDM excitation. Second, due to the reduced laser power, the excitation rate is lower and, thus, the effect of enhanced intensity for fast recombination processes is diminished. As a result, comparable photon emission (equal intensity) can be rationalized for both processes.

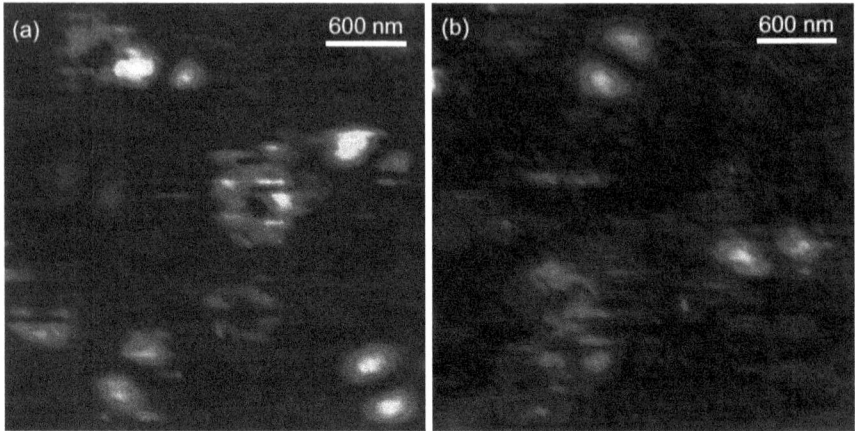

Fig. 21. Fluorescence microscopy images of free-standing Si nanocrystals. The fluorescence images (a) and (b) depict different sample areas, each probing 3.2×3.2 μm². The images were obtained by scanning through the focal region of an APDM. They reveal the excitation patterns of several individual, free-standing Si NCs deposited on a quartz cover slide by the cluster beam technique. Two kinds of excitation patterns (double lobes and ring-like structures) can be distinguished.

6.3. *Excitation isotropy of single CdSe/ZnS quantum dots*

In this section, we investigate the dimensionality of the excitation TDM of single CdSe/ZnS core-shell quantum dots using higher order laser modes. We studied commercially available CdSe/ZnS core-shall quantum dots, soluble in toluene or in water. This includes toluene soluble quantum dots QD1 (Evidot®, Evident Technologies, emission centered at 557, 580, 593 and 629 nm), QD2 (Lumidot™, Sigma Aldrich, at 610 nm) and QD3 (PlasmaChem at 610 nm) and water soluble quantum dots QD4 (Qdot® Invitrogen, emission centered at 565 nm). In the first part of the study we carried out imaging of CdSe/ZnS nanocrystals soluble in toluene (QD1, QD2 and QD3). According to the specification of the nanocrystals, they showed photoluminescence (PL) centered at five distinct spectral ranges. Each of the single-particle emission spectra had a single-band structure with full width at half maximum (FWHM) of about 15 nm, typical for PL of an individual CdSe/ZnS nanocrystal at room temperature. Figure 22 shows typical examples of excitation patterns for a single CdSe/ZnS quantum dot (QD3) obtained by scanning it through the focal region of a linearly (a), azimuthally (b) and radially (c) polarized laser beam. The images demonstrate fluorescence blinking, which may be caused by ionization of the nanocrystal.[115, 116] Such binary flickering between "on" and "off" states is evidence that the excitation pattern indeed originates from an individual quantum system. The pattern obtained upon excitation of the nanocrystal by a linearly polarized Gaussian beam has no characteristic features that enable determination of the dimensionality and orientation of the particle's excitation TDM. However, when the nanocrystal is scanned through the focal region of the APDM or the RPDM, it exhibits a pattern shape specific for a particular TDM orientation and dimensionality. Although strong intensity fluctuations make it difficult to precisely determine the shape of the excitation patterns, the images strongly resemble those obtained upon excitation of isotropic Nile Red fluorescent spheres (see insets of Figs. 22(b) and (c)). This observation indicates that the nanocrystals possess excitation isotropy. Remarkably, all studied CdSe/ZnS nanocrystals soluble in toluene (QD1, QD2 and QD3) revealed an identical shape of the excitation patterns regardless of the emission's spectral range. Since the PL tunability of the nanocrystal is imparted from its size and, partly, composition, this result suggests that excitation isotropy is not related to a specific size or composition of the nanocrystal, but is rather a common property of the CdSe/ZnS quantum dots.

Although the quantum dots QD1, QD2 and QD3 were produced by different suppliers, all the studied toluene soluble CdSe/ZnS nanocrystals possess a hydrophobic coating (organic molecules). In order to compare nanocrystals coated with different types of protection layers, we also studied water soluble quantum dots (QD4) spin-coated on the surface of a glass cover slide without a polymer matrix. Figures 23(b) and (e) show the measured excitation patterns of an individual CdSe/ZnS nanocrystal upon scanning through the focal region of APDM and RPDM, respectively. The images exhibit significantly shorter "off" states than the hydrophobic-coated nanoparticles investigated in the first part of the study. Excitation

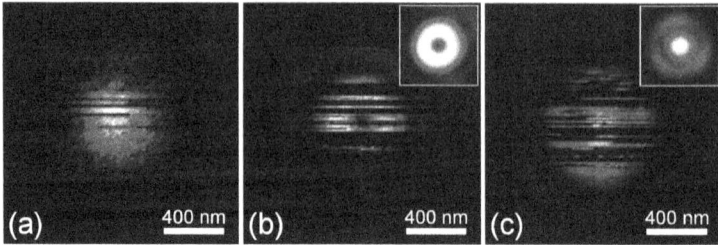

Fig. 22. Experimental excitation patterns created by scanning the same single CdSe/ZnS quantum dot (sample QD3, see experimental details) through the focal region of a linearly (a), azimuthally (b) and radially (c) polarized laser beam. The excitation patterns demonstrate fluorescence blinking, proving that the patterns originate from an individual nanocrystal (see main text for details). Insets in figures (b) and (c) show excitation patterns of Nile Red fluorescent spheres featuring excitation isotropy, excited with an azimuthally and radially polarized laser beam, respectively.

patterns revealed very good agreement with theoretical images, corresponding to a spherically degenerate i.e., isotropic excitation TDM (Figs. 23(c) and (f)). Comparing the cross sections through the centers of the experimental and theoretical patterns shows excellent agreement (Figs. 23(a) and (d)). This suggests that the studied single CdSe/ZnS quantum dots QD4 also have a spherically degenerate excitation TDM.

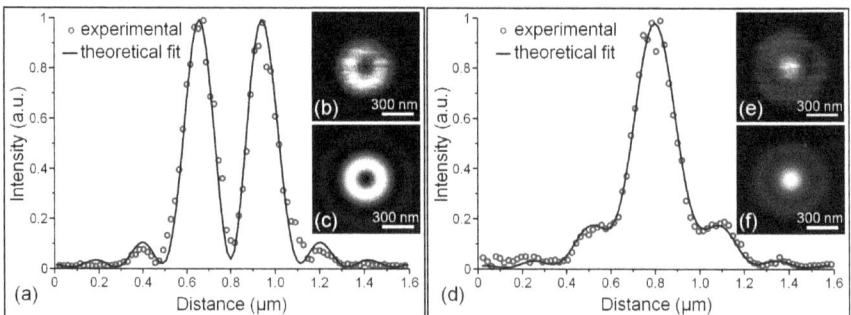

Fig. 23. Cross sections through the experimental ((b) and (e)) and simulated ((c) and (f)) patterns of an individual CdSe/ZnS quantum dot (sample QD4, see experimental detail), excited with an azimuthally (a) and radially (d) polarized laser beam.

6.4. *Optical characterization of single gold nanorods*

In the recent years the interest in gold nanoparticles has grown steadily and a variety of applications were developed such as biosensing,[117, 118] bio medical therapy[119] or efficiency enhancement of solar cells.[120] The optical properties of noble metal nanoparticles are dominated by collective oscillations of the conduction band electrons, named particle plasmon resonances. Among the many possible particle shapes, gold nanorods (GNR) are of special interest as their geometry allows tuning the particle plasmon resonance throughout nearly the whole visible spectrum. On resonance, GNRs exhibit a strong optical field enhancement at their ends, making them suitable as substrates for surface enhanced Raman scattering (SERS).[121] Furthermore, they are promising candidates for biosensing,[117] as they have a low toxicity, are photostable and do not show blinking. For many applications it is advantageous to determine the GNRs' orientation in a non-invasive and non destructive way. In principle, scanning probe techniques like scanning tunneling microscopy (STM), atomic force microscopy (AFM) or scanning near-field optical microscopy (SNOM) would be techniques of choice, but they are all limited to the surface of a structure, so lack the possibility to measure inside of samples. In recent years it was shown that certain far-field techniques, such as defocused imaging,[122] dark field imaging[123] or photothermal imaging,[124] are indeed able to determine the orientation of single GNRs by optical means. This goal can also be achieved by combining confocal microscopy with higher order laser modes, representing a simple and robust alternative to the techniques mentioned above. Moreover this approach can even give access to additional information such as the environment surrounding of the GNR. Already in 2006 Failla *et al.* showed that higher order laser modes, APDM or RPDM, are suited to determine the two dimensional orientation of single, isolated GNRs with high precision.[27, 51] Recently it was shown that this technique is also able to give a full optical three dimensional characterization.[28] Exploiting the field distribution of the APDM and the RPDM it is possible to determine the in-plane orientation with a precision of 1°, while the out-of-plane orientation can be determined with a precision of at least 5°. This can be achieved by using both the elastically scattered and the luminescence signal of single GNRs. As an example, Fig. 24 shows simultaneously acquired luminescence and scattering signals of single GNRs oriented at different angles θ with respect to the sample surface. These GNRs, with an aspect ratio (AR) of 2.1±0.5 and a short axis diameter of 16.5±1.2 nm, have been investigated upon excitation at 632.8 nm.

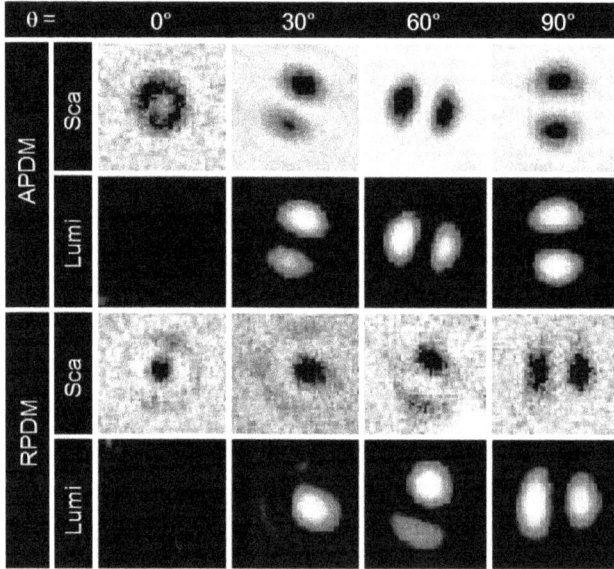

Fig. 24. Simultaneously recorded scattering and luminescence patterns of single gold nanorods. Each column shows a separate particle, the corresponding polar angle θ indicated on top. The two upper rows show patterns acquired using an APDM while the lower two rows show corresponding patterns recorded with an RPDM. As indicated on the left, both elastic scattering (Sca) and one-photon luminescence (Lumi) patterns are presented. The patterns acquired with the APDM exhibit a transition from a ring like pattern to a two lobbed pattern, making this mode sensitive for small polar angles ($\theta < 20°$). In contrast the RPDM has a higher sensitivity for larger polar angles ($\theta > 20°$), as the pattern shows a continuous transition from a spot-like to a double lobe pattern shape. This behavior makes the APDM and the RPDM complementary to each other and well suited to determine the three dimensional orientation of gold nanorods.

The luminescence pattern shape can be used to determine the three dimensional orientation of isolated, freely oriented GNRs in a simple and robust way as it is only weakly influenced by interfaces nearby. In contrast to the luminescence signal the elastically scattered signal is detected in an interferometric configuration and therefore strongly dependent on the phase relation of the elastically scattered and reflected light, thus it is strongly influenced by parameters such as the refractive index mismatch of nearby interfaces or the GNR's aspect ratio. This interferometric detection scheme allows the determination of the refractive index change of an interface, as shown in Ref. 125. In Ref. 27 we have shown that the scattering pattern is influenced by the refractive index mismatch at the focusing interface and that the GNRs do not necessarily need to be directly located on the interface to be sensitive to it. This could be applied to precisely monitor the growth of a dielectric layer by observing the change of the scattering pattern of a GNR. Another advantage of the interferometric detection scheme is that it allows determination of the orientation of a GNR spanning the full angular range, while the luminescence signal of standing GNRs is hardly detectable. Moreover, the interference of the elastically scattered and reflected light strongly increases the sensitivity, allowing detection of very small particles down to a few nanometers in size.[125] Furthermore, this technique can also be used to track dynamic changes of the orientation of GNRs[126] and is thus a promising candidate for bioimaging. As shown above, the combination of CVBs with confocal microscopy is a useful and simple tool to investigate single plasmonic nanoparticles down to a few nanometers in size and to gain additional information about dynamic changes or the particles' local environment. This technique can easily be implemented into commercial available microscopes and could be promoted to be a useful tool in various fields such as the investigation of biological samples or applications in nanophotonics.

7. Orientation and position determination of a single quantum Emitter inside an optical microcavity

Another interesting aspect of CVBs is their application in microcavities.[19, 58, 66] These structures can be used to tailor the optical properties of embedded quantum emitters[127-129] and offer the potential for developing new photonic devices. There are several excellent reviews that describe the application of microcavities in detail[20, 130, 131] (and references therein). For an efficient control of the optical properties of embedded quantum emitters, it is important to know their exact positions and orientations inside the microcavity. Here we describe a method to measure the position of a fluorescent sphere and the orientation of a single molecule in a microcavity using CVBs. The introduced method is based on the inhomogeneous distribution of in-plane and longitudinal field components in the focal volume of an RPDM inside the microcavity.

The tunable optical microresonator is sketched in Fig. 25. The cavity is formed by two silver mirrors, the upper and lower mirrors of 60 nm and 30 nm thick, respectively. The mirrors were fabricated by electron beam evaporation onto a standard glass cover slide and the curved surface of a plano-convex lens ($f = 150$ mm). An SiO_2-layer defines the spacing between the quantum emitters and the lower mirror and, hence, the z-position with respect to the optical resonator. A polymer film deposited onto the SiO_2 spacer contains either fluorescent spheres or single molecules (see Section 3.2 for preparation). This film provides a matrix to suppress translation and rotation of the quantum emitters and to maintain well-defined coupling conditions to the optical resonator during measurement. A refractive index matching liquid is injected between the dye-doped polymer layer and the upper cavity mirror and serves as an optically transparent, tunable intra-cavity medium. The microresonator assembly is placed in a mount that enables controllable tuning of the upper cavity mirror using piezoelectric elements. Due to the small curvature of the upper mirror, the microresonator can be seen as a planar resonator within the focal spot of our microscope objective lens (~500 nm). For the (x,y)-position of every molecule, the mirror spacing $L(x,y)$ and the corresponding quality factor Q of the cavity can be determined from the local white light transmission spectrum with high accuracy and reproducibility.[127, 128] The Q-factor is defined as the ratio of the transmission wavelength λ_0 and the peak width $\delta\lambda$ and has a value of about 50 for the used configuration.

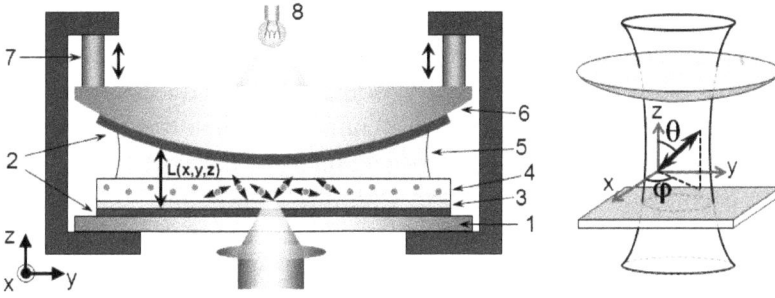

Fig. 25. Scheme of the tunable microresonator with (1) glass cover slide, (2) silver mirrors, (3) silica layer, (4) polymer (with QE), (5) immersion oil, (6) lens, (7) piezoelectric cell, (8) white light source. The coordinate system (right) defines the orientation of the molecular transition dipole μ (black arrow) with respect to the cavity mirrors.

In the following, we will briefly discuss how to model the field of an APDM and an RPDM in the microcavity. The focused field intensity distribution of the APDM is characterized by the in-plane components E_x and E_y of the incident laser beam and can be expressed as:

$$E_{APDM} = \frac{ikf^2 e^{-ikf}}{2w_0} \begin{pmatrix} 4iL_{1,0,1}(\rho,z)\sin(\varphi) \\ -4iL_{1,0,1}(\rho,z)\sin(\varphi) \\ 0 \end{pmatrix}. \tag{5}$$

Here, (ρ,z,φ) are the cylindrical coordinates with $z = 0$ corresponding to the focal point of a lens with a focal length f, $k = 2\pi/\lambda$ is the wavevector of the field with the wavelength λ, and w_0 is the beam waist of the laser beam. The functions $L_{n,m,l}$ characterize the in-plane and longitudinal components of the intra-cavity field.[66] For focusing the RPDM, which is characterized by the in-plane components E_x, E_y and the longitudinal component E_z, the modelling results in:

$$E_{RPDM} = \frac{ikf^2 e^{-ikf}}{2w_0} \begin{pmatrix} 4iL_{1,1,1}(\rho,z)\cos(\varphi) \\ 4iL_{1,1,1}(\rho,z)\sin(\varphi) \\ -4L_{0,0,2}(\rho,z) \end{pmatrix}. \tag{6}$$

On the basis of Eqs. (5) and (6), and taking the actual optical parameters of the resonator and the microscope objective lens into account, we can visualize the intensity distribution between the cavity mirrors. Figures 26(a)–(c) and (d)–(f) present experimental excitation patterns for the same single fluorescent sphere at three different resonator lengths L given in the figure using an APDM and an RPDM, respectively. The insets show the calculated (x,y)-field intensity distribution for the same L-values for a sphere situated 50 nm above the lower cavity mirror. The corresponding normalized (x,z)-field intensity distributions are shown above each experimental pattern with the white line indicating the position of the sphere. While the excitation pattern of the sphere always has a doughnut shape when excited with an APDM (see Figs. 26(a)–(c)), the sphere's excitation pattern changes its shape from a doughnut to a circular spot when excited with an RPDM (see Fig. 26(d)–(f)). By analyzing the excitation pattern, a precise determination of the sphere's longitudinal position inside the microcavity with an accuracy of a few nanometers[58] is possible.

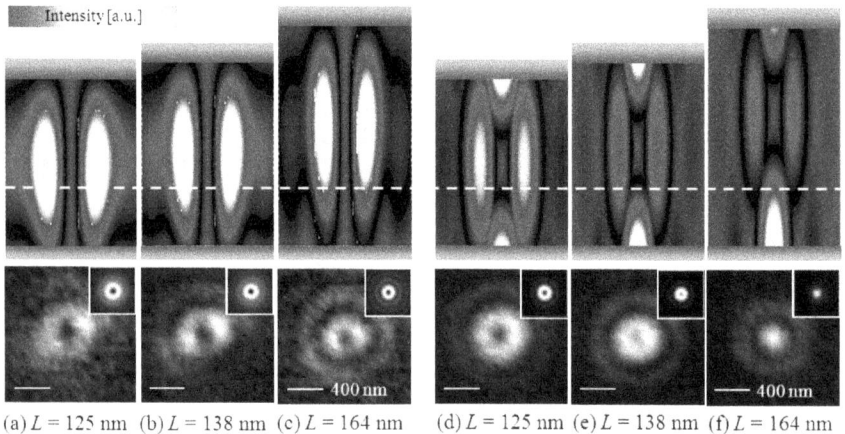

(a) L = 125 nm (b) L = 138 nm (c) L = 164 nm   (d) L = 125 nm (e) L = 138 nm (f) L = 164 nm

Fig. 26. Excitation patterns for the same single fluorescent sphere (lower row) for three different resonator lengths L resulting from excitation with an APDM (a)–(c) and an RPDM (d)–(f). The insets show the numerically predicted (x,y)-field intensity distributions and the corresponding normalized (x, z)-field intensity distribution is shown above the experimental patterns. The white line in the upper panels indicates the position of the sphere.

Moreover, the RPDM also allows the determination of the orientation of a single molecule in the microcavity.[19] In order to model the excitation pattern for a single molecule we calculate the projection of the intra-cavity field distribution on the molecular dipole moment, which results in:

$$E_{SM} = \frac{ikf^2 e^{-ikf}}{2w_0} \begin{pmatrix} 4iL_{1,1,1}(\rho, z)\cos(\varphi+\psi)\sin(\theta) \\ -4L_{0,0,2}(\rho, z)\cos(\theta) \end{pmatrix}. \tag{7}$$

The angle φ denotes the in-plane orientation of the molecular dipole, whereas θ describes the orientation towards the optical axis (see Fig. 25). If the actual mirror spacing L and the z-position of a single molecule are known, the intensity distribution of the fluorescence excitation pattern depends only on the dipole orientation (θ,φ). Hence, by comparing experimental and calculated excitation patterns of a single molecule, we can determine the orientation of a single molecule in the optical microresonator. This is shown in Fig. 27 for a series of recorded excitation patterns acquired from the same single molecule immobilized in the tunable microresonator. The SiO_2 spacer layer (see Fig. 9) with a thickness of 50 nm defines the z-position of the molecule and the mirror spacing L remains the only adjustable parameter. The black stripe in the second experimental scan image in Fig. 27 originates from fluorescence intensity blinking, a phenomenon assuring the observation of a single molecule.[132] Comparing the measured excitation patterns with the simulated patterns for different (θ, φ)-orientations of the molecular transition dipole, we find the best agreement between experiment and calculation for $\theta = 35° \pm10°$ and $\varphi = 255° \pm2°$. The accuracy of this method depends on the quality of the experimental pattern and can result in an error of just a few degrees.[19]

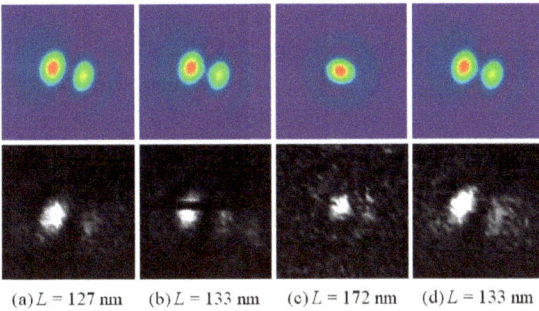

(a) $L = 127$ nm   (b) $L = 133$ nm   (c) $L = 172$ nm   (d) $L = 133$ nm

Fig. 27. A series of calculated (upper row) and experimental (lower row) fluorescence excitation patterns for the same single PI molecule in a tunable microresonator excited with an RPDM at different mirror separations L. The PI molecule is located 50 nm above the lower cavity mirror and the patterns are calculated for a transition dipole moment orientation of $\theta = 35°$ and $\varphi = 255°$. The size of each image is 1.5×1.5 μm^2.

8. Conclusions

This chapter has shown how the unique polarization properties of CVBs can be used to determine the three-dimensional orientation of the TDM of spatially isolated and immobilized single quantum emitters such as molecules and nanoparticles, with an inverted confocal microscope. Recorded excitation patterns obtained with RPDMs and APDMs have been compared with corresponding calculated patterns, thus revealing the orientation information. We have demonstrated the use of these modes for single molecules as well as for nanoparticles. We could determine not only the 3D orientation of the dipole moments, but also the dimensionality of the TDM and, by recording consecutive patterns, the orientational dynamics of the TDM, i.e., flipping processes.

Beyond the orientation measurements in non-confined media, we introduced the method to determine the three-dimensional orientation of single molecules inside a tunable $\lambda/2$-Fabry-Perot type microresonator. In this case, an unusual bimodal field distribution is formed when an RPDM is tightly focused into the cavity to probe the dipole orientation. This method has the potential to determine the TDM orientation in arbitrarily oriented multi-chromophoric molecular systems.[19]

In addition, the RPDM enables a smaller focal spot than a linear Gaussian mode.[38] Focusing with a PM gives rise to an even tighter spot.[56] The dominant longitudinal electric field component of the radial mode can not only be used to determine the three dimensional orientation of a molecule or nanoparticle, but also to more efficiently excite the probe of an apertureless SNOM.[40] Considering their widespread applications, it is quite convenient to use CVBs to investigate quantum systems when confocal microscopy is the method of choice.

Acknowledgments

The authors would like to thank Raphael Gutbrod for providing the images of the measurements inside a microresonator and for his contribution to this book chapter, as well as we thank Sebastian Bär.

Additionally we'd like to thank Sebastian Jäger and Andreas Kern for their support and stimulating discussions.

We gratefully acknowledge financial support from the "Kompetenznetz Funktionelle Nanostrukturen" of the Landesstiftung Baden-Württemberg, the European Commission through the Human Potential Program (Marie-Curie Research Training Network NANOMATCH, contract MRTN-CT-2006-035884), and the Deutsche Forschungsgemeinschaft DFG (ME 1600/6-1,2).

We also thank Coherent GmbH Germany for providing a SapphireTM 488-20 OPS laser. The WSxM software from Nanotec [133] was partly used for scan image processing.

References

1. W. E. Moerner and L. Kador, *Physical Review Letters* **62**, 2535 (1989).
2. M. Orrit and J. Bernard, *Physical Review Letters* **65**, 2716 (1990).
3. C. Zander, J. Enderlein, and R. A. Keller, *Single molecule detection in solution : methods and applications* (Wiley-VCH, Berlin, 2002).
4. T. Ha, T. Enderle, D. F. Ogletree, D. S. Chemla, P. R. Selvin, and S. Weiss, *Proc Natl Acad Sci U S A* **93**, 6264 (1996).
5. K. Kinosita, Jr., *FASEB journal : official publication of the Federation of American Societies for Experimental Biology* **13 Suppl 2**, S201 (1999).
6. K. Adachi, R. Yasuda, H. Noji, H. Itoh, Y. Harada, M. Yoshida, and K. Kinosita, *P Natl Acad Sci USA* **97**, 7243 (2000).
7. J. Jasny and J. Sepiol, *Chemical Physics Letters* **273**, 439 (1997).
8. M. Bohmer and J. Enderlein, *J Opt Soc Am B* **20**, 554 (2003).
9. A. P. Bartko and R. M. Dickson, *J Phys Chem B* **103**, 11237 (1999).
10. R. Schuster, M. Barth, A. Gruber, and F. Cichos, *Chemical Physics Letters* **413**, 280 (2005).
11. T. Ha, T. Enderle, D. S. Chemla, P. R. Selvin, and S. Weiss, *Physical Review Letters* **77**, 3979 (1996).
12. F. Guttler, J. Sepiol, T. Plakhotnik, A. Mitterdorfer, A. Renn, and U. P. Wild, *J Lumin* **56**, 29 (1993).
13. S. A. Empedocles, R. Neuhauser, and M. G. Bawendi, *Nature* **399**, 126 (1999).
14. P. Dedecker, B. Muls, J. Hofkens, J. Enderlein, and J. I. Hotta, *Optics Express* **15**, 3372 (2007).
15. B. Sick, B. Hecht, and L. Novotny, *Physical Review Letters* **85**, 4482 (2000).
16. K. S. Youngworth and T. G. Brown, *Optics Express* **7**, 77 (2000).
17. L. Novotny, M. R. Beversluis, K. S. Youngworth, and T. G. Brown, *Physical Review Letters* **86**, 5251 (2001).
18. H. Ishitobi, I. Nakamura, N. Hayazawa, Z. Sekkat, and S. Kawata, *J Phys Chem B* **114**, 2565 (2010).
19. R. Gutbrod, D. Khoptyar, M. Steiner, A. M. Chizhik, A. I. Chizhik, S. Bär, and A. J. Meixner, *Nano Lett* **10**, 504 (2010).

20. S. Bär, A. Chizhik, R. Gutbrod, F. Schleifenbaum, A. Chizhik, and A. J. Meixner, *Anal Bioanal Chem* **396**, 3 (2010).
21. A. I. Chizhik, A. M. Chizhik, D. Khoptyar, S. Bär, A. J. Meixner, and J. Enderlein, *Nano Lett* **11**, 1700 (2011).
22. A. M. Chizhik, A. I. Chizhik, R. Gutbrod, A. J. Meixner, T. Schmidt, J. Sommerfeld, and F. Hulsken, *Nano Lett* **9**, 3239 (2009).
23. A. M. Chizhik, T. Schmidt, A. I. Chizhik, F. Huisken, and A. Meixner, edited by S. Cabrini and T. Mokari (SPIE, San Diego, CA, USA, 2009), p. 739305.
24. A. I. Chizhik, T. Schmidt, A. M. Chizhik, F. Huisken, and A. J. Meixner, *Physics Procedia* **13**, 28 (2011).
25. A. M. Chizhik, A. I. Chizhik, A. J. Meixner, T. Schmidt, and F. Huisken, *AIP Conference Proceedings* **1275**, 63 (2010).
26. T. Schmidt, A. I. Chizhik, A. M. Chizhik, K. Potrick, A. J. Meixner, and F. Huisken, *Physical Review B* **86**, 125302 (2012).
27. A. V. Failla, H. Qian, H. Qian, A. Hartschuh, and A. J. Meixner, *Nano Lett* **6**, 1374 (2006).
28. F. Wackenhut, A. V. Failla, T. Züchner, M. Steiner, and A. J. Meixner, *Appl Phys Lett* **100** (2012).
29. A. I. Chizhik, A. M. Chizhik, D. Khoptyar, S. Baar, and A. J. Meixner, *Nano Lett* **11**, 1131 (2011).
30. D. Khoptyar, R. Gutbrod, A. Chizhik, J. Enderlein, F. Schleifenbaum, M. Steiner, and A. J. Meixner, *Optics Express* **16**, 9907 (2008).
31. H. Piwonski, C. Stupperich, A. Hartschuh, J. Sepiol, A. Meixner, and J. Waluk, *J Am Chem Soc* **127**, 5302 (2005).
32. H. Piwonski, A. Hartschuh, N. Urbanska, M. Pietraszkiewicz, J. Sepiol, A. J. Meixner, and J. Waluk, *J Phys Chem C* **113**, 11514 (2009).
33. A. M. Chizhik, R. Jäger, A. I. Chizhik, S. Bär, H. G. Mack, M. Sackrow, C. Stanciu, A. Lyubimtsev, M. Hanack, and A. J. Meixner, *Phys Chem Chem Phys* **13**, 1722 (2011).
34. R. Jäger, A. M. Chizhik, A. I. Chizhik, H.-G. Mack, A. Lyubimtsev, M. Hanack, and A. J. Meixner, edited by C. Silva (SPIE, San Diego, California, USA, 2011), p. 80980G.
35. Q. Zhan, *Optics Express* **12**, 3377 (2004).
36. Q. Zhan, *Advances in Optics and Photonics* **1**, 1 (2009).
37. A. Huss, A. M. Chizhik, R. Jäger, A. I. Chizhik, and A. J. Meixner, edited by K. Dholakia and G. C. Spalding (SPIE, San Diego, California, USA, 2011), p. 809720.
38. R. Dorn, S. Quabis, and G. Leuchs, *Physical Review Letters* **91** (2003).
39. J. Stadler, T. Schmid, and R. Zenobi, *Nano Lett* **10**, 4514 (2010).
40. L. Novotny, E. J. Sánchez, and X. Sunney Xie, *Ultramicroscopy* **71**, 21 (1998).
41. D. Zhang, U. Heinemeyer, C. Stanciu, M. Sackrow, K. Braun, L. E. Hennemann, X. Wang, R. Scholz, F. Schreiber, and A. J. Meixner, *Physical Review Letters* **104** (2010).
42. A. Boivin and E. Wolf, *Physical Review* **138**, B1561 (1965).
43. L. Novotny and B. Hecht, *Principles of nano-optics* (Cambridge University Press, Cambridge ; New York, 2006).

44. Q. Zhan and J. R. Leger, *Optics Express* **10**, 324 (2002).
45. K. Yonezawa, Y. Kozawa, and S. Sato, *Opt Lett* **31**, 2151 (2006).
46. G. Machavariani, Y. Lumer, I. Moshe, A. Meir, S. Jackel, and N. Davidson, *Appl Optics* **46**, 3304 (2007).
47. J. F. Bisson, J. Li, K. Ueda, and Y. Senatsky, *Optics Express* **14**, 3304 (2006).
48. M. A. Ahmed, A. Voss, M. M. Vogel, and T. Graf, *Opt Lett* **32**, 3272 (2007).
49. V. G. Niziev, R. S. Chang, and A. V. Nesterov, *Appl Optics* **45**, 8393 (2006).
50. M. Stalder and M. Schadt, *Opt Lett* **21**, 1948 (1996).
51. A. V. Failla, S. Jäger, T. Züchner, M. Steiner, and A. J. Meixner, *Optics Express* **15**, 8532 (2007).
52. G. Machavariani, Y. Lumer, I. Moshe, A. Meir, and S. Jacket, *Opt Lett* **32**, 1468 (2007).
53. E. Wolf, *Proceedings of the Royal Society of London. Series A. Mathematical and Physical Sciences* **253**, 349 (1959).
54. B. Richards and E. Wolf, *Proceedings of the Royal Society of London. Series A. Mathematical and Physical Sciences* **253**, 358 (1959).
55. M. Lieb and A. Meixner, *Opt. Express* **8**, 458 (2001).
56. J. Stadler, C. Stanciu, C. Stupperich, and A. J. Meixner, *Opt Lett* **33**, 681 (2008).
57. J. H. Callomon, T. M. Dunn, and I. M. Mills, *Philosophical Transactions of the Royal Society of London. Series A, Mathematical and Physical Sciences* **259**, 499 (1966).
58. R. Gutbrod, A. Chizhik, A. Chizhik, D. Khoptyar, and A. J. Meixner, *Opt Lett* **34**, 629 (2009).
59. A. G. Chizhik, R.; Chizhik, A.; Bär, S.; Meixner, A. J., *Proceedings SPIE* **7396** (2009).
60. F. Schleifenbaum, C. Blum, V. Subramaniam, and A. J. Meixner, *Mol Phys* **107**, 1923 (2009).
61. G. Calzaferri, S. Huber, H. Maas, and C. Minkowski, *Angewandte Chemie-International Edition* **42**, 3732 (2003).
62. A. Ajayaghosh, S. J. George, and V. K. Praveen, *Angewandte Chemie-International Edition* **42**, 332 (2003).
63. F. J. M. Hoeben, P. Jonkheijm, E. W. Meijer, and A. P. H. J. Schenning, *Chem Rev* **105**, 1491 (2005).
64. A. Del Guerzo, A. G. L. Olive, J. Reichwagen, H. Hopf, and J. P. Desvergne, *J Am Chem Soc* **127**, 17984 (2005).
65. J. S. Wilson, M. J. Frampton, J. J. Michels, L. Sardone, G. Marletta, R. H. Friend, P. Samori, H. L. Anderson, and F. Cacialli, *Adv Mater* **17**, 2659 (2005).
66. P. J. Langley and J. Hulliger, *Chem Soc Rev* **28**, 279 (1999).
67. G. Couderc and J. Hulliger, *Chem Soc Rev* **39**, 1545 (2010).
68. A. Lyubimtsev, M. N. Misir, M. Calvete, and M. Hanack, *Eur J Org Chem*, 3209 (2008).
69. M. M. Tsotsalas, K. Kopka, G. Luppi, S. Wagner, M. P. Law, M. Schafers, and L. De Cola, *Acs Nano* **4**, 342 (2010).
70. D. Bruhwiler, G. Calzaferri, T. Torres, J. H. Ramm, N. Gartmann, L. Q. Dieu, I. Lopez-Duarte, and M. V. Martinez-Diaz, *J Mater Chem* **19**, 8040 (2009).

71. K. Gierschner, L. Luer, D. Oelkrug, E. Musluoglu, B. Behnisch, and M. Hanack, *Adv Mater* **12**, 757 (2000).
72. J. Gierschner, H. J. Egelhaaf, H. G. Mack, D. Oelkrug, R. M. Alvarez, and M. Hanack, *Synthetic Met* **137**, 1449 (2003).
73. M. Aloshyna, B. Milian Medina, L. Poulsen, J. Moreau, D. Beljonne, J. Cornil, G. Di Silvestro, M. Cerminara, F. Meinardi, R. Tubino, H. Detert, S. Schrader, H. J. Egelhaaf, C. Botta, and J. Gierschner, *Advanced Functional Materials* **18**, 915 (2008).
74. G. Srinivasan, J. A. Villanueva-Garibay, K. Muller, D. Oelkrug, B. Milian Medina, D. Beljonne, J. Cornil, M. Wykes, L. Viani, J. Gierschner, R. Martinez-Alvarez, M. Jazdzyk, M. Hanack, and H. J. Egelhaaf, *Phys Chem Chem Phys* **11**, 4996 (2009).
75. L. Poulsen, M. Jazdzyk, J. E. Communal, J. C. Sancho-Garcia, A. Mura, G. Bongiovanni, D. Beljonne, J. Cornil, M. Hanack, H. J. Egelhaaf, and J. Gierschner, *J Am Chem Soc* **129**, 8585 (2007).
76. L. Viani, L. P. Tolbod, M. Jazdzyk, G. Patrinoiu, F. Cordella, A. Mura, G. Bongiovanni, C. Botta, D. Beljonne, J. Cornil, M. Hanack, H. J. Egelhaaf, and J. Gierschner, *J Phys Chem B* **113**, 10566 (2009).
77. C. Botta, G. Patrinoiu, P. Picouet, S. Yunus, J. E. Communal, F. Cordella, F. Quochi, A. Mura, G. Bongiovanni, M. Pasini, S. Destri, and G. Di Silvestro, *Adv Mater* **16**, 1716 (2004).
78. A. M. Chizhik, R. Berger, A. I. Chizhik, A. Lyubimtsev, L. Viani, J. Cornil, S. Bär, M. Hanack, J. Hulliger, A. J. Meixner, H. J. Egelhaaf, and J. Gierschner, *Chem Mater* **23**, 1088 (2011).
79. J. Hulliger, O. Konig, and R. Hoss, *Adv Mater* **7**, 719 (1995).
80. B. Wehrle, H. H. Limbach, M. Kocher, O. Ermer, and E. Vogel, *Angew Chem Int Edit* **26**, 934 (1987).
81. Y. D. Wu, K. W. K. Chan, C. P. Yip, E. Vogel, D. A. Plattner, and K. N. Houk, *J Org Chem* **62**, 9240 (1997).
82. H. H. Limbach, J. M. Lopez, and A. Kohen, *Philos T R Soc B* **361**, 1399 (2006).
83. M. J. Crossley, M. M. Harding, and S. Sternhell, *J Org Chem* **53**, 1132 (1988).
84. J. Braun, R. Schwesinger, P. G. Williams, H. Morimoto, D. E. Wemmer, and H. H. Limbach, *J Am Chem Soc* **118**, 11101 (1996).
85. J. Braun, H. H. Limbach, P. G. Williams, H. Morimoto, and D. E. Wemmer, *J Am Chem Soc* **118**, 7231 (1996).
86. M. J. Crossley, L. D. Field, M. M. Harding, and S. Sternhell, *J Am Chem Soc* **109**, 2335 (1987).
87. M. J. Crossley, M. M. Harding, and S. Sternhell, *J Am Chem Soc* **108**, 3608 (1986).
88. A. Vdovin, J. Waluk, B. Dick, and A. Slenczka, *Chemphyschem* **10**, 761 (2009).
89. P. Fita, N. Urbanska, C. Radzewicz, and J. Waluk, *Chem-Eur J* **15**, 4851 (2009).
90. J. Waluk and E. Vogel, *J Phys Chem-Us* **98**, 4530 (1994).
91. M. Drobizhev, N. S. Makarov, A. Rebane, G. de la Torre, and T. Torres, *J Phys Chem C* **112**, 848 (2008).
92. U. Even and J. Jortner, *The Journal of Chemical Physics* **77**, 4391 (1982).
93. P. Gregory, *J Porphyr Phthalocya* **4**, 432 (2000).
94. M. Ochsner, *J Photoch Photobio B* **39**, 1 (1997).

95. D. Phillips, *Pure Appl Chem* **67**, 117 (1995).
96. J. J. Schuitmaker, P. Bass, H. L. L. M. vanLeengoed, F. W. vanderMeulen, W. M. Star, and N. vanZandwijk, *J Photoch Photobio B* **34**, 3 (1996).
97. I. Lanzo, N. Russo, and E. Sicilia, *J Phys Chem B* **112**, 4123 (2008).
98. F. P. Zhi, X. Q. Lu, J. D. Yang, X. Y. Wang, H. Shang, S. H. Zhang, and Z. H. Xue, *J Phys Chem C* **113**, 13166 (2009).
99. T. Hasobe, S. Fukuzumi, and P. V. Kamat, *J Phys Chem B* **110**, 25477 (2006).
100. T. Hasobe, H. Imahori, P. V. Kamat, and S. Fukuzumi, *J Am Chem Soc* **125**, 14962 (2003).
101. A. Renn, U. P. Wild, and A. Rebane, *J Phys Chem A* **106**, 3045 (2002).
102. I. Thanopulos and E. Paspalakis, *Physical Review B* **76** (2007).
103. D. A. Muller, T. Sorsch, S. Moccio, F. H. Baumann, K. Evans-Lutterodt, and G. Timp, *Nature* **399**, 758 (1999).
104. J. B. Neaton, D. A. Muller, and N. W. Ashcroft, *Physical Review Letters* **85**, 1298 (2000).
105. C. Radloff and N. J. Halas, *Appl Phys Lett* **79**, 674 (2001).
106. S. T. Selvan, T. T. Tan, and J. Y. Ying, *Adv Mater* **17**, 1620 (2005).
107. E. P. O'Reilly and J. Robertson, *Physical Review B* **27**, 3780 (1983).
108. L. Skuja, *J Non-Cryst Solids* **239**, 16 (1998).
109. S. Munekuni, T. Yamanaka, Y. Shimogaichi, R. Tohmon, Y. Ohki, K. Nagasawa, and Y. Hama, *Journal of Applied Physics* **68**, 1212 (1990).
110. R. Tohmon, H. Mizuno, Y. Ohki, K. Sasagane, K. Nagasawa, and Y. Hama, *Physical Review B* **39**, 1337 (1989).
111. Y. D. Glinka, S. H. Lin, and Y. T. Chen, *Appl Phys Lett* **75**, 778 (1999).
112. A. Colder, F. Huisken, E. Trave, G. Ledoux, O. Guillois, C. Reynaud, H. Hofmeister, and E. Pippel, *Nanotechnology* **15**, L1 (2004).
113. J. Martin, F. Cichos, F. Huisken, and C. von Borczyskowski, *Nano Lett* **8**, 656 (2008).
114. L. E. Brus, P. F. Szajowski, W. L. Wilson, T. D. Harris, S. Schuppler, and P. H. Citrin, *J Am Chem Soc* **117**, 2915 (1995).
115. M. Nirmal, B. O. Dabbousi, M. G. Bawendi, J. J. Macklin, J. K. Trautman, T. D. Harris, and L. E. Brus, *Nature* **383**, 802 (1996).
116. K. T. Shimizu, R. G. Neuhauser, C. A. Leatherdale, S. A. Empedocles, W. K. Woo, and M. G. Bawendi, *Physical Review B* **63** (2001).
117. J. N. Anker, W. P. Hall, O. Lyandres, N. C. Shah, J. Zhao, and R. P. Van Duyne, *Nat Mater* **7**, 442 (2008).
118. G. Raschke, S. Kowarik, T. Franzl, C. Sonnichsen, T. A. Klar, J. Feldmann, A. Nichtl, and K. Kurzinger, *Nano Lett* **3**, 935 (2003).
119. H. Kang, B. H. Jia, J. L. Li, D. Morrish, and M. Gu, *Appl Phys Lett* **96** (2010).
120. H. A. Atwater and A. Polman, *Nat Mater* **9**, 205 (2010).
121. C. J. Orendorff, L. Gearheart, N. R. Jana, and C. J. Murphy, *Phys Chem Chem Phys* **8**, 165 (2006).
122. T. Li, Q. Li, Y. Xu, X. J. Chen, Q. F. Dai, H. Y. Liu, S. Lan, S. L. Tie, and L. J. Wu, *Acs Nano* **6**, 1268 (2012).

123. L. H. Xiao, Y. X. Qiao, Y. He, and E. S. Yeung, *Analytical Chemistry* **82**, 5268 (2010).
124. W. S. Chang, J. W. Ha, L. S. Slaughter, and S. Link, *P Natl Acad Sci USA* **107**, 2781 (2010).
125. T. Züchner, A. V. Failla, A. Hartschuh, and A. J. Meixner, *J Microsc-Oxford* **229**, 337 (2008).
126. T. Züchner, F. Wackenhut, A. V. Failla, and A. J. Meixner, *Appl Surf Sci* **255**, 5391 (2009).
127. M. Steiner, F. Schleifenbaum, C. Stupperich, A. V. Failla, A. Hartschuh, and A. J. Meixner, *Chemphyschem* **6**, 2190 (2005).
128. M. Steiner, A. V. Failla, A. Hartschuh, F. Schleifenbaum, C. Stupperich, and A. J. Meixner, *New J Phys* **10** (2008).
129. A. Chizhik, F. Schleifenbaum, R. Gutbrod, A. Chizhik, D. Khoptyar, A. J. Meixner, and J. Enderlein, *Physical Review Letters* **102** (2009).
130. K. J. Vahala, *Nature* **424**, 839 (2003).
131. V. N. Astratov, *Optics Express* **15**, 17171 (2007).
132. F. Kulzer and M. Orrit, *Annu Rev Phys Chem* **55**, 585 (2004).
133. I. Horcas, R. Fernandez, J. M. Gomez-Rodriguez, J. Colchero, J. Gomez-Herrero, and A. M. Baro, *Review of Scientific Instruments* **78** (2007).

CHAPTER 4

COMPREHENSIVE FOCAL FIELD ENGINEERING WITH VECTORIAL OPTICAL FIELDS

Weibin Chen[†,‡] and Qiwen Zhan[†,*]

†Electro-Optics Program, University of Dayton
300 College Park, Dayton, Ohio 45469, USA
‡Current Address: Seagate Technology,
7801 Computer Ave S, Edina, MN 55435
*E-mail: qzhan1@udayton.edu

The propagation and focusing properties of spatially variant polarization remain of continued research interest owing to their promising applications in physics, chemistry and biological sciences. The main challenges to these applications are the optimization of focus shape, size and the control of polarization distribution within the focal volume. In this chapter, systematic approaches that are suitable for comprehensive three-dimensional focal field engineering are reviewed. Three-dimensional flattop focusing with extended depth of focus and optical bubble is illustrated through the use of generalized cylindrical vector beam illumination and diffractive optical element (DOE). Combining the radiation pattern from an electric dipole and the Richards-Wolf vectorial diffraction method, the required input field at the pupil plane of a high numerical aperture objective lens for generating arbitrary three-dimensional polarization at the focal point with an optimal spot size can be found analytically by solving an inverse problem. Diffraction-limited spherical spots with three-dimensionally controllable polarization can be realized through applying this radiation reversal approach to 4-π microscopy. Several other exotic optical focal field distributions, such as high purity longitudinally polarized optical needle field with extended depth of focus, extremely long optical tube and uniform optical chain, are demonstrated through the reversal of the radiation pattern from linear electric and magnetic dipole arrays.

1. Introduction

Optical microscopy has been an indispensable tool in many scientific disciplines and industries owing to its nondestructive nature and the multi-dimensional information it provides. It utilizes a tightly focused optical field as a probe to interrogate the sample properties within the focal volume and generate the contrast for imaging. Hence, controlling the optical field properties within the focal volume plays a critical role in determining the function and performance of optical microscopy. For example, pupil plane apodization techniques have been developed to affect the spatial resolution of microscopy. The Gouy phase shift within the focal volume has been found to be important in determining the signal of coherent anti-Stokes Raman spectroscopy (CARS)[1] and third harmonic generation (THG) microscopy.[2-4] Spatial engineering of the focal intensity profile for a depletion pulse has been explored in stimulation emission depletion (STED) microscopy to provide a spatial resolution far beyond the diffraction limit.[5, 6]

With recent advances in the spatially variant polarization and particularly in cylindrical vector (CV) beams,[7] the vectorial distribution of focused electromagnetic fields has drawn significant amount of attentions. Focus shaping with vortex and elliptically symmetric polarized beams were studied numerically and experimentally.[8-11] For circularly polarized vortex beam, proper combination of vortex topological charge and handedness of circular polarization will generate focused field with flattop and strong longitudinal component.[8, 9] And elliptical symmetry of linear polarization states provides an additional degree of freedom in focus shaping to meet specific applications.[10, 11] Low numerical aperture (NA) element, such as axicon or axicon polarizer,[12-14] has been used to create radially or azimuthally polarized non-diffracting fields. Focused with a high NA objective lens, vector field distributions in the focal volume can be exploited for focal intensity shaping.[15-17] Furthermore, controllable polarization characteristics within the focal volume can be realized with spatial variant polarized illuminations.[18-21] These engineered optical vector focal fields pave the way to the applications in polarization sensitive orientation imaging,[20, 22] particle manipulation and acceleration,[23, 24] light-matter interaction on the nanoscale and so on.[25]

2. Three-dimensional focus shaping with CV beams

Focus engineering with CV beams has attracted increasing attentions for its many potential applications including optical microscopy, optical manipulations, optical micromachining and photolithography.[7] As reviewed in Chapter 1, radial and azimuthal polarizations are two basic modes of cylindrical polarization. Focus engineering using radially, azimuthally and cylindrically polarized beams have been studied through engineering the amplitude, phase and polarization within the pupil.[7] Many interesting focal field distributions such as optical needle,[26-30, 31] three-dimensional (3D) flattop focus,[16, 17, 32] optical bubble or cage enclosed by light field,[17, 33-38] 3D optical chain,[39] spherical spot[10, 41] or spot array[42] etc., have been created.

For high NA focusing, it has been demonstrated that a radially polarized light can be focused into an ultra-small focal spot with predominantly longitudinal polarization, while an azimuthally polarized beam can be focused into a hollow spot with purely transverse polarization.[15, 16] This observation led to a flattop focus shaping technique using a generalized CV polarization that can be regarded as the linear superposition of these two orthogonally polarized components.[16, 17] Advanced phase, amplitude and/or polarization filters at the pupil plane have also been applied to CV beams to achieve unique focal field distributions. The energy density pattern at the vicinity of the focus can be tailored in three dimensions by appropriately adjusting the parameters of the CV beam illumination, NA of the objective lens and the design of the filters.

The electric field of a generalized CV beam can be decomposed into a combination of radial polarization and azimuthal polarization as

$$\vec{E}(r,\varphi) = L(r)[\cos\phi_o\vec{e}_r + \sin\phi_o\vec{e}_\varphi], \tag{1}$$

where $\vec{e}_r$ is the unit vector in the radial direction and $\vec{e}_\varphi$ is the unit vector in the azimuthal direction. $L(r)$ is the pupil function denoting the relative amplitude of the field that only depends on radial position. ϕ_o is the polarization angle of each point from its radial direction as defined in Chapter 1.

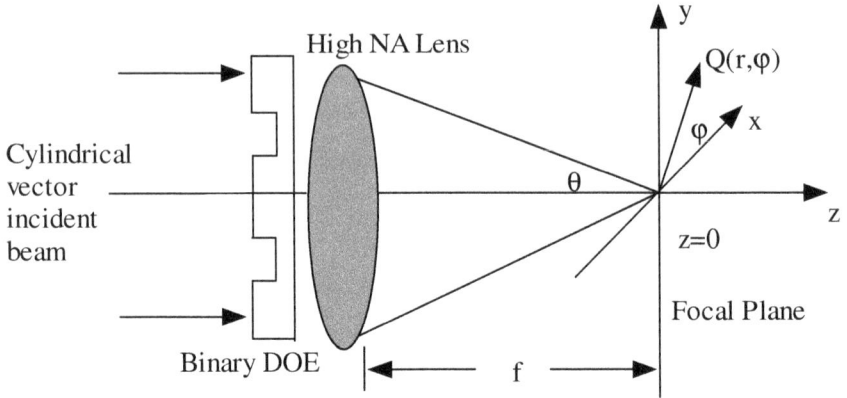

Fig. 1. Focusing of a generalized cylindrical vector beam with DOE.[17]

Figure 1 shows the schematic setup for 3D focus shaping.[17] The incident light is a generalized CV beam described by Eq. (1) with a planar wavefront over the input pupil. An aplanatic high NA objective produces a spherical wave converging to the focus of the lens. A binary DOE with three concentric regions is placed in front of the lens to modulate the wavefront of the generalized CV beams. Due to rotational symmetry and normal illumination, the phase modulation is identical for the radial and azimuthal polarization components. The NAs that correspond to the outer edge of the three regions are NA_1, NA_2 and NA respectively, where the local transmittances are $T_1 = 1$, $T_2 = -1$ and $T_3 = 1$. Thus the field distribution at the focus is given by

$$E = E_1 - E_2 + E_3, \qquad (2)$$

where E_1, E_2 and E_3 are the focal field contribution from each annular zone that can be calculated as follows.

The focusing property of a highly focused laser beam can be analyzed with the Richards and Wolf vectorial method.[15-17, 43-44] Due to the rotational symmetry of both the setup and illumination, the electric field distribution in the vicinity of the focal spot can be calculated in cylindrical coordinates. The focal field of a generalized CV beam contributed from each concentric region of the DOE can be written as

$$\vec{E}(r,\varphi,z) = E_z \vec{e}_z + E_r \vec{e}_r + E_\varphi \vec{e}_\varphi, \tag{3}$$

where $\vec{e}_z$ is the unit vector in the z direction, and [16, 17]

$$E_z(r,\varphi,z) = jA\cos\phi_o \int_{\theta_1}^{\theta_2} \sin^2\theta P(\theta)L(\theta)J_0(kr\sin\theta)e^{jkz\cos\theta}d\theta, \tag{4}$$

$$E_r(r,\varphi,z) = A\cos\phi_o \int_{\theta_1}^{\theta_2} \sin\theta\cos\theta P(\theta)L(\theta)J_1(kr\sin\theta)e^{jkz\cos\theta}d\theta, \tag{5}$$

$$E_\varphi(r,\varphi,z) = A\sin\phi_o \int_{\theta_1}^{\theta_2} \sin\theta P(\theta)L(\theta)J_1(kr\sin\theta)e^{jkz\cos\theta}d\theta, \tag{6}$$

where θ_1 and θ_2 is the angular transition points determined by the inner and outer transition points of the corresponding concentric region of the binary DOE. For example, for the concentric ring with π phase shift, the integration limits are given by $\theta_1 = \sin^{-1}(NA_1)$ and $\theta_2 = \sin^{-1}(NA_2)$, respectively. $P(\theta)$ is the pupil apodization function, k is the wave number and $J_n(x)$ is the Bessel function of the first kind with order n. $L(\theta)$ is the relative amplitude of the field, which is assumed to be dependent on the radial position only.[17]

In general, a focal spot with following two properties are often desirable: (a) flattop focusing both in longitudinal and transversal directions; (b) long depth of focus (DOF). From Eqs. (3)–(6), the field distribution functions of radial and azimuthal components include Bessel function $J_1(x)$, which is zero when $x = 0$. Therefore, they do not contribute to the on-axis field distribution. It is clear that the only non-zero contribution for on-axis field comes from the z-component. Thus, relative axial energy density distribution will not change with the rotation angle ϕ_o because only radial component of the generalized CV beams contributes to it. From these equations, it is easy to see that the focal fields created by radial polarization and azimuthal polarization are mutually orthogonal and spatially separated. If a generalized CV beam is used as illumination along with the binary DOE, one should be able to adjust the relative weighing of the radial and azimuthal components to create flattop profiles both in longitudinal and transversal directions.

(a)

(b)   (c)

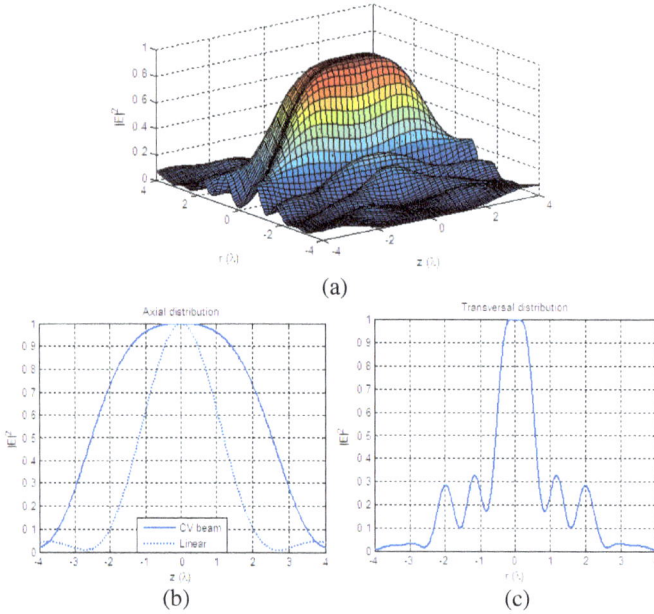

Fig. 2. Energy density distributions in the vicinity of focus with pupil apodization function under Sine condition. (a) 3D distribution; (b) Axial distribution; (c) Transversal distribution.[17]

To illustrate this, an aplanatic objective lens that satisfies the Sine condition with NA = 0.8 is used in the following example. Under the Sine condition, the pupil apodization function is $P(\theta) = \sqrt{\cos(\theta)}$.[45] The parameters for the binary DOE to obtain a long DOF and flattop focusing on the axis is found to be $NA_1 = 0.31$, $NA_2 = 0.5$. The rotation parameter for the CV beam is found to be $\phi_0 = 38°$ to achieve a flattop focusing in transversal direction. Figure 2(a) shows the calculated result of energy density distribution in the vicinity of the focus. The corresponding axial and transversal energy density distributions are shown in Fig. 2(b) and Fig. 2(c). For all of the calculations, the length unit is normalized to the wavelength λ in the medium, and the maximum energy density is normalized to unity. It can be seen that flattop profiles along both transversal and axial distribution have been obtained. The energy in the main lobe volume is about 59.14% of the total energy. The DOF defined by full width at half maximum (FWHM) of the $|E|^2$ along the axial direction is calculated to be 5λ. For comparison, the axial field distribution for a linearly polarized distribution is also calculated and shown in Fig. 2(b). The DOF is calculated to be 2.4λ. Clearly, an extended DOF is achieved with CV polarization. This extended DOF partly is due to the destructive interference from the middle ring of the DOE. In addition, the vector projection used in vectorial diffraction theory gives larger weighing to the higher spatial frequency components in the calculation of E_z, as seen in Eq. (4). This apodization effect also contributes to an extended DOF. Flattop focusing with longer DOF can be obtained with objective lenses under other pupil apodization functions.[17]

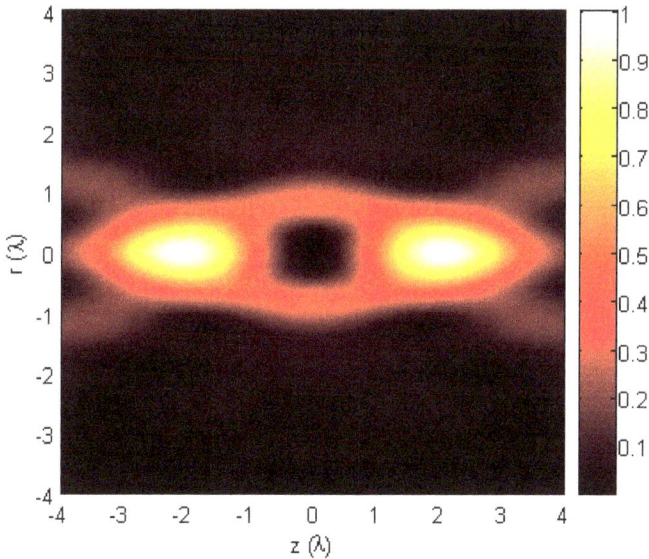

Fig. 3. Optical bubble created by radial polarization with DOE pupil plane modulation.[17]

The 3D flattop focusing is achieved by the feasibility of adjusting the destructive interference from the middle ring to "fill" the saddle. Through adjusting the destructive interference from the middle ring, it is possible to further increase the destructive interference to carve into the focus and generate an optical bubble which has a total dark volume surrounded by high field distributions.[17] In this case, radial polarization with $\phi_0 = 0°$ is used, since this may allow one to create an ultra-small dark volume. The parameter for the binary DOE are found to be

$NA_1 = 0.2$, $NA_2 = 0.65$ with $NA = 0.8$ and $NA_0 = 0.2$ (corresponding to the dark center size of a CV beam). Essentially, the DOE has two concentric rings, rather than three. Numerical simulation results for this design are shown in Fig. 3.[17] From this figure, it can be seen that a bubble-like focal field with a totally dark center is obtained. The optical bubble size defined by FWHM of the $|E|^2$ is calculated to be 2.01λ along the axial direction and 0.90λ along the transversal direction. A smaller optical bubble can be obtained with higher NA lens, or 4-π focusing.

Flattop focusing with extended DOF could find applications in laser cutting of metals, particle acceleration, materials processing and microlithography. Laser beams that can be focused into a flattop spot will allow faster, high quality laser cutting with lower operating costs. The laser machining efficiency of metals strongly depends on the polarization. Spatially homogeneous polarizations have substantial disadvantages. It has been pointed out that inhomogeneous types of polarization are much better in laser cutting.[46, 47] In the case of cutting metals with a large aspect ratio of sheet thickness to width, the laser cutting efficiency for a radially polarized beam is shown to be 1.5 ~ 2 times larger than for TM-polarized and circularly polarized beams.[46]

In order to increase the laser cutting efficiency and velocity, sharper focusing of the incident light is necessary for high-speed cutting of sheet steel. If a linear or circular polarized light is used in laser cutting, the radiation intensity decreases quickly along the focal axis, giving rise to a small ratio of cutting depth to cutting width. On the contrary, using CV beams with flattop focusing and long DOF, one may significantly increase the ratio of cutting depth to cutting width.

The optical bubble may find applications in particle trapping and fluorescence microscopy.[48, 49] For example, the resolution of fluorescence microscopy is mostly determined by the extent of the fluorescence spot. In order to improve the spatial resolution of fluorescence microscopy, STED technique has been developed.[48] In the STED technique, a depletion pulse following an excitation pulse is focused into a donut shape around the focus of the excitation beam. In the region where the focal field intensity of the depletion beam is above certain threshold intensity, fluorescence is inhibited. An optical bubble that has a total dark volume surrounded by high field distributions in three dimensions may be applied in the STED microscopy to improve the spatial resolution of fluorescence microscopy.

3. Three-dimensional polarization control within focal volume

Besides the intensity and phase distributions within the focal volume, polarization of the focal field is another important parameter that deserves attention. Many molecules and crystalline structures are anisotropic due to their specific spatial orientation, making polarization response a very sensitive contrast mechanism. For example, polarimetric imaging microscopy has been used to study the cellular organelle to infer nanoscale crystalline process as it occurs.[50] Magneto-optic (MO) Kerr imaging has been developed to study the domain structures and magnetization process of magnetic thin film.[51] A polarization dependent fluorescent pattern has been utilized for molecule orientation imaging.[22] Optical focus with prescribed linear polarization finds many applications in tip enhanced optical near field imaging.[52-54] In principle, full control of the state of polarization in the focal plane could provide much richer information in optical microscopy and significantly expand its functionality. In that aspect, vector point spread function (PSF) generation technique has been demonstrated to generate linear, radial or azimuthal state of polarization in the focal volume.[55]

Recent developments in nanofabrication and spatial light modulators (SLM) offer unprecedented level of light polarization engineering.[55-57] It becomes feasible to fully control the 3D polarization within the focal volume while maintaining the optimal diffraction-limited spot. By combining electric dipole radiation and Richards-Wolf vectorial diffraction method,[43, 44] the input field at the pupil plane of a high NA objective lens for generating arbitrary 3D oriented linear polarization at the focal point with an optimal spot size can be found analytically by solving an inverse problem. The corresponding input field to generate an arbitrary 3D elliptical state of polarization can also be found using two electric dipoles situated at the focus and polarized at orthogonal planes with different phase and amplitude.

In order to enable the 3D polarization control in practical optical microscopy, ideally one wants to control the 3D state of polarization while maintaining the focal spot to be diffraction-limited simultaneously. This requires a systematic approach to obtain the necessary input field at the pupil that can be focused into an optimal spot with arbitrary desired

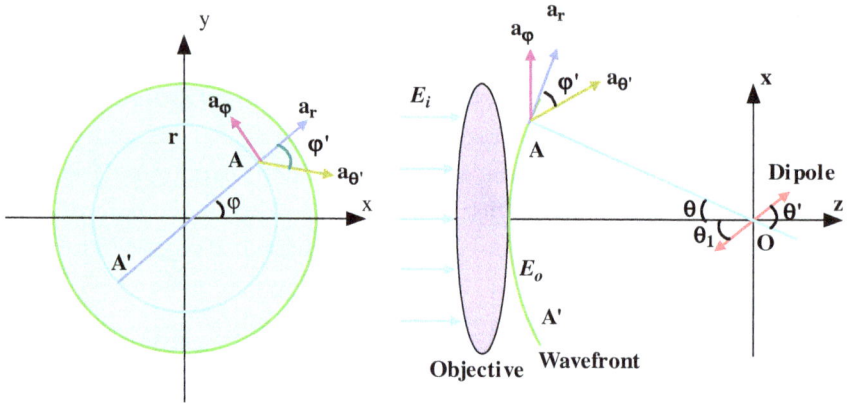

Fig. 4. Schematics setup for 3D polarization control. An electric dipole aligned in the direction of the desired 3D linear polarization at the focus is used to mimic a radiation source for the required pupil plane field calculation.[21]

state of polarization. Inspired by the observation of the connection between the electric dipole radiation and the focusing of radial polarization (see Fig. 13 in Chapter 1), such a systematic approach for the 3D linear polarization and field distribution control is illustrated in Fig. 4. An electric dipole situated at the focus of a high NA objective lens is aligned in the same direction as the desired linear polarization. The dipole radiation has well known angular patterns of field strength and local polarization distribution. The radiation from this dipole is collected and collimated by an aplanatic objective lens and the field distribution at the lens pupil can be found in conjunction with Richards-Wolf vectorial diffraction theory. The field at the lens pupil plane will be a vector field in general with nonuniform amplitude distribution. If this field distribution is used as illumination in the pupil plane and by reversing the propagation direction, the field of the electric dipole should be reconstructed up to the propagating components of the electric dipole radiation at the focal point, which should give the desired 3D polarization one began with. An optimal spot with a desired 3D linear polarization will be achieved. Hence, such method should provide a systematic way of obtaining the required field distribution at the lens pupil.

As shown in Fig. 4, an electric dipole oscillating in x-z plane with an angle θ_1 measured from the optical axis z, the dipole radiation angle to point A, which is also the angle between OA and dipole orientation, θ' can be expressed as

$$\theta' = 2\sin^{-1}[\sqrt{(1 - \sin\theta\sin\theta_1\cos\varphi - \cos\theta\cos\theta_1)/2}], \tag{7}$$

where θ is the focusing angle measured from the optical axis, and φ is the azimuthal angle. The dipole radiation field at point A is $\vec{R}(\theta') = C\sin\theta'\vec{a}_{\theta'}$, where $C = p_o k^2/4\pi\varepsilon_o\varepsilon f$ is a constant related to the dipole moment p_o, and $\vec{a}_{\theta'}$ is the unit vector of the radiation field at point A.[58] The angle φ' between $\vec{a}_{\theta'}$ and $\vec{a}_r$ is given by

$$\cos\varphi' = (-\cos\theta'\cos\theta + \cos\theta_1)/(\sin\theta\sin\theta'). \tag{8}$$

The dipole radiation field at point A can be decomposed into radial component $\vec{a}_r$ and azimuthal component $\vec{a}_\varphi$ on the spherical surface after the objective lens

$$\vec{E}_o(\theta,\varphi) = R(\theta')(\cos\varphi'\vec{a}_r - \sin\varphi'\vec{a}_\varphi). \tag{9}$$

Considering the bending effect, the field on the spherical surface can be projected to the Cartesian coordinates of the image space (dipole space) as

$$\vec{E}_o(\theta,\varphi) = R(\theta')[(\cos\varphi'\cos\theta\cos\varphi + \sin\varphi'\sin\varphi)\vec{i}$$
$$+ (\cos\varphi'\cos\theta\sin\varphi - \sin\varphi'\cos\varphi)\vec{j} + \cos\varphi'\sin\theta)\vec{k}]. \tag{10}$$

The input field $\vec{E}(\theta,\varphi)$ at the pupil plane can be found to be

$$\vec{E}_i(\theta,\varphi) = R(\theta')[(\cos\varphi'\cos\varphi + \sin\varphi'\sin\varphi)\vec{x}$$
$$+ (\cos\varphi'\sin\varphi - \sin\varphi'\cos\varphi)\vec{y}]. \tag{11}$$

Note that the field is expressed in terms of the refraction angle θ in the image space and the azimuthal angle φ. To express the input field in the pupil plane spatial coordinates (r,φ), the projection function of the objective lens needs to be considered.[45] For an objective lens that obeys sine condition $r = f \cdot \sin \theta$, where f is the focal length of the objective lens, the projection function from (r, φ) space to (θ, φ) space is $\sqrt{\cos \theta}$. Consequently, the input field at the lens pupil will be $\vec{E}_i(r, \varphi) = \vec{E}_i(\theta, \varphi) / \sqrt{\cos \theta}$, with $\theta = \sin^{-1}(r/f)$.

Finally the desired illumination field at the pupil to generate the prescribed 3D linear polarization at the focus is found to be

$$\vec{E}_i(r, \varphi) = R(\theta')[(\cos \varphi' \cos \varphi + \sin \varphi' \sin \varphi) \vec{x}$$
$$+ (\cos \varphi' \sin \varphi - \sin \varphi' \cos \varphi) \vec{y}] / \sqrt{\cos \theta}. \quad (12)$$

The Richards-Wolf vectorial diffraction method is applied to verify whether this input field will produce a focal spot with the desired polarization. The electric field $\vec{E}$ at any point $P(r_p, \psi, z_p)$ near the focus can be calculated as.[45]

$$\vec{E}(r_p, \psi, z_p) = \frac{i}{\lambda} \int_0^{\theta_{max}} \int_0^{2\pi} \vec{E}_o(\theta, \varphi) e^{-ikr_p \sin \theta \cos(\varphi - \psi) - ikz_p \cos \theta} \sin \theta d\theta d\varphi, \quad (13)$$

Where $\vec{E}(r_p, \psi, z_p)$ is the electric field vector at point P; λ is the wavelength; $\theta_{max} = \sin^{-1}(NA)$ with NA as the numerical aperture of the aplanatic lens; $\vec{E}_o(\theta, \varphi)$ is the field distribution after the refraction of the lens given by Eq. (10).

Arbitrary 3D linear polarization can be obtained through changing the dipole orientation. For example, to produce a $45°$ linear polarization in x-z plane, one can simply set $\theta_1 = 45°$. The input field to produce the desired 3D linear polarization at the pupil plane $\vec{E}_i(r, \varphi)$ is shown in Fig. 5(a). The input field is a vector beam with spatially variant amplitude and polarization. Figure 5(b) illustrates the 3D slice projection of the focal spot in three orthogonal planes using the calculated input pupil field. A homogeneous spot with uniform polarization is obtained around the focal point. The FWHMs of the focused spot are 0.513λ along the x-axis, 0.414λ along the y-axis, and 0.966λ along the z-axis (depth of focus), respectively. The conversion efficiency from the input power to the main focal spot is 60.6%.

To generate 3D elliptical polarization states, one can introduce a second electric dipole polarized in the y-z plane with certain field strength ratio η and phase shift ϕ with respect to the dipole oscillating in the x-z plane. For this second dipole, the azimuthal angle is $\varphi_2 = \varphi - \pi/2$. Radiation angle θ_2' and φ_2' can be expressed as

$$\theta_2' = 2\sin^{-1}[\sqrt{(1 - \sin\theta\sin\theta_2\cos\varphi_2 - \cos\theta\cos\theta_2)/2}] , \tag{14}$$

$$\cos\varphi_2' = (-\cos\theta_2'\cos\theta + \cos\theta_2)/(\sin\theta\sin\theta_2') . \tag{15}$$

Following the same procedure, the illumination field at the lens pupil plane due to the second dipole can be found as

$$\vec{E}_{i2}(r,\varphi) = \eta e^{j\phi} R(\theta_2')[(\cos\varphi_2'\cos\varphi_2 + \sin\varphi_2'\sin\varphi_2)\vec{x}$$

$$+ (\cos\varphi_2'\sin\varphi_2 - \sin\varphi_2'\cos\varphi_2)\vec{y}]/\sqrt{\cos\theta} . \tag{16}$$

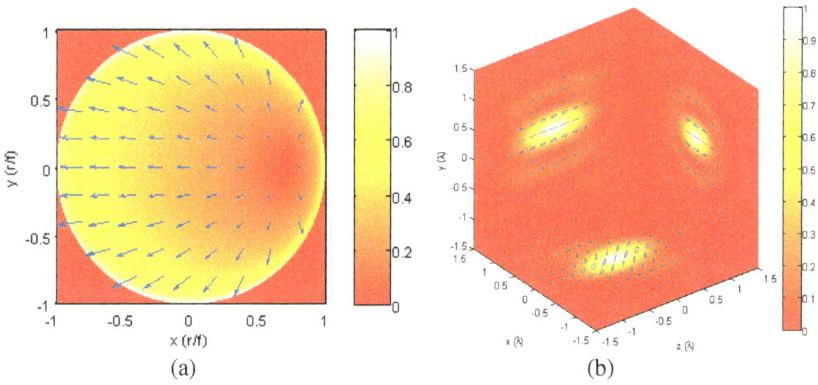

Fig. 5. (a) Pupil plane field for generating desired 3D linear focal polarization. (b) Projection of intensity and state of polarization distributions on three orthogonal planes in the focal region. The field at focus is 45° linearly polarized in x-z plane.[21]

Coherent superposition of this field with the field computed with Eq. (12) gives the required pupil field that can create a desired 3D elliptical focal field polarization. Figure 6 shows an example for achieving a 3D polarization state with projected polarization 45° linear in x-z plane, circular in both the x-y and y-z planes. The pupil illumination (shown in Fig. 6(a)) can be found by setting $\theta_1 = 45°$, $\theta_2 = 90°$, the field strength ratio $\eta = \sin(\theta_1)$ and phase shift $\Phi = \pi/2$. The intensity and the state of polarization distributions of the focused spot are projected to three orthogonal planes shown in Fig. 6(b). The FWHMs of the focused spot are 0.468λ along x-axis, 0.470λ along y-axis, and 0.938λ along z-axis, respectively. The conversion efficiency from the input power to the main spot is 72.3%.

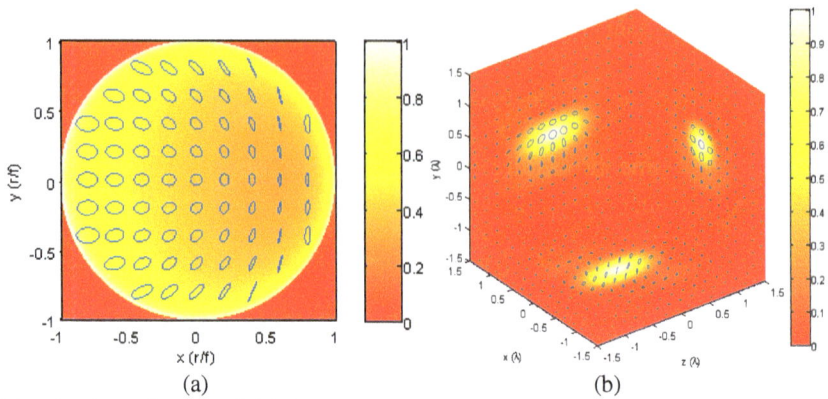

Fig. 6. (a) Pupil plane field for generating desired 3D elliptical focal polarization. (b) Projections of intensity and state of polarization distributions in the focal region onto three orthogonal planes. The projected polarization of the field in the focus is circular in the x-y plane, 45° linear in the x-z plane and circular in the y-z plane.[21]

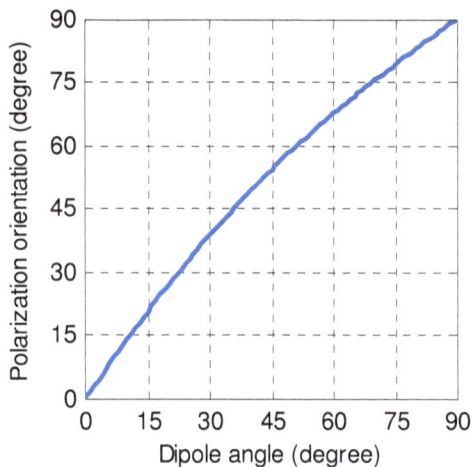

Fig. 7. Deviation from linear relationship with lower NA lens.[21]

Numerical aperture equal to 1 is used in the previous calculation. Such high NA can be achieved with reflective type of parabolic objectives.[59] The method also works for refractive objective with lower collection angle. For an electric dipole polarized in the x-z plane with an angle θ_1 measured from the optical axis z, if the radiation field is collected by an objective lens of NA equal to 1, the illumination field found at the input field plane can be focused into a 3D linear polarization with a same angle measured from the optical axis (equal to θ_1). However, if an objective lens of NA less than 1 is used, the obtained 3D linear polarization at the focus will be slightly off from the starting angle θ_1. Figure 7 shows the relationship between the dipole orientation and the polarization direction of the focused field with an aplanatic lens of NA = 0.95. A small deviation from the linear relationship is found due to the cutoff of the contribution from the illumination edge. Therefore, to obtain a focused field with desired 3D polarization using a lower NA objective, one needs to choose a dipole with orientation angle slightly smaller than the desired inclination angle as the start to calculate the required pupil field.

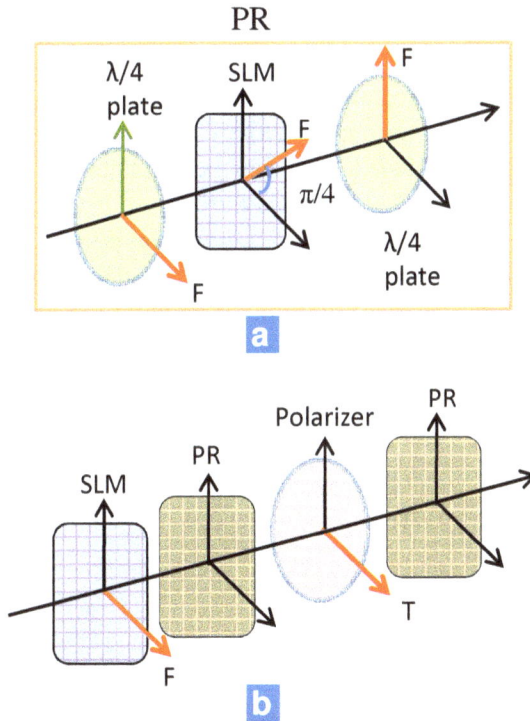

Fig. 8. Schematic setup for generating vector beams with amplitude distribution. (a) Pure polarization rotator (PR) using SLM sandwiched between two orthogonally oriented $\lambda/4$ waveplates. F: fast axis. (b) Generation of arbitrary polarization and amplitude distribution by using a SLM and two pure polarization rotators.[21]

In principle, the field necessary for 3D polarization control has nonuniform spatial distributions in polarization, magnitude and phase. Such a complicated field requires the full control of the spatial distribution of light field. This is becoming possible with the advances in nanofabrication and spatial light modulators. As shown in Fig.8(a), a pure polarization rotator has been demonstrated by sandwiching a SLM between two orthogonally oriented λ/4 waveplates.[55, 60, 61] The fast axis of the SLM has an angle of π/4 with respect to the two waveplates. The Jones matrix of this rotator is

$$
T = R\left(-\frac{\pi}{2}\right)\begin{bmatrix} 1 & 0 \\ 0 & -j \end{bmatrix} R\left(\frac{\pi}{2}\right) R\left(-\frac{\pi}{4}\right)\begin{bmatrix} e^{j\varphi_{xy}/2} & 0 \\ 0 & e^{-j\varphi_{xy}/2} \end{bmatrix} R\left(\frac{\pi}{4}\right)\begin{bmatrix} 1 & 0 \\ 0 & -j \end{bmatrix} \tag{17}
$$

$$
= -je^{-j\varphi_{xy}/2} R\left(\frac{\varphi_{xy}}{2}\right),
$$

where φ_{xy} is the spatially distributed retardation from the SLM. This polarization rotator is independent of the initial state of polarization. The amount of rotation is determined by the retardation of the SLM. Vector beams with amplitude distribution can be obtained using three SLMs (Fig. 8(b)). The first SLM generates a pure phase pattern of $\delta_{xy}/2$ for the whole aperture. The combination of the first pure polarization rotator and a linear polarizer provides the necessary amplitude modulation. Linearly polarized beam with desired amplitude distribution is obtained after the linear polarizer. The second pure polarization rotator then rotates the local polarization to the desired vector field direction. The Jones matrix of the system can be found as

$$
T = -je^{-j\varphi'_{xy}/2} R\left(\frac{\varphi'_{xy}}{2}\right)\begin{bmatrix} 1 & 0 \\ 0 & 0 \end{bmatrix}\left(-je^{-j\varphi_{xy}/2}\right) R\left(\frac{\varphi_{xy}}{2}\right)\begin{bmatrix} e^{-j\delta_{xy}/2} & 0 \\ 0 & e^{-j\delta_{xy}/2} \end{bmatrix}\begin{bmatrix} 1 & 0 \\ 0 & 0 \end{bmatrix} \tag{18}
$$

$$
= -j\cos\left(\frac{\varphi_{xy}}{2}\right) e^{-j(\varphi_{xy}+\varphi'_{xy}+\delta_{xy})/2} R\left(\frac{\varphi'_{xy}}{2}\right).
$$

The phase retardations $\varphi_{xy}/2$ and $\varphi'_{xy}/2$ due to the two pure polarization rotators can be compensated by the first SLM.

By the inverse dipole radiation method, the focused field has maximum intensity in the focal point with a diffraction-limit spot size. This is very valuable for applications such as particle trapping and manipulation. For example, for a dielectric particle with higher dielectric constant than the ambient, the gradient force of the focused field tends to pull and trap the particle to the highest intensity region of the focal spot.[62] Recently, there is a strong interest in manipulating and trapping anisotropic particles, either geometrical or optical anisotropy, such as carbon nanotube, nanorods, ferroelectric materials or birefringent crystals. If the tightly focused beam is linearly polarized, the nano- or micro-particles will be trapped and aligned in the plane of polarization. Arbitrary 3D linear polarization control makes it possible to manipulate these particles by changing the polarization of the tightly focused beam. Real time polarization control with SLMs enables real time control of the trapping force, and manipulates the particles in desired trajectory. If the focused beam is 3D elliptically polarized, the anisotropic particles will experience a spinning torque.[24] The tightly focused beam with desired 3D polarization provides strong alignment force or spinning torque for particle manipulation.

4. Spherical spot with controllable 3D polarization

A spherical spot with equal 3D spatial resolutions has important applications in optical microscopy, single molecule fluorescence spectroscopy,[63] optical data storage, particle trapping and optical tweezers.[64] For conventional optical microscope with single objective lens, the axial extent of the focused spot is always several times larger than the transversal extent, resulting in lower axial resolution. A tightly focused spherical spot with diffraction-limited size that provides equal axial and transversal resolutions is strongly desirable. In order to improve the axial resolution, 4-π microscopy that uses two opposing high NA objective lenses has been developed.[65, 66] The axial resolution is improved by the coherent interference of the two counter propagating focused wavefronts. However, the axial resolution generally is still worse than the transverse resolution, leading to an elongated focal spot.

In this section we introduce a new approach to obtain spherical spot with 4-π focusing using aplanatic lenses.[41] Adopting an approach similar to the radiation reversal method described in Section 3, the input field at the pupil plane of high NA objective lenses in a 4-π microscopy for

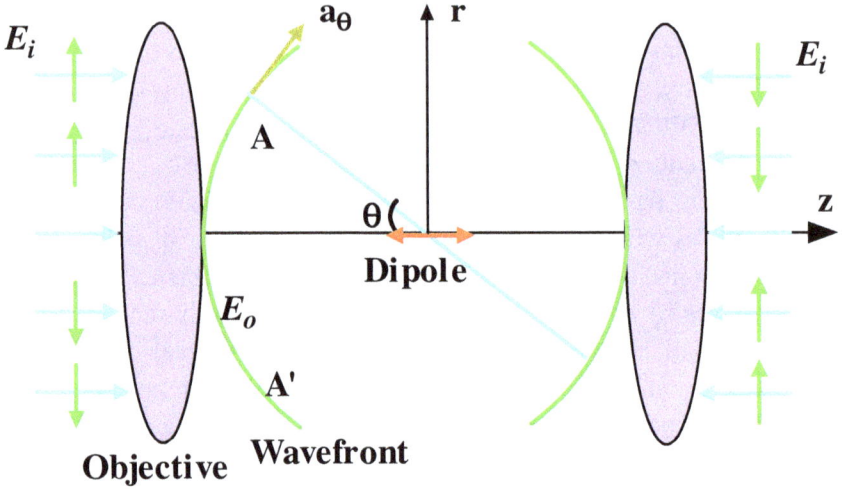

Fig. 9. Schematic setup for generating a spherical spot. A dipole antenna situated at the foci of two high NA objective lenses is aligned along the optical axis. The radiation field is collected and collimated by the 4-π focusing system, and the input field at the pupil planes for creating spherical focal spot is found.[41]

generating spherical focal spot with diffraction-limited size can be found analytically by solving the inverse problem through combining the dipole antenna radiation pattern and the Richards-Wolf vectorial diffraction method. A spherical spot can be obtained by choosing an appropriate dipole antenna length with the calculated field at the pupil plane being used as illumination. Even more importantly, we show that it is also possible to control the 3D state of polarization within the spherical focal spot.

The schematic for the generation of a spherical focal spot is illustrated in Fig. 9. A dipole antenna situated at the foci of two high NA objective lenses is aligned along the optical axis. The dipole antenna has a length L that is comparable with the wavelength λ of the radiation field. The field at the pupil plane is radially polarized owing to the symmetry of the setup. The angular field radiation pattern $R(\theta)$ of the dipole antenna can be written as

$$\vec{R}(\theta) = C\left[\cos\left(\frac{kL}{2}\cos\theta\right) - \cos\left(\frac{kL}{2}\right)\right]\bigg/\sin\theta\,\vec{a}_\theta\,, \qquad (19)$$

where $\vec{a}_\theta$ is the unit vector of the radiation field in the image space after the objective lens. Then the field distribution calculated above is used as illumination in the pupil plane. By reversing the optical path, the electric fields in the vicinity of the focal spot for radial polarization can be calculated by Richards-Wolf vectorial diffraction method.[15, 16]

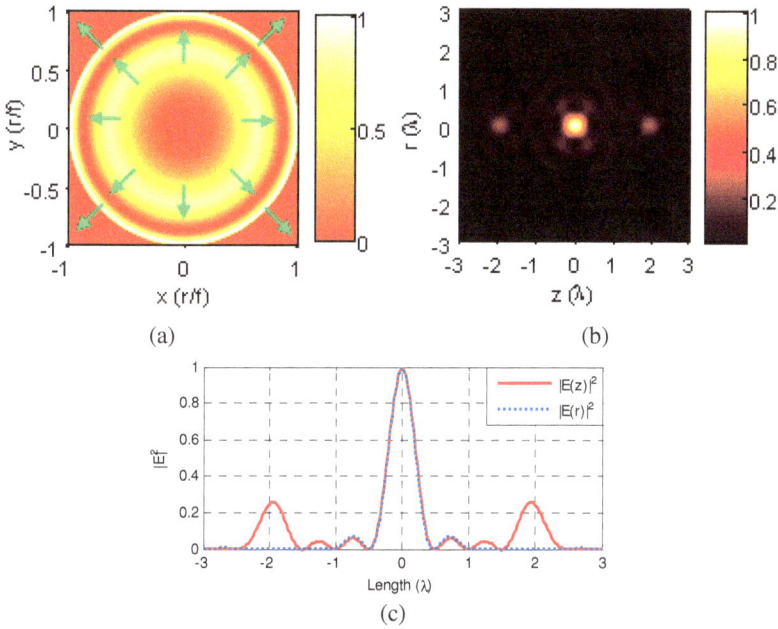

Fig. 10. Generation of spherical spot with the use of $L = 4\lambda$ dipole antenna. (a) Input field at the pupil plane for producing a spherical spot. (b) Calculated electric energy density $|E|^2$ distribution in the vicinity of focal point. (c) Linescans of corresponding axial and transversal energy density distributions.[41]

For $L = 0$ or very small dipole length, the problem is reduced to the inverse point dipole situation that has been dealt with in Ref. 40. And indeed an elongated spot has been resulted. However, it is found that a spherical spot can be obtained by adjusting the dipole antenna length to be $L = m\lambda$ ($m = 3, 4, 5 \dots$). For example, for dipole antenna length $L = 4\lambda$, the input field $\vec{E}_i(r)$ at the pupil plane to produce a spherical spot is shown in Fig. 10(a). The input field is a radial polarization with spatial amplitude modulation. Figure 10(b) shows the energy density $|E|^2$ distribution. The corresponding linescans of the axial and transversal energy density distributions are shown in Fig. 10(c). A spherical spot with nearly equal axial and transversal focal spot sizes with size mismatch less than 0.012λ has been obtained. The FWHM spot size is calculated to be 0.442λ, corresponding to a focal volume of $0.045\ \lambda^3$.

In the above example, the maximum sidelobes are about 2λ away from the mainlobe, which is half of the dipole antenna length. The sidelobes are far enough from the mainlobe so they do not affect the performance of optical microscopy in most cases including nonlinear or confocal optical microscopy. In addition, the maximum sidelobes can be shifted further away by manipulating the dipole antenna length L. For example, if L is set to be 6λ, the sidelobes are further pushed away from the main focal spot. The general trend is that a larger integer m for dipole antenna length leads to farther separation of the maximum sidelobes from the main focus and better spherical focal spot. However, the price

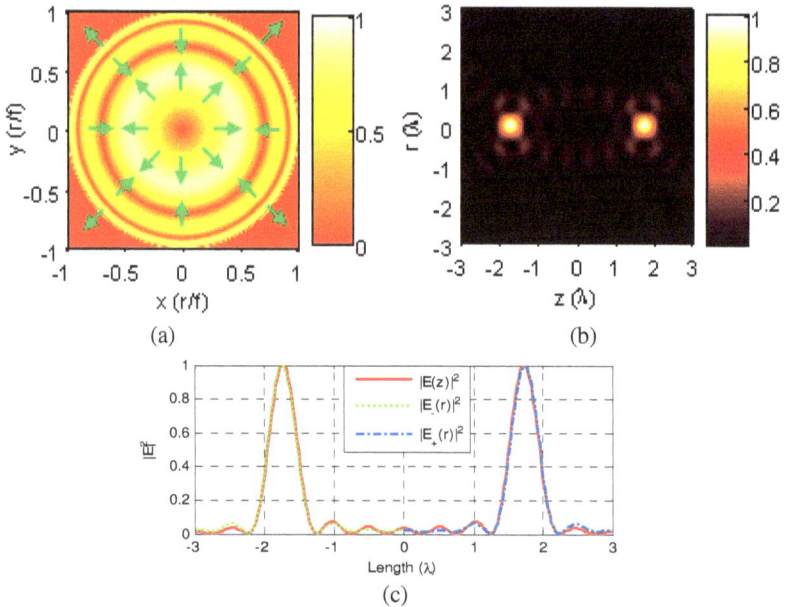

Fig. 11. Generation of two diffraction-limited spherical spots with the use of $L = 3.5\lambda$ dipole antenna. (a) Input field at the pupil plane for producing two spherical spots. (b) Calculated electric energy density $|E|^2$ distribution in the vicinity of focal point. (c) Linescans of corresponding axial and transversal energy density distributions, $E_-(r)$ and $E_+(r)$ are transversal linescans for the two spherical spots respectively overlaid on the axial linescan.[41]

one has to pay is a more complicated pupil plane spatial amplitude modulation of the radially polarized illumination.

In addition, two identical spherical spots with diffraction-limited sizes along the optical axis can also be achieved by setting the dipole antenna length to be an odd integer number times a half wavelength, i.e., $L = (2m+1)\lambda/2$, where m is an integer. For example, for $L = 3.5\lambda$, the input field $\vec{E}_i(r)$ at the pupil plane to produce two spherical spots is found and shown in Fig. 11(a). The input field is a radially polarized beam with spatial phase modulation. The local polarization alternates directions in different annulus caused by π phase jump between the adjacent zones. Figure 11(b) shows the energy density $|E|^2$ distribution in the vicinity of the focal point. The corresponding axial energy distributions are shown in Fig. 11(c). Two diffraction-limited spherical spots have been obtained. The sidelobes are very low comparing to the two main spots. The distance of the two spots is equal to the dipole antenna length L. The creation of the two identical spots is due to the interference from the multiple annular illumination.

A diffraction-limited spherical spot with controllable arbitrary 3D polarization can be realized through incorporating 3D state of polarization control technique into the spherical spot generation in 4-π microscopy.[67] Spherical spot with arbitrary 3D linear polarization can be obtained through changing the dipole antenna orientation. For example, in order to produce a spherical spot with 45° linear polarization in the x-z plane, one can simply set the dipole orientation angle $\theta_1 = 45°$. The amplitude and polarization patterns at the lens pupil plane are shown in Fig. 12(a). The input field is vector beam with spatially variant amplitude and polarization distributions. Figure 12(b) illustrates the 3D slice projection of the focal spot in three orthogonal planes. Intensity and local polarization distributions are both shown. Figure 12(c) shows the state of polarization distribution within the focal volume. The focal spot maintains linearly polarized, with a maximum polarization angle deviation of 8.77°, and a root mean square (RMS) deviation of 2.03°. Thus, a spherical spot with nearly uniform polarization is obtained in the vicinity of the focal point. The corresponding linescans of the axial and transversal energy density distributions are shown in Fig. 12(d). A spherical spot with nearly equal axial and transversal focal spot sizes with size mismatch less than 0.016λ has been obtained. The FWHM spot size is calculated to be 0.450λ, corresponding to a focal volume of 0.048 λ^3.

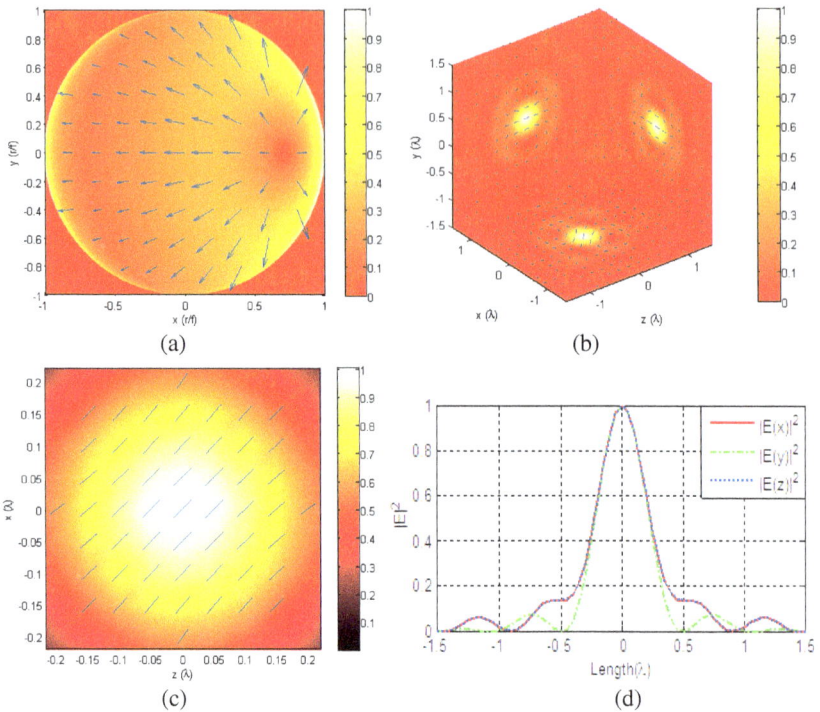

(a)

(b)

(c)

(d)

Fig. 12. (a) Input field at the pupil plane for creating a spherical spot with 3D linear polarization. (b) Projection of intensity and state of polarization distributions onto three orthogonal planes in the focal region. The field at focus is 45° linearly polarized in x-z plane. (c) State of polarization distribution within the FWHM focal volume. (d) Linescans of energy density distributions along three orthogonal axes.[67]

5. Focus shaping through inverse dipole array radiation

The radiation reversal approach described above can be further extended to dipole array (including electric dipole array, magnetic dipole array and a mixture of electric and magnetic dipole arrays) for systematically engineering even more exotic optical focal fields. In this section we illustrate the capability and flexibility of this approach through demonstrating the creation of ultra-long diffraction-limited optical needle field, 3D optical tube, 3D flattop focus with extended DOF, and uniform optical chain, etc.

5.1. *High purity optical needle field*

Creation of optical needle field has attracted lots of research interests recently. An optical field is a special focal field distribution that is substantially polarized along the axial direction. Some of the desired characteristics of an optical needle field include high purity of the axial polarization, small transverse size and long DOF. Through reversing the radiation field from an electric dipole array situated near the focus of a high NA lens, the required incident field distribution in the pupil plane for producing an ultra-long optical needle field can be found analytically.[31] As shown in Fig. 13, the angular radiation of a dipole array with $2N$ electric dipoles oscillating along optical axis can be expressed as[68]

$$\bar{E}_0(\theta) = E_{DA}(\theta)\bar{a}_\theta = C\sin\theta AF_N\bar{a}_\theta, \qquad (20)$$

where AF_N is the array factor (AF) related to the phase delay caused by the spacing distance d_n and initial phase difference β_n of each pair of the dipole that are mirror symmetric with respect to the focal plane

$$AF_N = \sum_{n=1}^{N} A_n \left[e^{j(kd_n\cos\theta+\beta_n)/2} + e^{-j(kd_n\cos\theta+\beta_n)/2} \right], \qquad (21)$$

where A_n is the ratio of the radiation amplitude between the nth dipole pair and the standard dipole pair with normalized amplitude. The DOF is mainly determined by the total number of dipole elements $2N$. Higher N generally leads to longer DOF. However, the corresponding pupil plane

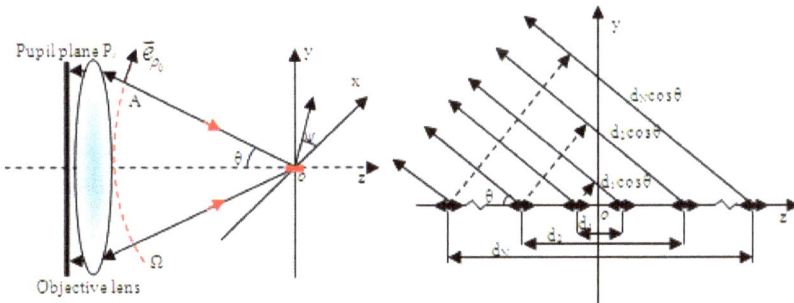

Fig. 13. Schematic of the method of pupil plane field synthesis in order to achieve specific focal field characteristics through the reversing of the radiation pattern of a dipole array. (a) System layout with dipole array oscillating along optical axis in the focal volume and the coordinates used in the calculation; and (b) Far-field geometry and conceptual diagram of the dipole array with $2N$ dipole elements.[31]

distribution may become increasingly complicated and the number of parameters for optimization will increase accordingly.

Examples of the optical needle fields generated with this method are illustrated in Fig. 14. The DOF defined as the axial full width of above 80% maximum intensity can reach 5λ and 8λ for $N = 2$ and $N = 3$, respectively. The beam purity η defined as the percentage of the longitudinal component intensity of total field can be expressed as,[26]

$$\eta = \Phi_z / (\Phi_z + \Phi_r), \Phi_{z(r)} = \int_0^{r_0} \left| E_{z(r)}(r, z) \right|^2 r dr , \tag{22}$$

where r_0 is the first zero point in the distribution of radial component intensity. In the focal plane (z = 0), it has been calculated as $\eta_{N=2} = 86\%$ and $\eta_{N=3} = 87\%$. A very important advantage of this method is that the η remains nearly constant throughout the entire DOF (>81% for both cases). This demonstrates that very high longitudinal polarization component purity has been achieved in the entire DOF. Nearly flattop axial distribution in focal volume can be achieved. With an increase of the DOF from 5λ to 8λ, the intensity near the edge will have a shaper transition, a very desirable property for focal fields with extended DOF. A FWHM spot size of the total field intensity of 0.408λ and 0.405λ is realized for $N = 2$ and $N = 3$, respectively. The later case leads to a spot size of $0.129\lambda^2$ and the FWHM is restricted below 0.43λ for $-3.5\lambda < z < 3.5\lambda$.

The corresponding distributions of the required incident field in the pupil plane P_i are shown in Figs. 14 (e) and (f). For dipole arrays with $2N$ elements, the computed incident distributions can be divided into $2N$ annular bright zones separated by dark rings. The characteristic of this distribution can be expressed as $P(\rho_i)\exp(-j\psi(\rho_i))$, where $P(\rho_i)$ denotes a continuous amplitude distribution. The innermost core of the pupil field is dark, indicating a polarization singularity point. The peak amplitude of the bright zones grows monotonically from the center to the outermost zone. The phase $\psi(\rho_i)$ takes binary values that alternates between 0 and π from one zone to another. The pupil field is radially polarized. Due to the alternating phase, the polarization actually flips its directions between adjacent zones.

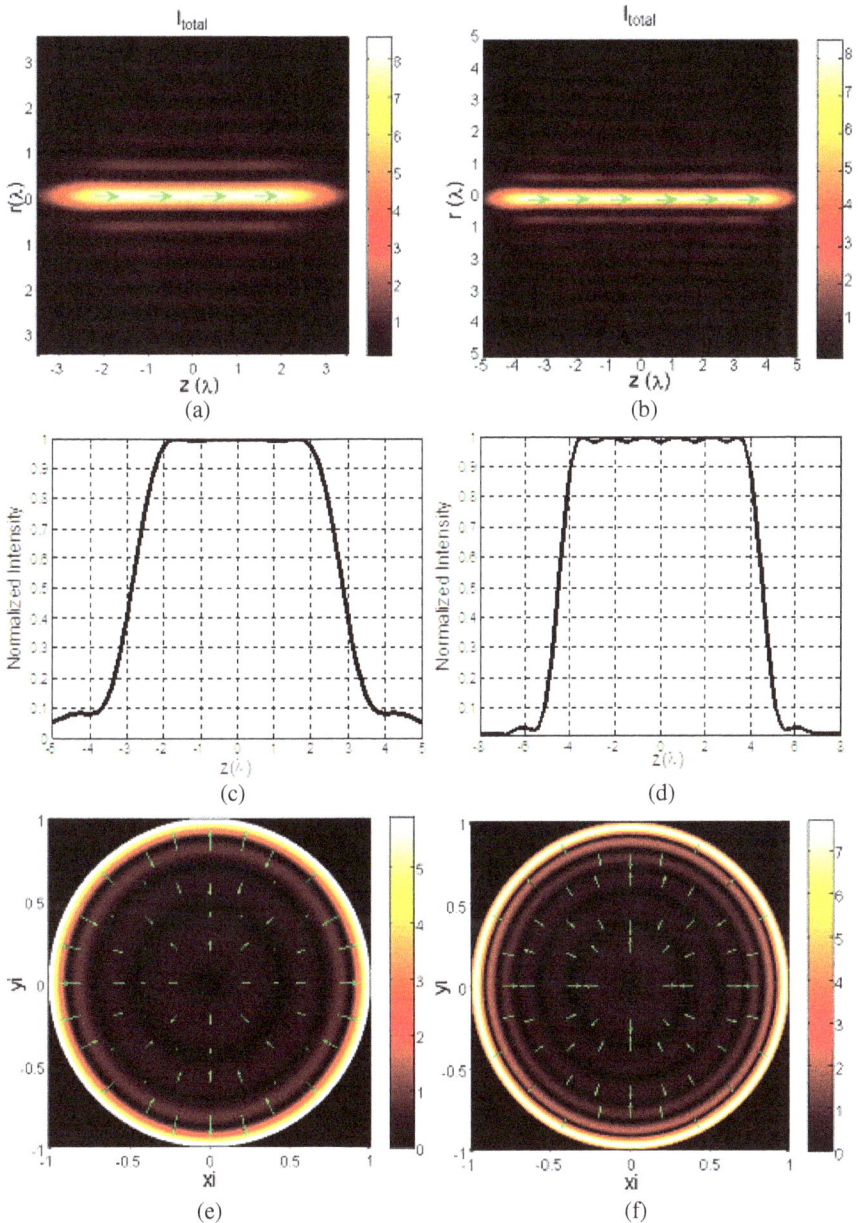

Fig. 14. Total intensity distribution in the transverse r-z plane and axial intensity distribution in the focal volume for the obtained optical needle field with extended DOF for (a) (c) N = 2; and (b) (d) N = 3. The corresponding required incident field in the pupil plane Pi are illustrated for (e) N = 2 and (f) N = 3.[31]

The desired incident pupil field can be considered a radially polarized field filtered by a complex pupil filter, where $P(\rho_i)$ is the amplitude transmittance and $\psi(\rho_i)$ is the phase of the complex filter. In the above design, $P(\rho_i)$ is a continuous function while $\psi(\rho_i)$ takes binary values. Liquid crystal spatial light modulators (SLM) have been demonstrated for the control of the spatial distribution of the light field in the objective's pupil.[21, 55, 57] However, precise control of continuous transmittance is very challenging due to the needs of extremely careful microscale adjustment and the pixelated format of the SLM, especially when rapid amplitude transmittance spatial variation occurs. In order to simplify the pupil filter design and draw a comparison with the conventional binary DOE design,[26, 27, 30, 69] a discrete complex filter design with non-continuous gray scale amplitude transmittance was introduced.

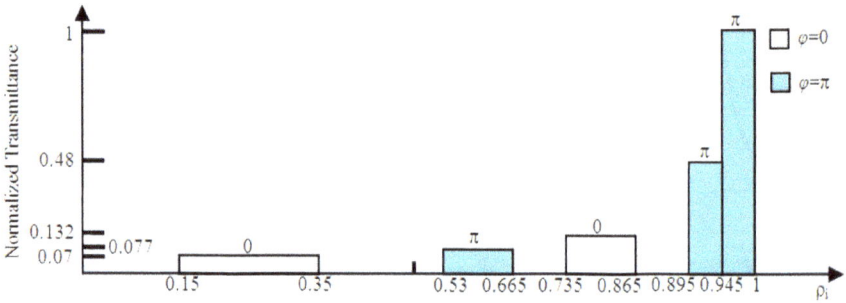

Fig. 15. Structure and parameters of discrete complex pupil filter.[31]

An example of this new type of filter is designed based on the continuous pupil field distribution of a four-dipole array ($N = 2$) shown in Fig. 14(e). Zones with zero transmittance are designed according to the dark central core and the three dark rings in Fig. 14(e). An amplitude threshold of 5% of the maximal amplitude in the pupil plane shown in Fig. 14(e) is chosen. The transmittances for those regions with amplitude lower than this threshold are set to zero. The continuous amplitude transmittance was reduced to discrete transmittance through averaging. Five annular zones with discrete transmittance ranging from 0.07 to 1 and alternating phase 0 or π are shown in Fig. 15. The transition points and radial widths of these annular transmitted zones are optimized and given in Fig. 15. Each transmitted zone corresponds to one bright ring in Fig. 14(e) except for the outermost zone. Owing to the sharp amplitude variation in the outermost zone, it was divided into two zones with the same phase and different average transmittance. Note that the parameters of the outermost annular zone strongly influence the transverse distribution of the focal field while the inner zones help to flatten the axial field intensity distribution to achieve an extended DOF.

The corresponding total field intensity distributions in the transverse plane and along the z-axis are illustrated in Figs. 16(a) and (b) respectively. Similar to the continuous pupil filer, this discrete filter also gives rise to longitudinally polarized field with extended DOF. The results are comparable with those shown in Figs. 14(a) and (c). Thus such a discrete complex pupil filter can be used as a proper substitute of the continuous pupil filter. The performance of this discrete filter demonstrates significant advantages over binary DOE based methods. The optical needle obtained with the discrete pupil filter has higher beam purity throughout the focal volume, tighter lateral confinement and smaller sidelobes.

In practical applications, imperfections of the designed filter will influence the quality of the generated optical needle field. In general, the deviation of radius, transmitted amplitude and phase in each zone will

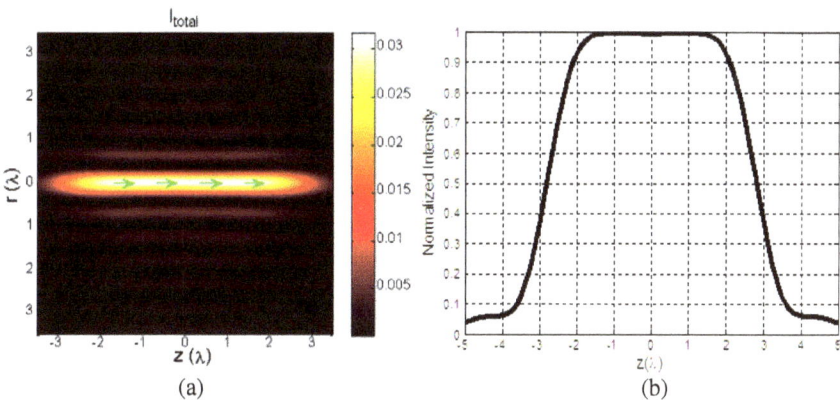

(a)   (b)

Fig. 16. Focal intensity distribution using the discrete complex pupil filter. (a) Total intensity distribution in the transverse r-z plane; (b) Axial total intensity distribution.[31]

affect the optical needle field quality to some extent. For ± 2.5% error in transition radius or ± 10% error in transmittance of the outmost zone, it is found that the axial intensity peak-to-valley fluctuations can be controlled below 5% of the peak intensity with the DOF greater than 4λ. The corresponding polarization purity still remains above 80% through the total DOF. For ± 0.2π phase deviations from π, axial intensity fluctuations can be controlled below 2% of the peak intensity while maintaining the long DOF (> 4λ) and high polarization purity (> 80%) throughout the DOF.

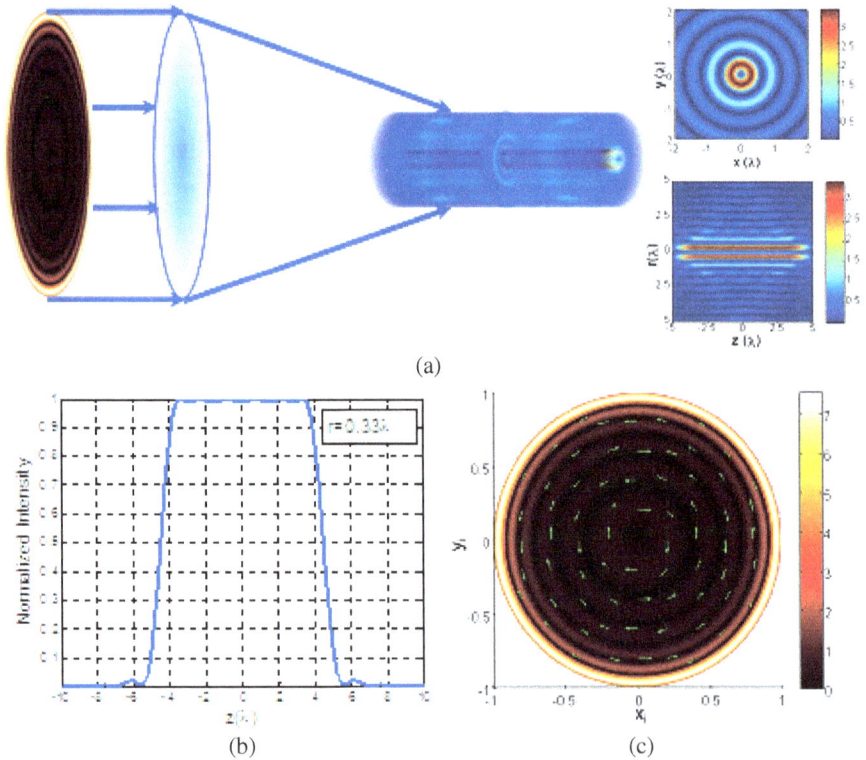

Fig. 17. (a) Concept scheme of construction of 3D optical tube field with the reversal radiation of a magnetic dipole array (NA = 0.95); (b) Axial field intensity along wall of the pipe when r = 0.33λ; (c) Required pupil field distribution for pipe field.[32]

5.2. *3D optical tube, flattop focus and optical chain*

The inverse method can be further explored by including the radiation patterns from magnetic dipole array. An ultra-long diffraction-limited 3D optical tube with maximal uniformity can be created with modulated azimuthal polarization illumination. Combination of co-located electric dipole array and magnetic dipole array allows the realization of ultra-long 3D homogenized focus that has flattop profiles in both transversal and longitudinal directions.[32]

By using magnetic dipole array with $2N = 6$ elements, an ultra-long (~8λ) sub-wavelength-wide 3D optical tube with azimuthal polarization spiraling around z-axis is created.[32] The transverse FWHM of this optical tube is about 0.312λ in the focal plane and remains nearly constant throughout the DOF. From the axial intensity distribution in Fig. 17(b) at $r = 0.33\lambda$, excellent intensity uniformity of the tube wall without obvious axial sidelobes can be observed. The required illumination field at the normalized pupil plane to create such optical tube is given in Fig. 17(c). The illumination features six continuous amplitude zones with alternating azimuthal polarization (π phase shifted) in neighboring zones. Incident light with this pupil distribution will be focused into the optical tube with local azimuthal polarization.

More complex focus engineering can be achieved with the combination of electric and magnetic dipole arrays. Proper choice of amplitude ratio tgα between magnetic and electric dipoles enables creation of 3D flattop focus with extremely long DOF.[32] Figures 18(a) and (b) show the obtained 3D flattop focus with 8λ DOF and required incident pupil field when $\alpha = 52°$. Radiated amplitude ratio is $\text{tg}\alpha = 0.74$. Corresponding axial and transverse intensity distributions are shown in Fig. 18(c) and (d). In the main lobe ($r < 0.3\lambda, |z| < 4\lambda$), both axial flattop (shown for $r = 0$, 0.1λ, 0.2λ) and transverse flattop (shown for $|z| = 0, \lambda, 2\lambda, 3\lambda$) can be retained very well. From Fig. 18(e), analysis of polarization structure in the focal plane denotes that in central zone longitudinal polarization is dominant and in marginal zone azimuthal polarization is stronger than other components. Radial polarization is always very weak in the entire DOF.

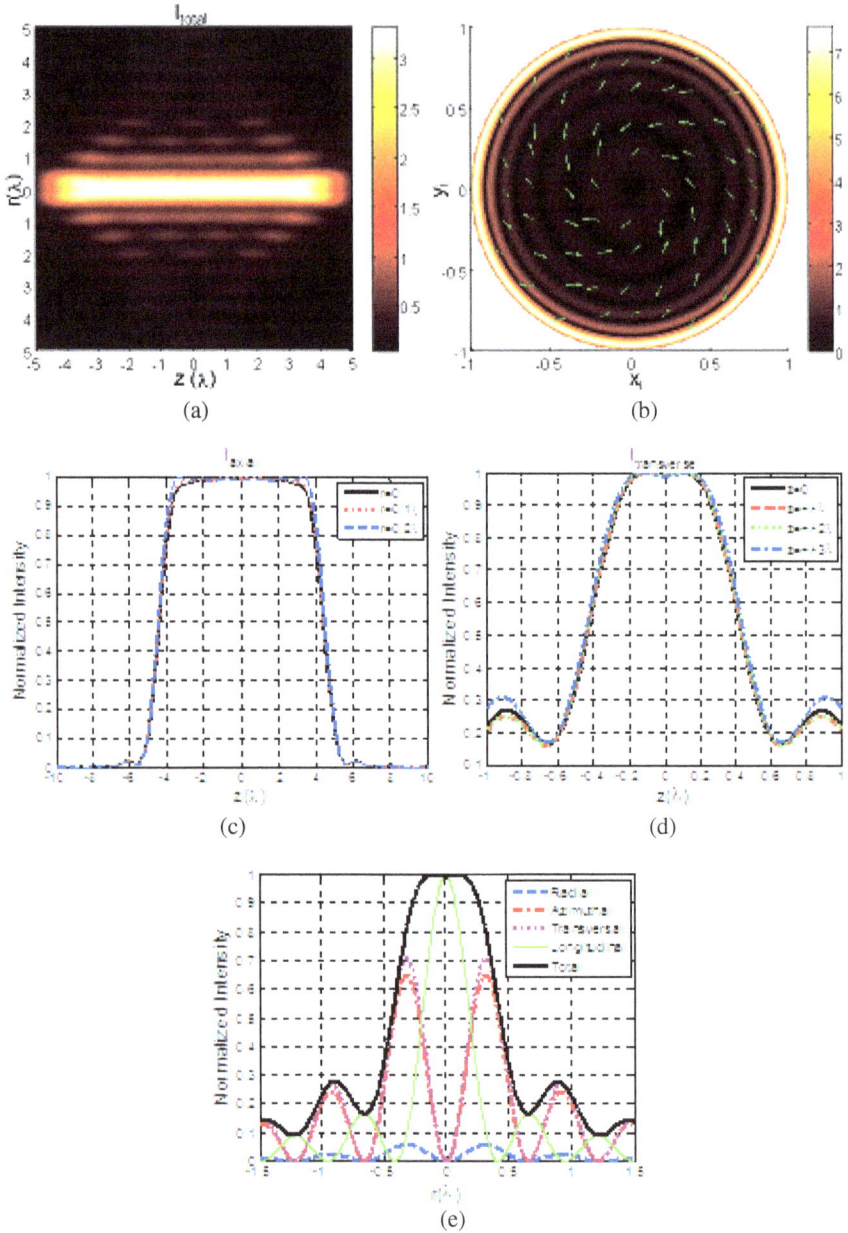

Fig. 18. Creation of 3D flattop focus (NA = 0.95). (a) Total intensity distribution in r-z plane with long DOF (~8λ); (b)Required incident pupil field; (c) Axial intensity along r = 0 and r = 0.2λ; (d)Transverse intensity along z = 0, ±λ, ±2λ,±3λ; (e) Polarization structure at the focal volume.[32]

In addition to optical needle and optical tube, 3D hollow focus has attracted attentions for its potential applications in the dark optical trapping and fluorescence microscopy as erasing beam. It has been demonstrated that filtered radially polarized beam can be focused into a bubble with a higher optical barrier along the axial direction than radial direction,[17, 34, 38, 39] while azimuthally polarized beam can provide a bubble with higher optical barrier in the radial direction than axial direction (unclosed hollow spot).[15, 16] Therefore, for 3D optical chain created through solely filtering and focusing of radially polarized beam, the potential barrier of the bubbles array is not uniform.[39] Optimized linear superposition of azimuthal and radial components with double belts has been investigated to construct single optical cage with uniform optical barrier.[36, 38] But in order to meet the need of multiple site optical trapping or multiple spot optical microscopies, it is necessary to construct interconnected bubble array (optical chain) with uniform barrier or spherical spot array.

Through a combination of a mix of electric and magnetic dipole arrays with appropriate relative strength and locations, very complicated focal field distributions can be created. Due to the orthogonality of the reversed radiations from electric and magnetic dipoles, the contributions from these two arrays can be considered separately. Through controlling the relative amplitude and phase of electric dipoles, optical chain can be achieved.[33] To construct uniform 3D optical chain, the magnetic dipole array is introduced to provide better radial barrier. A 3D tube field (see Fig. 17) from magnetic dipole array can be created to surround the previously constructed field by electric dipole array. If these two fields are superposed with each other, 3D chain-like field with uniform barriers along both the axial and radial directions can be constructed.[33]

Figure 19(a) shows this composite field and the required pupil field is given in Fig. 19(b). The 3D optical chain with equal barriers sounding the dark spot (optical bubble) is actually achieved in the focal volume. The barrier ratios (I_r/I_z) between barrier peaks along radial and axial direction for each of the bubbles are close to 1 (above 93%). Such an optical chain distribution can be useful in optical trapping application with multiple trapping sites. The normalized intensity gradient of this optical chain is shown in Fig. 19(c) (along z-axis when r = 0) and Fig. 19(d) (along radial direction of the four bubbles with location z = ±0.6λ and z = ±1.8λ). From the gradient plots, 3D stable trapping and manipulating of particle array with different index of refractions along z-axis can be realized. Particles with refractive index higher than the ambient can be trapped at the bright spots while particles with refractive index lower than the ambient can be trapped within the bubbles. The radial trapping strengths are almost identical for the four bubbles shown in this example. The corresponding barrier width ratio is about 0.34λ/0.6λ.

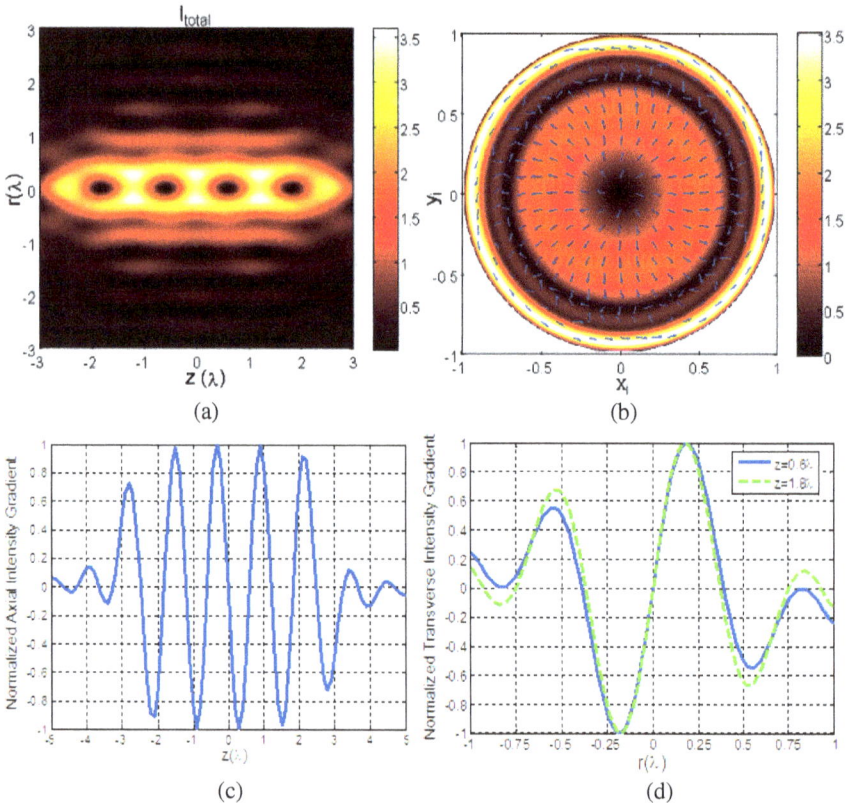

Fig. 19. Reconstructed field in the focal volume (a) and corresponding input pupil field (b) from electric-magnetic dipole array. Normalized intensity gradient are shown in (c) for axial intensity gradient when r = 0 and (d) for transverse intensity gradient when z = ±0.6λ and ±1.8λ.[33]

Longer chain with more bubbles can be achieved as long as enough electric and magnetic dipoles are used in the design procedure. Not surprisingly, the required input pupil field becomes increasingly complicated. The constructed optical chain may be used as an effective tool for multiple particles trapping, delivering and self-assembling.[39, 70-74]

6. Conclusions

In this chapter, we demonstrated that a wide variety of special focal field with specifically engineered intensity and polarization distributions can be created with proper vectorial optical fields as illumination at the pupil plane. Inspired by the intrinsic spatially variant polarization distributions from various antenna radiations, systematic approaches to find the required vectorial optical fields at the pupil of objective lens for the creation of focal field with various desired characteristics are developed and validated with numerical calculation with the Richards-Wolf vectorial diffraction method. These generalized methods allow us to take advantage of the extensive body of the existing literatures in the microwave and radio frequency antenna theory. Various exotic optical focal fields, including but not limited to, 3D flattop focus, 3D optical bubble, 3D spherical focal spots with arbitrary polarization, extremely long optical needles, optical tube and optical chain etc., have been demonstrated to illustrate the capability and flexibility of the proposed methods. These approaches enable us to find the analytical expressions of the required vectorial optical field for the generation of these comprehensively engineered optical focal fields. With the recent advances in nanofabrication and SLM, generation of these required complex vectorial optical fields become practical and increasingly accessible. The comprehensive focal field engineering techniques presented in this chapter may find important applications in single molecular imaging, tip enhanced Raman spectroscopy, high resolution optical microscopy, and particle trapping and manipulation.

References

1. J. Cheng, A. Volkmer and X. S. Xie, *J. Opt. Soc. Am. B*, **19**, 1363 (2002).
2. T. Y. F. Tsang, *Phys. Rev. A*, **52**, 4116 (1995).

3. R. W. Boyd, *Nonlinear Optics* (Academic Press, Boston, 1992).
4. J. Cheng and X. S. Xie, *J. Opt. Soc. Am. B*, **19**, 1604 (2002).
5. S. W. Hell and J. Wichmann, *Opt. Lett.*, **19**, 780 (1994).
6. S. W. Hell, *Science*, **316**, 1153 (2007).
7. Q. Zhan, *Advances in Optics and Photonics*, **1**, 1 (2009).
8. Q. Zhan, *Opt. Lett.*, **31**, 867 (2006).
9. S. Yang, P. E. Powers and Q. Zhan, *Opt. Commun.*, **282**, 4657 (2009).
10. G. M. Lerman and U. Levy, *Opt. Lett.*, **32**, 2194 (2007).
11. G. M. Lerman, Y. Lilach, and U. Levy, *Opt. Lett.*, **34**, 1669 (2009).
12. L. Cicchitelli, H. Hora, and R. Postle, *Phys. Rev. A*, **41**, 3727 (1990).
13. R. D. Romea and W. D. *Kimura, Phys. Rev. D*, **42**, 1807 (1990).
14. Z. Bouchal and M. Olivik, *J. Modern Opt.*, **42**, 1555 (1995).
15. K. S. Youngworth and T. G. Brown, *Opt. Express*, **7**, 77 (2000).
16. Q. Zhan and J. R. Leger, *Opt. Express*, **10**, 324 (2002).
17. W. Chen and Q. Zhan, *Opt. Commun.*, **265**, 411 (2006).
18. H. Kang, B. H. Jia and M. Gu, *Opt. Express*, **18**, 10813 (2010).
19. I. Iglesias and B. Vohnsen, *Opt. Commun.*, **271**, 40 (2007).
20. F. Abouraddy and K. C. Toussaint, Jr., *Phys. Rev. Lett.*, **96**, 153901 (2006).
21. W. Chen and Q. Zhan, *J. Opt.*, **12**, 045707 (2010).
22. L. Novotny, M. R. Beversluis, K. S. Youngworth and T. G. Brown, *Phys. Rev. Lett.*, **86**, 5251 (2001).
23. S. Takeuchi, R. Sugihara and K. Shimoda, *J. Phys. Soc. Jpn.*, **63**, 1186 (1994).
24. M. E. J. Friese, T. A. Nieminen, N. R. Heckenberg and H. Rubinsztein-Dunlop, *Nature*, **394**, 348 (1998).
25. P. Banzer, U. Peschel, S. Quabis and G. Leuchs, *Opt. Express*, **18**, 10905 (2010).
26. H. F. Wang, L. P. Shi, B. Lukyanchuk, C. Sheppard and C. T. Chong, *Nat. Photonics*, **2**, 501 (2008).
27. E. Karimi, G. Zito, B. Piccirillo, L. Marrucci and E. Santamato, *Opt. Express*, **16**, 21069 (2008).
28. S. Yang and Q. Zhan, *J. Opt. A*, **10**, 125103 (2008).
29. K. Kitamura, K. Sakai and S. Noda, *Opt. Express*, **18**, 4518 (2010).
30. K. Huang, P. Shi, X. Kang, X. Zhang and Y.P. Li, *Opt. Lett.*, **35**, 965 (2010).
31. J. Wang, W. Chen and Q. Zhan, *Opt. Express*, **18**, 21965 (2010).
32. J. Wang, W. Chen and Q. Zhan, *Opt. Commun.*, **284**, 2668 (2011).
33. J. Wang, W. Chen and Q. Zhan, *J. Opt.*, **14**, 055004 (2012).
34. Y. Kozawa and S. Sato, *Opt. Lett.*, **31**, 820 (2006).
35. N. Bokor and N. Davidson, *Opt. Lett.*, **31**, 149 (2006).
36. N. Bokor, N. Davidson, *Opt. Commun.*, **279**, 229 (2007).
37. Y. Kozawa and S. Sato, *Opt. Lett.*, **33**, 2326 (2008).
38. X. L. Wang, J. Ding, J. Q. Qin, J. Chen, Y. X. Fan and H. T. Wang, Opt. Commun., **282**, 3421 (2009).
39. Y. Zhao, Q. Zhan, Y. Zhang and Y. P. Li, *Opt. Lett.*, **30**, 848 (2005).

40. N. Bokor and N. Davidson, *Opt. Lett.*, **29**, 1968 (2004).
41. W. Chen and Q. Zhan, *Opt. Lett.* , **34**, 2444 (2009).
42. S. Yan, B. Yao, W. Zhao and M. Lei, *J. Opt. Soc. Am. A*, **27**, 2033 (2010).
43. E. Wolf, *Proc. R. Soc. London Ser. A*, **253**, 349 (1959).
44. B. Richards and E. Wolf, *Proc. R. Soc. London Ser. A*, **253**, 358 (1959).
45. M. Gu, *Advanced optical imaging theory* (Springer-Verlag, New York, 1999).
46. V. G. Niziev and A. V. Nesterov, *J. Phys. D*, **32**, 1455 (1999).
47. A. V. Nesterov and V. G. Niziev, *J. Phys. D.*, **33**, 1817 (2000).
48. T. A. Klar, S. Jakobs, M. Dyba, A. Egner and S. W. Hell, *Proc. Natl. Acad. Sci.*, **97**, 8206 (2000).
49. P. Torok and P. R. T. Munro, *Opt. Express*, **12**, 3605 (2004).
50. R. Oldenbourg and G. Mei, *J. Microsc.*, **180**, 140 (1995).
51. W. K. Hiebert, A. Stankiewicz and M. R. Freeman, *Phys. Rev. Lett.*, **79**, 1134 (1997).
52. E. J. Sanchez, L. Novotny and X. S. Xie, *Phys. Rev. Lett.*, **82**, 4014 (1999).
53. D. Mehtani, N. Lee, R.D. Hartschuh, A. Kisliuk, M. D. Foster, A. P. Sokolov and J. F Maguire, *J. Raman Spectrosc.*, **36**, 1068 (2005).
54. Y. Saito, M. Motohashi, N. Hayazawa, M. Iyoki and S. Kawata, *Appl. Phys. Lett.*, **88**, 143109 (2006).
55. M. R. Beversluis, L. Novotny and S. J. Stranick, *Opt. Express*, **14**, 2650 (2006).
56. Z. Bomzon, G. Biener, V. Kleiner and E. Hasman, Opt. Lett., **27**, 285 (2002).
57. X. Wang, J. Ding, W. Ni, C. Guo and H. Wang, *Opt. Lett.*, **32**, 3549 (2007).
58. M. Born and E. Wolf, *Principle of optics* (Pergamon Press, Oxford, 1970).
59. J. Stadler, C. Stanciu, C. Stupperich and A. J. Meixner, *Opt. Lett.*, **33**, 681 (2008).
60. C. Ye, *Opt. Eng.*, **34**, 3031 (1995).
61. Q. Zhan and J. R. Leger, *Appl. Opt.*, **41**, 4630 (2002).
62. A. Shakin, *IEEE J. Sel. Top. Quantum Electron.*, **6**, 841 (2000).
63. S. Weiss, *Science*, **283**, 1676 (1999).
64. A. Ashikin and J. Z. Dziedzic, *Science*, **235**, 1517 (1987).
65. S. Hell and E. Stelzer, *J. Opt. Soc. Am. A*, **9**, 2159 (1992).
66. M. C. Lang, T. Staudt, J. Engelhardt and S. Hell, *New J. Phys.*, **10**, 043041 (2008).
67. W. Chen, Q. Zhan, *Opt. Commun.*, **284**, 41 (2011).
68. A. Balanis, *Antenna Theory: Analysis and Design, 3rd Edition* (Wiley-Interscience, 2005).
69. Y. S. Xu, S. Janak, C. J. R. Sheppard and N. G. Chen, *Opt. Express*, **15**, 6409 (2007).
70. M. P. MacDonald, L. Paterson, K. Volke-Sepulveda, J. Arlt, W. Sibbett, K. Dholakia., *Science*, **296**, 1101 (2002).
71. J. E. Curtis, B. A. Koss, D. G. Grier, *Opt. Commun.*, **207**, 169 (2002).
72. D. McGlin, G. C. Spalding, H. Melville, W. Sibbett, and K. Dholakia, *Opt. Commun.*, **225**, 215 (2003).

73. J. R. Moffitt, Y. R. Chemla, S. B. Smith, and C. Bustamante, *Annu. Rev. Biochem.*, **77**, 205 (2008).

74. T Čižmár, L C Dávila Romero, K Dholakia and D L Andrews, *J. Phys. B: At. Mol. Opt. Phys.*, **43**, 102001 (2010).

CHAPTER 5

PLASMONICS WITH VECTORIAL OPTICAL FIELDS

Guanghao Rui[*] and Qiwen Zhan

Electro-Optics Program, University of Dayton
300 College Park, Dayton, Ohio 45469, USA
[*]E-mail: grui01@udayton.edu

An overview of the recent developments in the field of plasmonics using vectorial optical fields as the excitation is presented. Surface plasmon polaritons are collective oscillation of free electrons at metal/dielectric interface. As a wave phenomenon, surface plasmon polaritons can be focused by using appropriate excitation geometry and metallic/dielectric structure. The strong spatial confinement and high field enhancement make surface plasmon polaritons very attractive for near-field optical imaging and sensing applications. As one class of spatially variant polarization, cylindrical vector beam are the axially symmetric beam solution to the full vector electromagnetic wave equations. Due to its rotational spatial polarization symmetry, cylindrical vector beam has been discovered to be the ideal source for surface plasmon polaritons excitation with axially symmetric metal/dielectric structures. In this chapter, we review the interaction of vectorial optical fields with various plasmonic structures, including planar thin film, bull's eye, nanoantenna, conical tip and spiral antenna. Compared with excitation using conventional spatially homogeneous state of polarization, plasmon excitation with the use of matched vectorial optical fields can lead to higher coupling efficiency and stronger field enhancement with focal spot beyond the diffraction limit.

1. Surface plasmon polaritons

Surface plasmon polaritons (SPP) are free electron oscillations near metal/dielectric interface due to the interactions between incident photons and conduction electrons of the metal. The resonant interaction gives rise to many unique properties of SPP (e.g. short effective wavelength, high spatial confinement and strong field enhancement). In the field of optics, one of the most attractive aspects of SPP is its capability of concentrating and channeling light with subwavelength structures, enabling miniaturized photonic circuits with dimensions much smaller than those are currently available.[1, 2] The spatial confinement leads to a local electric field enhancement that can be used to manipulate light-matter interactions and boost the efficiency of optical nonlinear effects.[3, 4] Therefore, efficient SPP excitation is an interesting and important topic in the area of near-field optics, especially for applications in sensing, imaging, and lithography.[5-7]

As illustrated in Fig. 1, the electric field of SPP propagating on a metal interface ($z = 0$) in the x direction can be expressed as[8]

$$E_{spp}\left(x, z\right) = E_0 e^{ik_{spp}x - k_z|z|}. \tag{1}$$

This solution corresponds to a surface mode propagating along the interface with transverse wavevector $k_{//} = k_{spp}$ and exponentially decaying from the surface with decay constant k_z. The dispersion relationship of SPP on a planar metal/dielectric interface is given by[8]

$$k_{spp}^2 = \left(\frac{\omega}{c}\right)^2 \frac{\varepsilon_i \varepsilon_m}{\varepsilon_i + \varepsilon_m}, \tag{2}$$

where ω is the radial frequency of the SPP wave, ε_m is the complex dielectric constant of the metal, and ε_i is the dielectric constant of the surrounding dielectric medium. An example of this ω-k dispersion curve

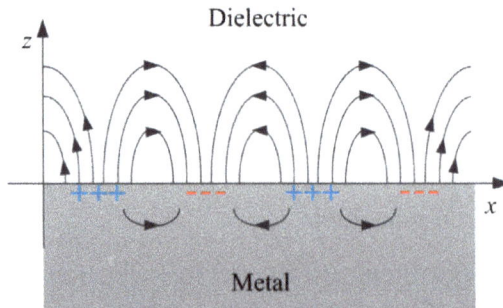

Fig. 1. Surface plasmon polaritons.

is illustrated in Fig. 2. From this figure, it can be seen that the SPP wavenumber is larger than that of photons in the adjacent dielectric medium. Thus, the light illuminating a smooth surface cannot be directly coupled to SPP. In order to excite SPP optically, transverse wavenumber of the excitation light wave needs to be equal to the wavenumber of the excited SPP. This linear momentum match can be realized by a variety of techniques such as prism coupling (Kretschmann and Otto configurations), grating coupling, and scattering by subwavelength structures. The Kretschmann configuration is widely used to excite the

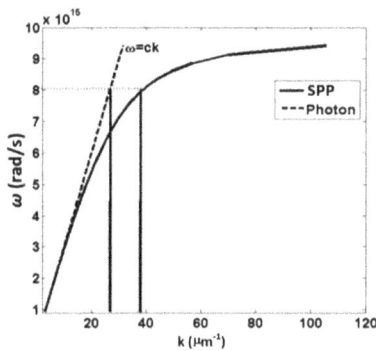

Fig. 2. Dispersion relationship of SPP.

SPP on the surface of thin film. This configuration comprises of a metal film and a dielectric prism that is in contact with the metal. The metal film is illuminated through the dielectric prism at an incident angle larger than the critical angle. At the angle of incidence where the transverse component of photon wavenumber in the prism coincides with the wavenumber of the SPP on the air/metal surface, resonant light tunneling through the film occurs and light is coupled into the surface polaritons.

As a wave phenomenon, SPP can be focused by appropriate metallic structures, i.e., plasmonic lens. Efficient excitation, focusing, coupling and guiding SPP remain continued research interest. A major challenge related to the SPP focusing is to achieve tight focusing with high energetic efficiency at the focus. Required by the boundary conditions, SPP can only be excited by light field that is *p*-polarized with respect to the interface. Thus the polarization of excitation source plays a major role in the excitation and manipulation of SPP. The availability of vectorial optical fields with spatially engineered polarization offers tremendous opportunities in plasmonic focusing and manipulation. As special cases of the vectorial optical field, radially and azimuthally polarized beams with spatially variant states of polarization have attracted much interest as a SPP excitation sources. Radial polarization, whose local electric field is linearly polarized along the radial directions, was discovered to be the ideal source for SPP excitation with axially symmetric metal/dielectric structures. Recently, it was found that azimuthally polarized beam could also be coupled to SPP efficiently by matching its polarization to spatially arranged triangular apertures. Compared with their spatially homogeneous counterparts (linear, elliptical, and circular polarizations), vectorial beams offer higher coupling efficiency, smaller and homogeneous focusing spot, and stronger local field strength at the focus. These potential advantages continuously motivate researches and scientists to further exploit the feasibility of improving the performance of plasmonic devices with the use of vectorial beams illumination. In this chapter, we will review the recent progresses on the interactions of vectorial optical fields with various plasmonic structures.

2. Interaction of vectorial fields with plasmonic structures

2.1. *Planar metallic thin film*

The polarization selective property of SPR leads to important application of radial polarization in plasmonic focusing. It is found that efficient plasmonic excitation and focusing can be achieved through matching the axially symmetric structure to the polarization symmetry of radially polarized illumination.[9-11] When a radially polarized beam is focused onto a planar plasmonic interface by an aplanatic lens (Fig. 3), the entire beam is *p*-polarized with respect to the interface. For objective lens with sufficiently high numerical aperture (NA), light within an annular ring with narrow angular width in the pupil can satisfy the resonant condition and be coupled into the SPP modes to produce plasmonic focus at the center (see inset of Fig. 3(a)). Owing to the rotational polarization symmetry, the SPP excited from all directions interfere constructively

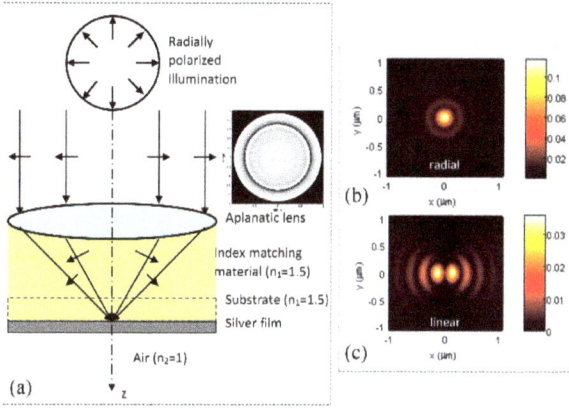

Fig. 3. (a) Diagram of the setup for SPR excitation under highly focused radially-polarized illumination. An aplanatic lens focuses the radially polarized beam onto the dielectric/silver interface. The space between the substrate and the lens is filled with index-matching material with the same refractive index as the substrate. The entire focused beam is *p*-polarized with respect to the silver/dielectric interface. Inset: Simulated intensity distribution at the back focal plane of the objective lens after reflection. Calculated total field strengths at the bottom of the silver layer are shown in the insets for (b) radial polarization and (c) linear polarization illumination (adapted from Ref. 11).

and create an enhanced local field in the center (Fig. 3(b)). In contrary, for linearly polarized illumination, only part of the focused optical beam is *p*-polarized with respect to the interface and can be coupled to SPP. More severely, the plasmonic focal field at the geometrical focus splits into two lobes due to a destructive interference between counter propagating SPP waves (shown in Fig. 3(c)). Compared with radial polarization excitation, the plasmonic focal field is inhomogeneous and much weaker under linearly polarized illumination. Clearly, different polarizations have remarkable influences on the plasmonic focal field produced by the same structure.

Richard-Wolf vectorial diffraction theory is used to calculate the plasmonic focal fields in Fig. 3. For radial polarization excitation, the plasmonic focal field can be expressed as[11]

$$E_r(r,\varphi,z) = 2A \int_{\theta_{min}}^{\theta_{max}} \cos^{1/2}(\theta)P(\theta)t_p(\theta)\sin\theta\cos\theta$$
$$\times J_1(k_1 r\sin\theta)\exp\left[iz(k_2^2 - k_1^2\sin^2\theta)^{1/2}\right]d\theta, \qquad (3)$$

$$E_z(r,\varphi,z) = i2A \int_{\theta_{min}}^{\theta_{max}} \cos^{1/2}(\theta)P(\theta)t_p(\theta)\sin^2\theta$$

$$\times J_0(k_1 r\sin\theta)\exp\left[iz(k_2^2 - k_1^2\sin^2\theta)^{1/2}\right]d\theta, \qquad (4)$$

where $t_p(\theta)$ is the transmission coefficient of p-polarization at the incident angle of θ, θ_{max} and θ_{min} are the maximal and minimal incident angles corresponding to the annular illumination, $P(\theta)$ is the pupil apodization function, $J_m(x)$ is the mth order Bessel function of the first kind, k_1 and k_2 are the wavenumbers in the glass and air, respectively. Numerical simulations using the vectorial diffraction model given above are shown in Fig. 4. The total plasmonic field consists of a radial component and a longitudinal component normal to the surface. Under a highly focused radial polarization excitation, the longitudinal component is much stronger and dominates the total field distribution.

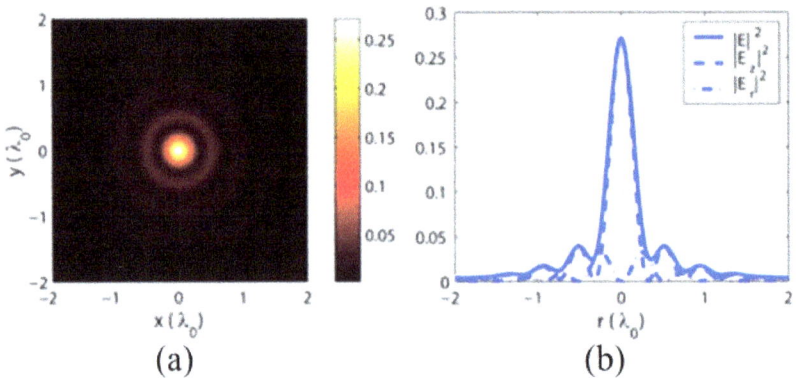

Fig. 4. Numerical simulation results using vectorial diffraction theory. (a) Total intensity at the bottom of the silver layer, (b) Linescan of (a) through the center. The longitudinal and transverse components are also shown as dashed curve and dashed-dotted curve, respectively (adapted from Ref. 11).

The plasmonic focusing characteristics excited by focused radial polarization has been confirmed by the experimental setup shown in Fig. 5.[12] The radially polarized beam is generated by coupling a charge +1 vortex beam into a few-mode fiber. An aplanatic oil immersion lens (NA = 1.25) is used to focus the radially polarized beam onto a glass/silver (50 nm) interface. A circular photomask is placed after the

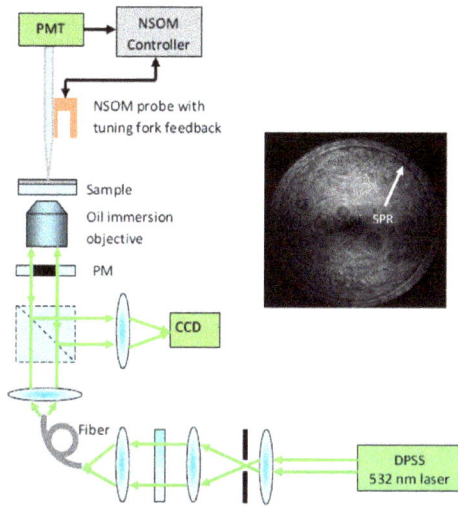

Fig. 5. Experimental setup for generating and detection of plasmonic focusing field with highly focused radially polarized illumination. Inset: The intensity distribution at the back focal plane of the objective lens after reflection captured by a CCD camera (adapted from Ref. 12).

collimated radially polarized beam to block the center of the illumination corresponding to the incident angle below the angle satisfying the SPR condition. A collection-mode near-field scanning optical microscope (NSOM) with a metal-coated tapered fiber probe (nominal aperture size ~ 50-100 nm) was used to image the plasmonic focal field distribution near the silver-air interface. It is known that a NSOM probe with very small aperture is more sensitive to the electric field component parallel to the surface,[13, 14] and the signal detected by the aperture NSOM fiber probe is proportional to $\left| \nabla_{\perp} E_z \right|^2$, predicting to a donut pattern with a dark center. The measured near-field intensity distribution of the focused plasmonic field is shown in Fig. 6(a). Logarithmic scale is used for better visualization of the outer rings. Focusing of SPP waves generated from all directions is clearly observed. Very good agreement is obtained between the measured and calculated transverse profiles (Fig. 6(b)).

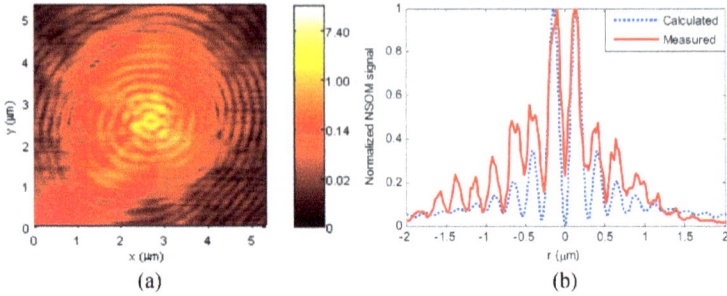

(a)   (b)

Fig. 6. (a) Measured near-field intensity distribution of the focusing SPP field. (b) Comparison of measured and calculated transverse profiles of the intensity distribution near the air/silver interface (adapted from Ref. 12).

More interestingly, the generated plasmonic field shows the properties of an evanescent Bessel beam. Bessel beam is one kind of non-diffracting beam solution of the Maxwell's wave equations. An ideal Bessel beam can be regarded as the superposition of a set of plane waves with wavevectors along the surface of a cone, which can be generated using an axicon or conical lens.[15] For the plasmonic focal field excited by a radial polarization, the SPR condition of the silver film (shown as inset of Fig. 5) performs an angular filter function to the transmitted field that mimics an axicon.[12] Figure 7(a) shows the normalized transverse profiles of the intensities at different distances from the surface of the silver layer measured in the experiment. It can be seen that the shape of the lobes remains almost constant for different planes, showing the non-diffracting nature of the evanescent Bessel beam. The evanescent decaying nature is illustrated in Fig. 7(b).

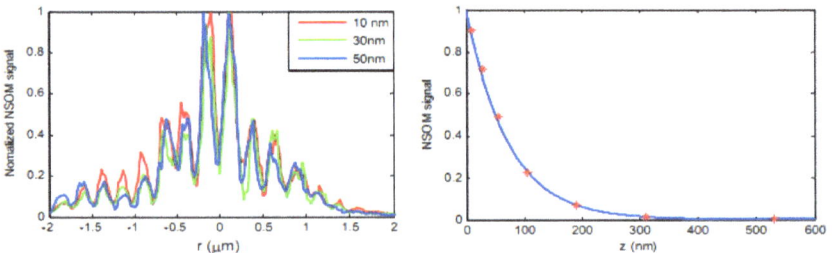

Fig. 7. (a) Measurement at different planes to reveal the non-diffraction nature of the evanescent Bessel beam. (b) Measured evanescent decaying property of the evanescent Bessel beam (adapted from Ref. 12).

This experiment successfully demonstrates the generation of evanescent Bessel beam via SPR excitation with a tightly focused radially polarized beam. Compare with previous methods using conical devices, this setup greatly simplifies the alignment procedure. This principle can also be extended to other resonant structures that exhibit a narrow angular filter function for *p*- and/or *s*-polarization. For examples, J_0 and J_1 type evanescent Bessel beams can be generated by the band-edge mode[16] or the defect mode[17] of a one-dimensional photonic band gap structure excited by highly focused radially and azimuthally polarized beam, respectively. These evanescent Bessel beam could be used as a virtual probe for near-field imaging, sensing, as well as stable trapping of metallic nanoparticles.

2.2. *Bull's eye structures*

2.2.1. *Plasmonic lens made of single circular slot*

Other than the Kretschmann and Otto configuration, subwavelength structures such as subwavelength surface defects have also been widely used to excite SPP.[18] With proper spatial arrangement, the SPP generated by these subwavelength structures can interfere with each other and produce a plasmonic focus. Hence these spatially arranged subwavelength structures form a plasmonic lens. One of the most popular plasmonic lenses consists of a circular slot milled into a metallic film.[19] When the slit width is smaller than half of the excitation wavelength, part of the diffracted light whose wavenumber coincides with the SPP wavenumber will be coupled to SPP. The energy is guided toward the focal point of the circular slot with the propagation direction normal to the slit.

Similar to the plasmonic focusing produced on planar metallic thin film described in Section 2.1, the plasmonic focusing properties of a circular slot plasmonic lens strongly depends on the excitation polarization. When linearly polarized beam is used as the excitation, only the portions of the circle where the incident electric field is locally perpendicular to the circle edge could excite SPP. The dominant E_z component would interference destructively (Fig. 8), leading to an inhomogeneous focal field with split two lobes.[19, 20] On the contrary, radial polarization is the ideal SPP excitation source for plasmonic structures with axial symmetry such as the circular slot plasmonic lens. When a radially polarized beam illuminates the circular slot, the entire beam is p-polarized with respect to the annular edge of the slit, providing an efficient way to generate highly focused SPP through constructive interference and creating an enhanced local field compared with linearly polarized illumination. Under radial polarization excitation, the in-plane electric field components (E_R) vanish at the center of the plasmonic lens due to the fact that for points on the opposite side of the circle E_R are excited in antiphase, leading to destructive interferences. However, the out-of-plane electric field components (E_z) are generated in phase and accumulate the same phase along their propagation toward the center, leading to constructive interferences and a solid and homogeneous focal spot (Fig. 8).

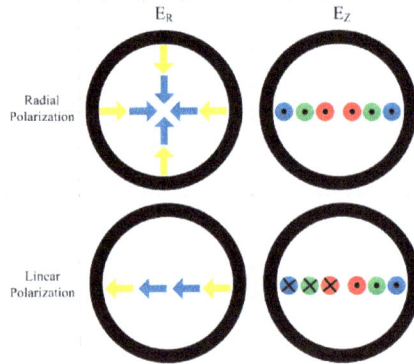

Fig. 8. Schematic diagram of the annular ring plasmonic lens and the orientation of the field under radially and linearly polarized illumination.

Experiment has been conducted to confirm the focusing properties of single circular slot plasmonic lens under radial polarization excitaiton.[21] The plasmonic lens is made of an circular slot etched into an silver thin film by focused ion beam (FIB) milling. A NSOM with probe of 300 nm diameter aperture was used to map the plasmonic focal field. Fig. 9(a) shows the measured near-field intensity distribution at a constant height of 2 μm over the silver/air interface. Clearly SPP are excited from all the azimuthal directions, propagate toward the geometric center of the annular ring and interfere constructively. It should be noticed that, at the operating wavelength (1.064 μm) of this work, the intensity of E_z is much larger than that of E_R. Thus, the measured field distribution corresponds mostly to the out-of-plane electric component despite of the fact that the aperture NSOM probe is more sensitive to E_R component (as we discussed in Section 2.1). The normalized line scan through the center of plasmonic lens (Fig. 9(b)) shows a profile of $(J_0)^2$ function with a FWHM beyond the diffraction limit.

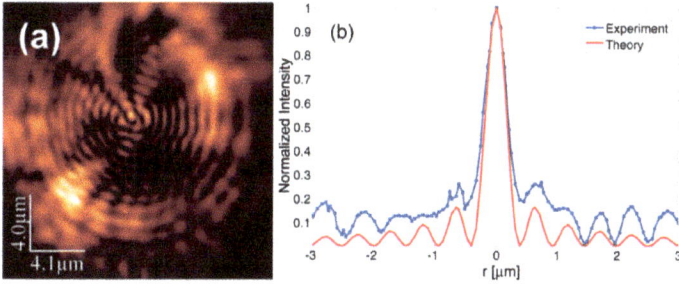

Fig. 9. (a) Measured NSOM signal showing SPP focusing in the annular ring plasmonic lens under radially polarized illumination. (b) Normalized experimental and theoretical line-scan through the center of the plasmonic lens (adapted from Ref. 21).

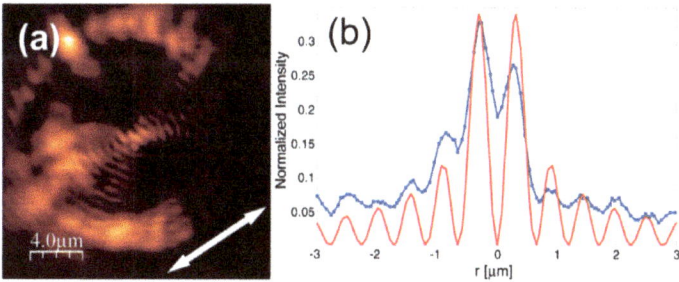

Fig. 10. (a) Measured NSOM signal showing SPP focusing in the annular ring plasmonic lens under linearly polarized illumination. (b) Normalized experimental and theoretical line-scan through the center of the plasmonic lens (adapted from Ref. 21).

To demonstrate the polarization selectivity of the annular ring plasmonic lens, experiment with linear polarization excitation was also conducted (Fig. 10).[19] The polarization direction of the illumination is indicated by the white arrow in Fig. 10(a). SPP are only generated by the section of the slot that the polarization is mainly p-polarized locally with respect to the slot. Line scans of the focal field in Fig. 10(b) show good agreements between experimental results and numerical predictions. In contrary to the case of radially polarized illumination that obtains a solid spot in the focus, a pattern of two separate lobes is observed in the center of the plasmonic lens due to the destructive interference of the E_z.

2.2.2. Plasmonic lens made of multiple concentric circular slots

Although the single circular slot has the ability to focus SPP into a small solid spot with enhanced intensity, the coupling efficiency and intensity at the focus are low for many potential applications. This can be improved by Bull's eye structure with multiple concentric circular slots.[22] The function of the multiple concentric circular slots is to enlarge the collecting area of incident photons so as to boost the SPP generation efficiency. To obtain an enhanced local field at the center of the bull's eye structure, the period of the concentric rings need to match the SPP wavelength such that the SPP excited at different slits have a propagation phase differences of $2\pi \cdot n$ ($n = 1, 2, 3...$). The focusing property of such a bull's eye plasmonic lens under radially polarized illumination has been explored experimentally.[23] A 200 nm silver film was deposited onto a glass substrate by e-beam evaporation. This thickness was chosen to prevent far field transmission of the incident laser beam through the silver layer directly. Bull's eye structure with single ring, 5-ring, and 9-

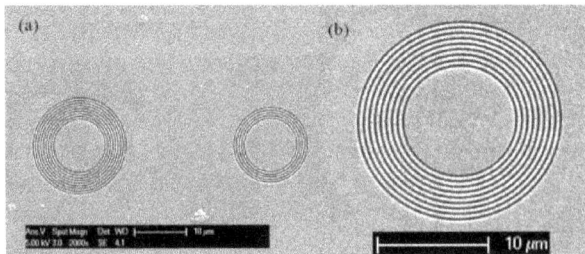

Fig. 11. (a) SEM image of the bull's eye plasmonic lens with 5-ring and 9-ring. (b) Zoom-in of the 9-ring plasmonic lens (adapted from Ref. 23).

ring were fabricated with FIB milling (SEM images are shown in Fig. 11). To meet the phase matching requirement, the singularity center of radially polarized beam need to be carefully aligned with the geometrical center of bull's eye structure. The SPP intensity distribution near the silver/air interface was directly mapped by a collection mode NSOM.

Compared with the single circular slot experiment reported in Ref. 21, the collected NSOM signals are rather different. In this work, the measured intensity distribution near the silver/air interface of single circular slot plasmonic lens shows a donut shape with a very small dark center (shown in Fig. 12(a)), while a solid spot at the focus was obtained (Fig. 9) in Ref. 21. The difference is mainly due to the fact that a much smaller aperture is used (50 ~ 80 nm as opposed to 300 nm). As mentioned in Section 2.1, for very small aperture, the NSOM detected signal is proportional to $\left|\nabla_{\perp}E_z\right|^2$, leading to the small hole in the center.

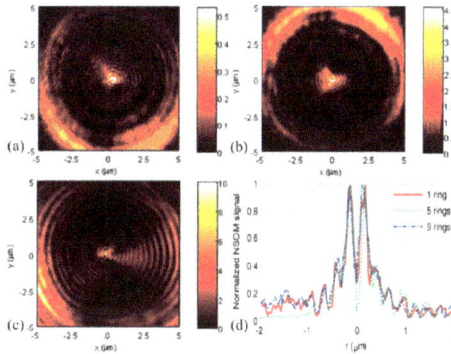

Fig. 12. Measured near-field energy density distribution at the silver/air interface for (a) single ring (b) 5-ring and (c) 9-ring plasmonic lens. (d) Measured normalized transverse profiles of energy density distribution through the center of bull's eye plasmonic lens with single ring, 5-ring and 9-ring (adapted from Ref. 23).

The focal field can be strengthened by adding more concentric rings satisfying the Bragg condition for the SPP wavelength. Figures 12(b) and (c) show the measured 2D near field intensity distributions of bull's eye plasmonic lens with 5-ring and 9-ring respectively. The SPPs are excited from all azimuthal directions around the entire concentric rings and propagate toward the center. The focusing effect can be clearly seen as the SPP interference fringes getting stronger when they are closer to the geometrical center. The peak intensity increases with more rings. A comparison of the transverse profiles of the measured plasmonic focal field for with single ring, 5-ring and 7-ring is shown in Fig. 12(d). The main lobes of the three curves almost overlap with each other, indicating that the focal spot remains the same size as more rings are added to the Bull's eye structure.

In order to optimize the field strength at the focus, the parameters of the Bull's eye plasmonic lens need to be adjusted to match the illumination condition. Radially polarized illumination has a donut shape field distribution of $E_{in} = r \cdot \exp(-r^2 / w^2)$, where r is the radial coordinate and w is the beam waist of illumination. The intensity of SPP has a linear dependence on r because the SPP are excited along a circle with perimeter of $2\pi \cdot r$, so the field strength is proportional to $r^{0.5}$. Due to Ohm losses in metal, the SPP field decays exponentially during propagating along the plasmonic lens with a propagation loss of exp[-Im(k_{spp})·r] where k_{spp} is the wavenumber of the SPP. Therefore, the focal field strength E_{fe} is proportional to $r^{0.5} \times \exp[-\text{Im}(k_{spp}) \cdot r]$. Consider the spatial distribution and spot size of the incident radial polarization, the overall field strength can be expressed as $E_p = E_{in} \times E_{fe}$.

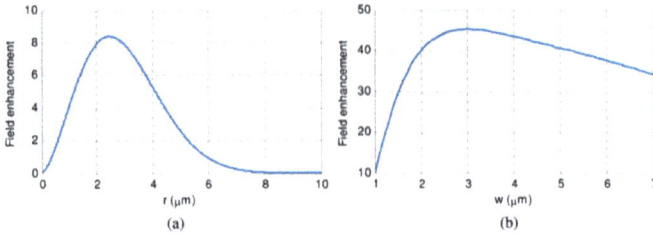

Fig. 13. (a) Field enhancement curve for single ring plasmonic lens with different radius. (b) Field enhancement curve for 11-ring plasmonic lens optimized at $w = 3$ μm under radially polarized illumination with different beam waist (adapted from Ref. 23).

Figure 13(a) shows the calculated field strength for single rings structure with different radius under excitation wavelength of 532 nm. One can see that the field strength is maximal when the single ring locates at $r = 2.48\ \mu$m. Due to the propagation loss of SPP, the intensity of the focal field cannot keep increasing by adding more rings to the bull's eye plasmonic lens. From Fig. 13(a), it can be seen that the benefit of additional ring with radius larger than 8 μm is negligible. The condition of illumination also changes the field enhancement. Fig. 13(b) shows the relationship between the beam waist w and the field strength for the 11-ring plasmonic lens optimized at $w = 3\ \mu$m. It indicates that the field enhancement factor of a bull's eye plasmonic lens can also be modified by adjusting the spot size of the illumination, which could be achieved by controlling the focus of the objective lens.

2.2.3. *Single ring plasmonic lens with circular Bragg gratings*

Although the Bull's eye plasmonic lens provides tight focusing with high energetic efficiency at the focus, the scattering field by the fully etched slots is strong. A possible way to solve the dilemma between the field enhancement and the scattering is to introduce circular gratings on both

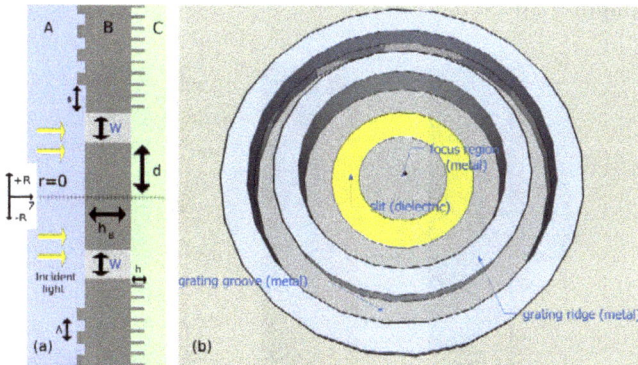

Fig. 14. Diagram of the plasmonic lens design. (a) Cross-section view. (b) Top view (adapted from Ref. 24).

sides of the single ring plasmonic lens (Fig. 14).[24] Radially polarized beam is used as the excitation source because the plasmonic lens having the circular symmetry. Surface relief circular gratings with a period equal to the SPP wavelength are added in region A to enhance the coupling of incident light to SPP. An annular ring is fully etched into the metal film with the same center of the circular gratings. This vertical annular slit can be regarded as a Fabry-Perot (F-P) cavity. Hence, to build up the field inside the slit constructively, the thickness of the silver film should satisfy the F-P resonant condition. SPP being coupled from the slit can propagate towards the center (in-focus) or the edges (counter-focus) of the plasmonic lens on the back side (region C). The counter-focus propagation leads to the scattering that lowers the SPP focusing efficiency.

The scattering can be minimized by adding a circular Bragg reflector made of circular gratings with a period of half the SPP wavelength. The distance between the innermost Bragg reflector and the outer edge of the annular slit strongly influences the performance of the Bragg reflector (Fig. 15). When this shift equals $\lambda_{spp}/4$, transverse field component E_R interfere destructively, while the longitudinal component E_z interfere constructively at the slit. Therefore, the scattering from the slit is negligible and efficient coupling to SPP is obtained. The results are totally the opposite when the shift distance is $\lambda_{spp}/2$. In this case, E_R is in phase, while the E_z is π out of phase, leading to a strong scattering and low SPP coupling efficiency.

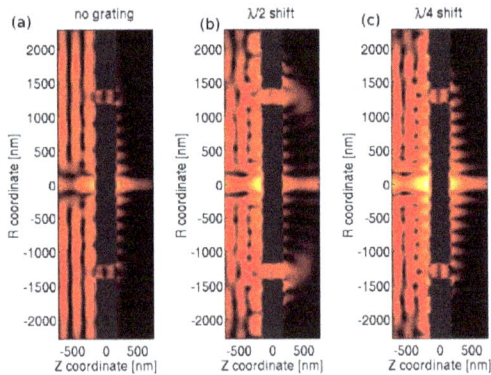

Fig. 15. Logarithmic plot of the 2D electric power density for the plasmonic lens with (a) no Bragg reflector, (b) Bragg gratings shifted from the slit with $\lambda_{spp}/2$, (c) Bragg gratings shifted from the slit with $\lambda_{spp}/4$ (adapted from Ref. 24).

2.2.4. *Applications in compact radial polarizer design*

The strong polarization selectivity of the bull's eye plasmonic structure described above can be used in the design of a compact radial polarizer.

Its feasibility has been demonstrated both by simulation and experiment.[25] The experimental results show that the bull's eye plasmonic lens strongly reflects azimuthal polarization and allows radial polarization to transmit through. This kind of subwavelength concentric metallic grating structure can also be fabricated on the core region of a cleaved few-mode fiber to facilitate the launch of cylindrical vector beams into optical fiber (see Fig. 8 in Chapter 1).[26] For the gold grating used in the experiment, the extinction ratio between the radial and azimuthal component is around 50, leading to the generation of a radially polarized beam with relatively high purity. Such a miniature device is suitable for the end mirror coupler in an all-fiber CV beam laser design.

2.3. *Extraordinary optical transmission with radial polarization*

Since the discovery of the extraordinary optical transmission (EOT) by Ebbesen,[27] this anomalous optical phenomenon arising from various subwavelength structures has been extensively studied due to their broad potential applications. The characteristics of EOT can be explained by the excitation of electromagnetic resonance modes in these structures. Usually, two types of resonances are associated with the EOT phenomenon: the SPP modes and the cavity modes.[28] Both mechanisms may lead to almost perfect transmittance. Owing to its unique polarization symmetry, radial polarization has been applied to various axially symmetric metallic structures for the investigation of their EOT performances. A brief review of this area is given in the following.

2.3.1. *Coaxial aperture*

The transmission properties of circular coaxial aperture under highly focused CV beam have been investigated experimentally.[29] For an on-axis radially polarized illumination, the EOT is up to 4 times higher than the transmission through a hollow aperture of the same diameter. This EOT phenomenon is attributed to the excitation of a TEM mode inside the coaxial aperture without a cut-off diameter. In the contrary, for the hollow aperture, a strongly focused radially polarized beam can couple to the TM_{01} mode that exhibits a cut-off diameter. In addition, a strong polarization contrast is observed between the transmission of coaxial aperture for radially and azimuthally polarized illumination (shown in Fig. 16).

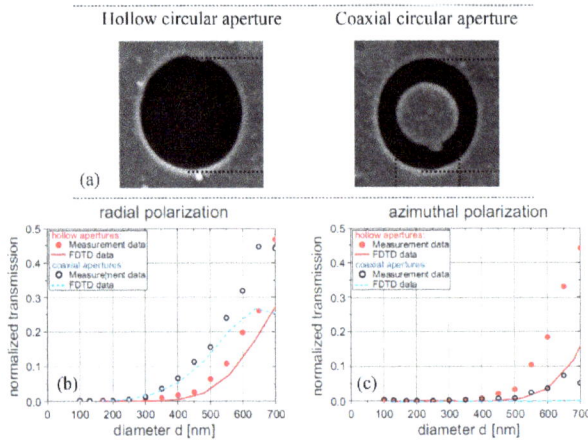

Fig. 16. (a) SEM images showing the aperture type. Experimental and numerical results for on-axis illumination with strongly focused (b) radially and (b) azimuthally polarized light (adapted from Ref. 29).

2.3.2. *Circular nanoantenna*

The SPP excitation efficiency is fairly low for an individual hollow or coaxial circular aperture, which limits its EOT performance. In order to excite SPP more efficiently, a widely adopted method is to introduce periodical corrugations around the circular aperture. However, optical antenna is another powerful that can be used to tremendously boost the SPP coupling efficiency. Besides, the effect of polarization was not considered in the previous EOT design. Recently, a plasmonic structure capable of generating extremely high EOT through combining circular nanoantenna and an annular nanoslit (Fig. 17) has been investigated.[30] A circular gold nanoplate is placed over a vertical annular nanoslit etched into a gold film. A SiO_2 thin film is sandwiched between the nanoplate and the metal film as supporting layer. Radially polarized beam that matches this axially symmetric structure illuminates the structure from the nanoplate side.

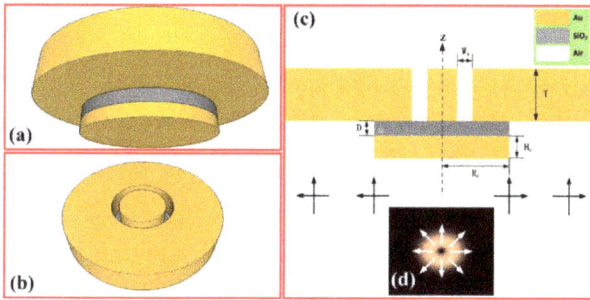

Fig. 17. Diagram of an EOT setup. The circular nanoplate over the opening of a vertical annular slit etched in a gold film forming a nanoantenna. (a) Bottom view. (b) Top view. (c) Cross-section view. (d) Spatial distribution of instantaneous electric field for the radially polarized illumination (adapted from Ref. 30).

The metallic nanoplate plays an important role in this EOT setup. It acts like a nanoantenna that contributes to a more efficient collection of the incident photon energy (Fig. 18(a)). It can be clearly observed that the incoming fields are collected by the nanoantenna and a standing wave pattern in the shape of concentric annular fringes is formed in the horizontal F-P cavity. To couple the localized energy in the horizontal F-P cavity, an annular slit is etched thoroughly into the gold film at the exact position where the magnetic field in the dielectric film is maximal. The annular slit can be regarded as a vertical F-P cavity. Therefore, the thickness of the gold film is another key factor to enhance the transmission. Figure 18(b) shows the transmission spectrum for the sole vertical F-P cavity, from which one can find a resonant thickness of 410 nm and anti-resonant thickness of 270 nm. The efficiency of coupling from the horizontal circular cavity to the vertical annular cavity strongly depends on the gold film thickness. Figure 18(c) and (d) show the $|H_y|$ distributions of the EOT setup with resonate and non-resonant thickness of the annular nanoslit, respectively. For the resonate nanoslit, the field intensity in the horizontal F-P cavity is very weak compared with the field in the vertical nanoslit, indicating that the majority of the energy collected by the circular nanoantenna is coupled to the vertical F-P cavity.

To evaluate the EOT design, transmission efficiency η is defined as the ratio between the integration of the z-component of Poynting vector over the output opening of the annular nanoslit and the integration of the

z-component of Poynting vector over the input opening of the bare annular slit. Transmission efficiency η of 114 can be obtained for the resonate structure. For the case of anti-resonant thickness, a large portion of energy is still localized in the horizontal F-P cavity, leading to a fairly low transmission efficiency ($\eta = 10.45$).

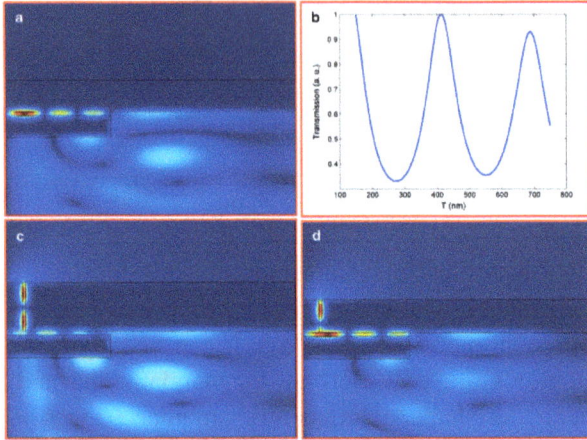

Fig. 18. (a) Distribution of the magnetic field for circular nanoantenna over a gold film. (b) Transmission curve of the vertical F-P cavity versus thickness of the gold film. Distribution of the magnetic field for circular nanoantenna over a gold film with a (c) resonant and (d) non-resonant nanoslit (adapted from Ref. 30).

The EOT performance can be further improved by adjusting the parameters of circular nanoantenna. Figure 19 shows the transmission efficiency profiles versus the radius of nanoantenna with different thicknesses for both resonant and anti-resonant vertical annular nanoslit. The curves show a similar trend for both cases except that the peak values are generally smaller for the anti-resonant vertical annular cavity. Higher EOT can be obtained with thinner circular nanoantenna. As the thickness of the nanoantenna approaches the skin depth, part of incident light transmitting directly into the horizontal cavity will provide addition source in the dielectric layer, leading to a higher transmission. In addition, the radius of the nanoantenna plays an important role. The fluctuating transmission curves versus the radius of the nanoantenna show a tradeoff between the collection capability and propagating loss of SPP.

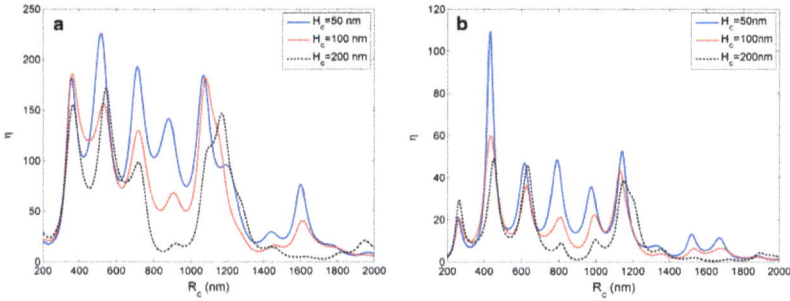

Fig. 19. Transmission curves versus radius of the circular nanoantenna with different thickness for (a) resonant and (b) non-resonant nanoslit (adapted from Ref. 30).

This novel EOT design can be integrated onto the end of an optical fiber.[26] When a radial polarized beam is generated and propagating in the fiber, a strong enhanced transmitted field could be created by the combination of the circular nanoantenna and the annular nanoslit. This type of device may be used to detect weak signals generated by events or analytes within the annular nanoslit as a highly sensitive optical sensor.

2.4. *Polarization mode matching and optimal plasmonic focusing*

The examples discussed in Sections 2.1-2.3 show that optimal plasmonic effects can be achieved through matching the plasmonic structure to specific spatially engineered polarization distribution of the illumination (e.g., radial polarization in these examples). This can be understood within the context of the antenna radiation theory. In the antenna theory, the receiving antenna should match to the polarization mode pattern from the emitting antenna to maximize the received signal. This is also true for plasmonic focusing devices. Essentially, the plasmonic focusing device can be regarded as a receiving antenna for the incoming optical signal. In order to optimize the received signal, the illumination polarization pattern needs to match the corresponding polarization mode pattern that would radiate from the same plasmonic structure as if it were used as a transmitter.

Taking the circular slot plasmonic structures as an example, from the antenna theory, it is known that the radiation pattern of a circular slot antenna in the upper half space is similar to an electric dipole (illustrated in Fig. 20). The radiation mode has its polarization aligned in the radial direction if it is viewed at the direction normal to the slot antenna. Thus, for a plasmonic lens made of circular slot etched in the metal film, the desired illumination should be radially polarized. This is confirmed by the circular slot plasmonic lens presented in the sections above. Such an observation can be further extended to many other plasmonic structures with their own specific spatially engineered polarization modes for optimal plasmonic excitation and localization. Furthermore, these studies and their connections to the polarization mode matching of receiving antenna point out that the extensive body of the existing literatures in microwave and radio frequency antenna can be applied to the design and optimization of functional plasmonic devices for better plasmonic energy localization and manipulation. One of such examples is discussed in the following section.

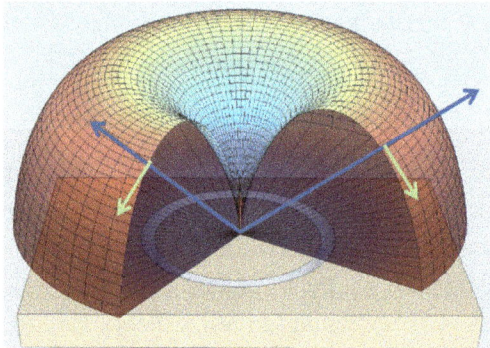

Fig. 20. Radiation pattern of a circular slot antenna.

2.5. *Archimedes' spiral plasmonic lens*

In the previous sections, optimal SPP focusing has been demonstrated both theoretically and experimentally through matching the radial polarized illumination to axially symmetric plasmonic structures. However, it requires stringent alignment between the singularity center of the spatially variant polarized beam and the center of the axially symmetric plasmonic lens. This requirement necessitates a scanning mechanism for imaging applications. Utilizing the polarization mode matching concept presented in Section 2.4, plasmonic lens made of Archimedes' spiral slot etched into metallic thin film has been developed to eliminate this alignment requirement. From the antenna theory, an Archimedes spiral slot transmitting antenna radiates circularly polarized mode (Fig. 21). The handedness of the circular polarization depends on the handedness of the spiral. Consequently, the performance of a spiral plasmonic lens as a receiver will exhibit spin dependence on the illumination. A left-handed spiral (LHS) plasmonic lens produces strongly focused plasmonic field under right-handed circular (RHC) polarization illumination while defocuses the left-handed circular (LHC) polarization. The use of circularly polarized illumination eliminates the stringent requirement of center alignment. Meanwhile, the spin dependence adds additional functionality to the spiral plasmonic devices.

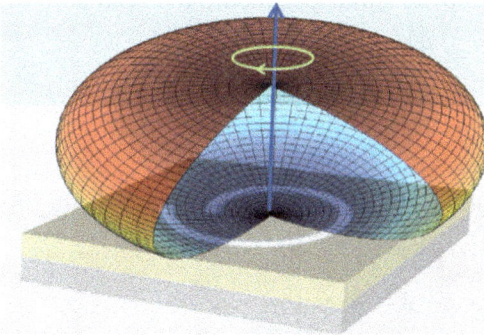

Fig. 21. Radiation pattern of a spiral slot antenna.

2.5.1. *Single spiral plasmonic lens*

Circular polarization, which can be decomposed into radial and azimuthal polarization components with a spiral phase wavefront,[31] has been used as an alternative to radial polarization in SPP excitation and focusing. It has been shown that right-handed circular polarization can be re-written in the cylindrical coordinates as:

$$\vec{E}_{RHC} = \frac{1}{\sqrt{2}}(\vec{e}_x + i\vec{e}_y) = \frac{1}{\sqrt{2}}e^{i\varphi}(\vec{e}_r + i\vec{e}_\varphi), \tag{5}$$

and the left-handed circular polarization can be expressed as:

$$\vec{E}_{LHC} = \frac{1}{\sqrt{2}}(\vec{e}_x - i\vec{e}_y) = \frac{1}{\sqrt{2}}e^{-i\varphi}(\vec{e}_r - i\vec{e}_\varphi).\tag{6}$$

Using this simple decomposition, it has been shown that a spiral plasmonic lens can focus circular polarization with the same handedness into a solid spot and defocus circular polarization of the opposite chirality into a donut shape with a dark center.[32, 33] In the cylindrical coordinates, a left-hand single Archimedes' spiral (LHS) (Fig. 22) can be described as:

$$r = r_0 - \Lambda\phi/2\pi,\tag{7}$$

where r_0 is a constant and Λ equals the SPP wavelength. Circularly polarized beam illuminates the spiral structure normally from the glass substrate side (along the z direction pointing out of the page in Fig. 22). Under a RHC polarized illumination, the plasmonic field near the geometric center of the spiral plasmonic lens can be derived analytically as:[32]

$$\vec{E}_{spp}(R,\theta) = \vec{e}_z 2\pi E_{0z} r_0 e^{-k_z z} e^{ik_r r_0} J_0(k_r R),\tag{8}$$

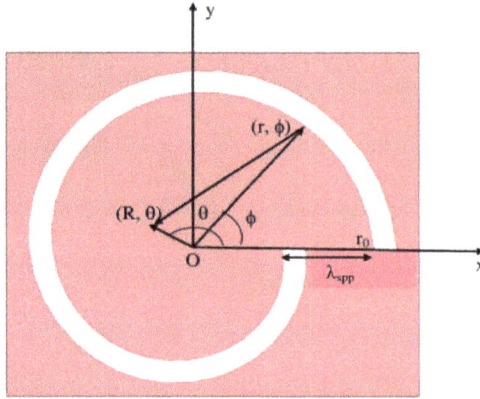

Fig. 22. Diagram of a left-turn single spiral and the coordinates used in the calculation (adapted from Ref. 32).

where E_{0z} is a constant and k_r is the wavenumber of the SPP. From this derivation, a LHS plasmonic lens focuses RHC polarization into a 0th-order evanescent Bessel beam with a central peak. It also indicates that a larger spiral size (r_0) leads to stronger field strength at the focus. On the other hand, LHC would be defocused by a LHS with the expression as:

$$\vec{E}_{spp}(R,\theta) = \vec{e}_z 2\pi E_{0z} r_0 e^{-k_z z} e^{ik_r r_0} e^{-2i\theta} J_2(k_r R).$$ (8)

This represents a second-order evanescent Bessel beam with a donut shape and topological charge of 2. This phenomenon can be explained by the spin-orbit interaction manifested by the geometric Berry's phase. The spin of the incident photon is defined as $\sigma_i = 1$ for RHC and $\sigma_i = -1$ for LHC. Similarly, the chirality of the spiral structure σ_s is 1 for right-hand spiral (RHS) and -1 for LHS structure. The total orbital angular momentum of the plasmonic field is $l = \sigma_i + \sigma_s$. Therefore, a LHS plasmonic lens focuses RHC into a solid spot ($l = 0$) and defocuses LHC into a donut with a dark center and a spiral wave front ($l = -2$). Field distributions corresponding to the RHC and LHC illuminations are shown in Fig. 23.

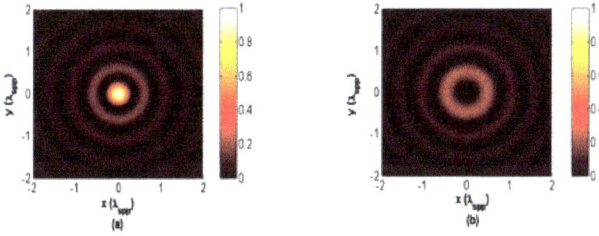

Fig. 23. Analytical results for intensity distributions of (a) RHC and (b) LHC polarizations illuminating a LHS plasmonic lens (adapted from Ref. 32).

The analytical expressions given by Eqs. (7) and (8) are only suitable for a single turn spiral structure that is relatively large compared with the SPP wavelength. The propagation loss and transverse components of SPP are also ignored. In order to take all these effects into consideration, finite element method numerical modeling (COMSOL) is performed. Simulation results with RHC illuminating on the RHS and LHS plasmonic lenses are shown in Fig. 24. Clearly, a RHS structure focuses RHC polarization into a donut shape with a dark center, while a solid spot with a central peak could be obtained for a LHS structure. Thus, the analytical expressions provided excellent insight of the focusing behavior of the spiral plasmonic lens. The numerically modeled spiral plasmonic lens has a footprint on the order of $4\lambda_{spp}$, making it an extremely small polarization sensitive device. The spin dependence of the spiral plasmonic structure shows that a pair of LHS and RHS structures with a small detector integrated in the center may be used as miniature circular polarization analyzers to analyze the circular polarization content of the incident field.

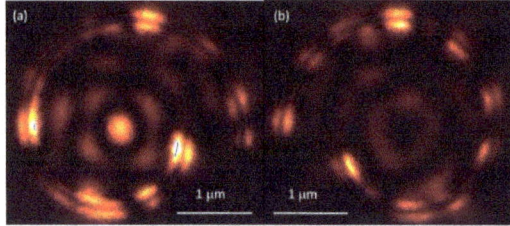

Fig. 24. Simulation results of the intensity distributions on (a) LHS and (b) RHS under RHC polarization illumination (adapted from Ref. 32).

The spin-dependent behavior of spiral plasmonic lens has been observed experimentally.[33] Single-turn LHS and RHS spiral structures are etched into a gold film with FIB milling (Fig. 25). Circularly polarized light with wavelength of 808 nm illuminates the spiral from the glass substrate side and the SPP distribution is imaged by a collection mode NSOM (aperture size of 50-80 nm). Figure 26(a) and (b) show the 2D NSOM images of the SPP distributions at the air/gold interface for a LHS plasmonic lens under LHC and RHC illuminations, respectively. As expected, a donut shaped spot with a dark center is observed for RHC illumination due to the weak coupling efficiency of the longitudinal electric component into apertures NSOM probe. However, the peak

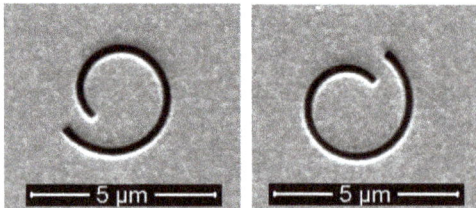

Fig. 25. SEM images of (a) a LHS and (b) a RHS plasmonic lens etched into gold film deposited on a glass substrate (adapted from Ref. 33).

NSOM signal in the vicinity of focus with RHC illumination is 2 times higher than that of with LHC illumination. This intensity difference indicates the different plasmonic focusing behaviors for different spin angular momentums, even though the shapes of NSOM images look similar. The near-field images of focusing properties of RHS structure are shown in Fig. 26(c) and (d) and similar phenomenon is observed for the LHC illumination that shows the higher peak intensity.

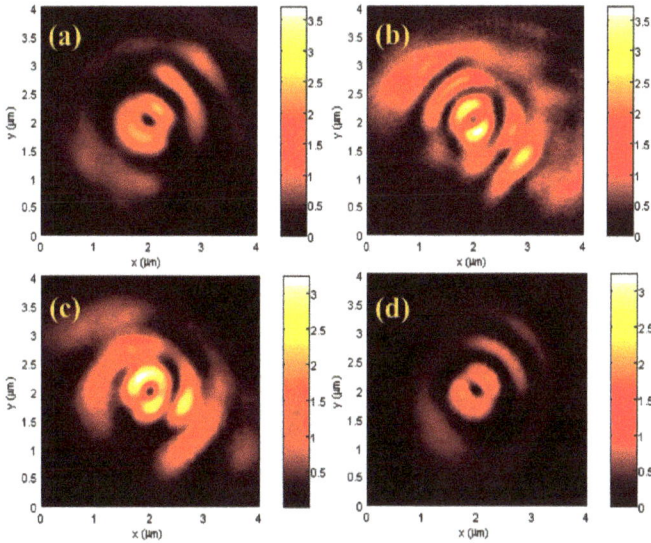

Fig. 26. Measured NSOM images at the gold/air interface for a LHS and RHS plasmonic lens under (a) (c) LHC and (b) (d) RHC polarized illuminations, respectively (adapted from Ref. 33).

2.5.2. *Spiral plasmonic lens array*

As mentioned above, the focusing effect of spiral plasmonic lens does not require alignment between the centers of illumination and a spiral structure, enabling its applications in parallel imaging with an array format. A 10×10 array of LHS plasmonic lens (Fig. 27) was experimentally characterized with two-photon fluorescence (TPF) microscopy.[34] The field intensity distributions of SPP in the vicinity of spiral surface can be visualized by TPF in the far field (see Fig. 28). For the RHC polarized illumination, a bright spot at the center of each LHS element is observed, while a donut

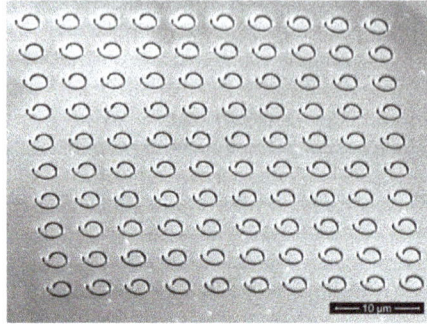

Fig. 27. SEM image of a 10×10 LHS plasmonic lens array (adapted from Ref. 34).

shape with dark center is obtained under LHC illumination. To better reveal the distinction of intensity distributions of collected fluorescence signals of RHC and LHC illuminations, line scans across the center of one spiral are shown in Fig. 28(c).

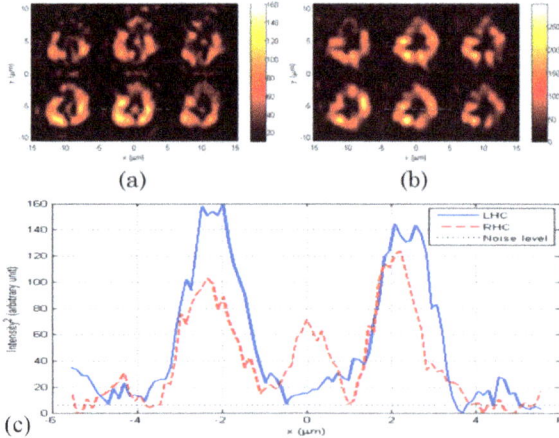

Fig. 28. TPF microscopy images of intensity distributions of LHS array under (a) RHC and (b) LHC polarized illuminations. (c) Line-scan across the center of one spiral (indicated by the dashed lines in (a) and (b)). The estimated noise level is indicated by the black dashed line (adapted from Ref. 34).

The focusing characteristics of each element in the spiral array obtained by the TPF microscopy agrees with that of the single spiral structure measured by NSOM.[33] The circular polarization extinction ratio is calculated by the ratio of integrated fluorescence signals within a central circular area in the vicinity of the focus under RHC and LHC illuminations. An estimated circular polarization extinction ratio of a TPF signal larger than 200 can be realized with a detector diameter up to $0.3\lambda_{spp}$. More importantly, the fabrication process of the spiral plasmonic lens is compatible with the conventional wire grid micropolarizers for linear polarization detection, making them suitable for full Stokes parameter polarimetric imaging application.[35]

2.5.3. *Hybrid spiral plasmonic lens*

For the spiral slot plasmonic lens described above, only the radial component of the circularly polarized excitation can be efficiently coupled to SPP as this component is locally p-polarized with respect to the spiral slot. The azimuthal component is not collected because it is *s*-polarized with respect to the spiral slot, leading to the loss of half of the incident power. Recently, plasmonic focusing under azimuthal polarization excitation has been demonstrated with a plasmonic lens consists of spatially arranged triangular apertures.[36] The focusing property of this plasmonic lens exhibits strong dependence on the orientation of the triangular apertures because of a geometric phase effect.[36] If the triangle array was arranged along a spiral line, it also can couple the azimuthal component of a circularly polarized illumination into SPP.[37] Similar to the spiral slot structure, the spiral triangle array focuses the RHC and LHC illuminations into spatially separated plasmonic fields and may serve as a miniature circular polarization analyzer (Fig. 29).

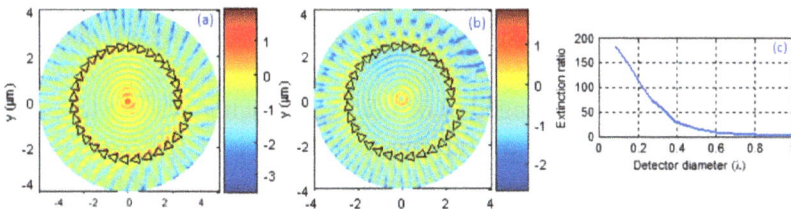

Fig. 29. Logarithmic electric density distributions at the silver/air interface of a spiral triangle array plasmonic lens under (a) LHC and (b) RHC polarized illuminations. (c) Circular polarization extinction ratio curve versus the detector diameter (adapted from Ref. 36).

The characteristics of the spiral triangle array offers pathway to a high efficiency circular polarization analyzer design. A hybrid plasmonic lens (Fig. 30), which consists of both a spiral slot and a spiral triangle array, has been demonstrated for much higher field enhancement and power conversion efficiency compared to the pure spiral slot lens.[37] Unlike the spiral plasmonic lens consisting of either spiral slot or spiral triangle array, which can efficiently couple only one component of the circularly polarized excitation, this hybrid structure enables the coupling of both radial and azimuthal components into SPP.

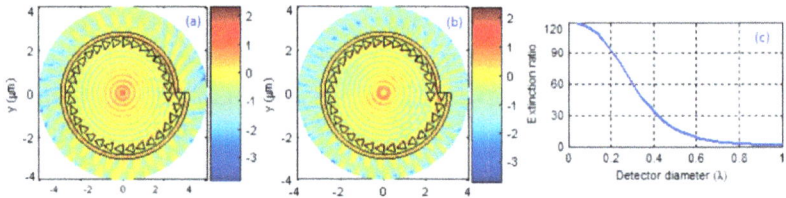

Fig. 30. Logarithmic electric density distributions at the silver/air interface of a hybrid plasmonic lens consisting of both spiral slot and spiral triangle array under (a) LHC and (b) RHC polarized illuminations. (c) Circular polarization extinction ratio curve versus the detector diameter (adapted from Ref. 37).

The intensity of the plasmonic field at the focus is sensitive to the distance of radial displacement between the spiral slot and the triangle array. To obtain the maximal field enhancement at the focus, SPP excited at the spiral slot and spiral triangle array need to be in phase to achieve constructive interference. For circularly polarized beam, there is a $\pi/2$ phase difference between the radial and azimuthal components (see Eqs. (5) and (6)).[31] Numerical study shows that the intensity of plasmonic field at the focus is highest when the spacing between the spiral slot and spiral triangle array is arranged to be $0.75\lambda_{spp}$, which gives rise to a phase difference of $3\pi/2$ and an overall phase difference of 2π. Figures 30(a) and (b) illustrate the logarithmic energy density distribution of this hybrid spiral plasmonic lens under RHC and LHC illumination, clearly showing a spin-dependent behavior. Compared with a two-turn spiral slot, the field enhancement and power conversion efficiency of this hybrid plasmonic lens are increased by 39.53% and 94.69%, respectively. Such a hybrid plasmonic lens can be used as a high efficiency circular polarization analyzer, with a circular polarization extinction ratio better than 61.5 for a detector with diameter less than 0.3λ (Fig. 30(c)).

2.6. *Applications in near-field optical probe designs*

Optical microscopy has been widely used for the observation of small structures due to their non-contact and nondestructive natures. Owing to the diffraction of light, the smallest feature can be resolved with a traditional optical microscopy is on the order of half wavelength of the incident light. In order to improve the spatial resolution and beat the diffraction limit, NSOM that exploit the evanescent wave components containing high spatial frequency information about the sample have been developed. Different from the conventional optical microscopy, the image formation by NSOM is a result of electromagnetic interaction between the near-field optical probe and the sample via evanescent photon coupling. Therefore the NSOM probe needs to be placed very close to the sample surface. The most commonly used NSOM probe is made of metallic coated tapered glass fiber with a nanometric aperture opening at the fiber end. With this technology, the resolution of the NSOM image is primarily determined by the size of the aperture. However, in order to achieve good image quality with reasonable signal-to-noise-ratio (SNR), a compromise has to be made between the aperture size and optical throughput. Typical aperture size of NSOM in practice is limited to around 50 nm.

To overcome the drawbacks of the aperture NSOM, apertureless techniques utilizing the field enhancement effect of SPP have been developed. Compared with aperture NSOM, apertureless NSOM may use a broader variety of tip material to achieve better spatial resolution and wider spectral response range. Usually an apertureless NSOM uses a sharp metallic tip that can effectively scatter the evanescent field localized around the tip apex. The apex of the tip functions as an efficient scatterer with strongly enhanced local electric field. Hence the performance of an apertureless NSOM strongly depends on the local field enhancement produced at the tip apex, which in turn is heavily influenced by the probe design as well as the optical illumination geometry. The discussions in Sections 2.1-2.5 demonstrated that plasmonic focusing and localization for a specific plasmonic structure can be optimized with its matched polarization mode excitation. Naturally, these findings should find important applications in efficient apertureless NSOM probe designs by producing highly enhanced local plasmonic fields.

2.6.1. *Metal coated conical dielectric tip*

When a radially polarized beam illuminates an axially symmetric fully coated apertureless NSOM probe (Fig. 31), the induced SPP will converge toward the tip apex and interference constructively because of the rotational symmetry of the input polarization and the probe.[38, 39] In contrast, if a linear or circular polarization is coupled to the probe, the field at the end of the tip will cancel out because the opposite sides on the probe surface have opposite charges. A COMSOL model has been developed to investigate the localized field distribution in the vicinity of a silver coated tip (shown in Fig. 32).[39] The tip has a half cone angle of 16.4° and a radius of curvature of 20 nm. The entire tip is coated with 50 nm silver, and the radius of curvature of the silver film at the apex is set to be 5 nm. Radially polarized beam illuminates the tip from the bottom with wavelength of 632.8 nm. Figure 32 shows the logarithmic 2D and 3D plots of electric energy density distribution in the vicinity of the tip. Clearly, a strongly enhanced localized field is observed at the tip apex. A FWHM of less than 10 nm with an electric field enhancement of about

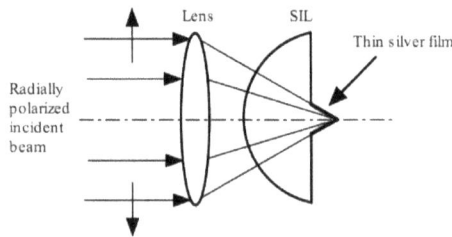

Fig. 31. Diagram of an apertureless probe design with radially polarized illumination (adapted from Ref. 39).

320 can be obtained for this probe design. For specific applications such as near-field Raman spectroscopy, high sensitivity at nanometer scale spatial resolution is hard to achieve due to the extremely weak Raman scattering cross-sections. High Raman enhancement factor, which is proportional to the fourth power of the localized field enhancement, is crucial. This metal-coated probe with such a high field enhancement is extremely instrumental to meet the challenges facing in near-field Raman microscopy.

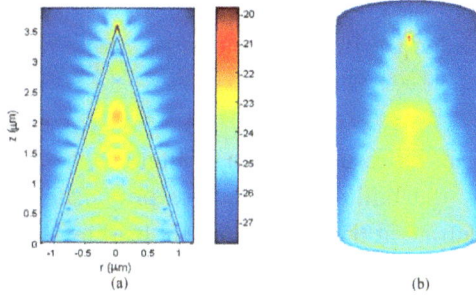

Fig. 32. Logarithmic (a) 2D and (b) 3D plot of electric energy density distribution at the end of probe tip (adapted from Ref. 39).

2.6.2. *Metal tip integrated with bull's eye plasmonic lens*

The field enhancement factor produced by an apertureless probe is crucial for the performance of the apertureless NSOM. It is possible to further increase the field enhancement factor by cascading several plasmonic focusing and localization structures together. A conceptual design to acquire much larger field enhancement by combining the concentration effects from both bull's eye structure and a sharp conical tip is illustrated in Fig. 33.[40] A silver tip is integrated at the center of a 5-ring bull's eye plasmonic lens. Under a radially polarized excitation, the concentric annular rings focus the SPPs to the geometrical center of the structure where the conical tip further localized the SPP field and induces higher field enhancement at the tip apex.

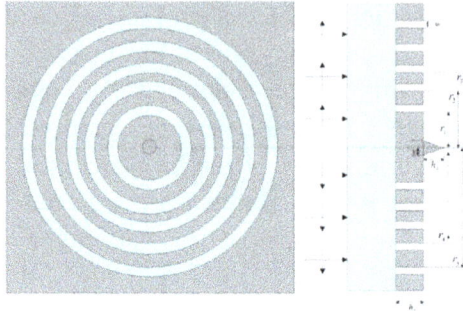

Fig. 33. Diagram of the probe design that combines concentric annular rings and a conical tip under radially polarized illumination. The quartz substrate and metal are indicated by light blue and gray zone, respectively (adapted from Ref. 40).

From the simulation results shown in Fig. 34, it can be seen the SPP waves focused by the bull's eye structure propagate toward the center and further concentrate at the end of the tip, leading to a strong localized field enhancement. For this silver probe design, intensity distribution with an FWHM of about 4 nm and electric field enhancement of about 467 can be achieved with 632.8 nm optical excitation. Furthermore, the longitudinal plasmonic field drives the tip as an oscillating dipole, leading to a strong local enhancement at the tip apex. This phenomenon is similar to the lighting rod effect. Therefore the enhancement factor does not strongly depend on the tip cone angle and the excitation wavelength, enabling easier fabrication and broader spectroscopic applications.

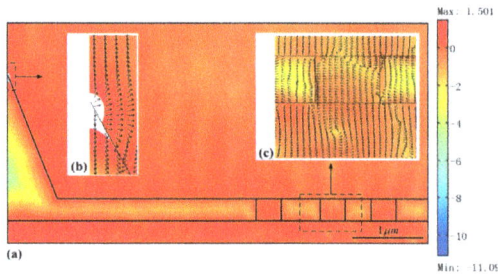

Fig. 34. (a) Logarithmic 2-D plot the electric energy density distribution of the probe. Electric field distribution around the end of the tip and slit are shown in insets (b) and (c) (adapted from Ref. 40).

This type of near-field probe structure can be regarded as a tandem of two optical antennas. The bull's eye structure functions as a receiving antenna aimed at converting incoming beam to focused cylindrical plasmons on the backside. Subsequently, the focused plasmonic field will be coupled to the vertical tip working as a transmitting antenna that assists in retransmitting to the far field. To investigate the efficiency of coupled plasmonic antennas and their resonance interactions, a device comprised of single ring and a short vertical gold nanowire has been fabrication with FIB milling and low current electron beam-assisted local deposition (Fig. 35).[41]

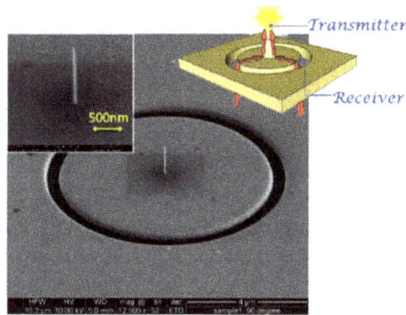

Fig. 35. SEM image of the plasmonic wire grown at the center of a single ring structure. Right inset: scheme of the proposed structure. Left inset: zoom-in of the wire antenna (adapted from Ref. 41).

Transmission spectra of the fabricated device were measured using a radially polarized tunable near-infrared laser. Figure 36(a) shows both the measured and calculated spectral transmission of the sample with different dimensions of the nanowire antenna embedded in air. The transmission spectrum for the same receiving configuration with various dimensions of the plasmonic nanowire is normalized by the ring antenna transmission. The longest wire antenna shows a significant transmission enhancement of 30%, indicating 3 orders of magnitude enhancement of the nanowire antenna cross section. To study the influence of environment to the transmission properties of the device, similar experiment was performed with the same device but embedding the wire antenna in polyamide (shown in Fig. 36(b)). It is found that the transmission peaks were blue shifted and the peak value of the transmission enhancement remains to be similar. The transmission peaks corresponds to the first F-P resonance of the plasmonic wire antenna.[42] The role of the nanowire antenna as part of the transmitting antenna has also been demonstrated experimentally. When the direction of the excitation flipped, the nanowire antenna becomes a part of receiving antenna that can be used to produce strongly enhancement local field for near-field optical imaging and sensing.

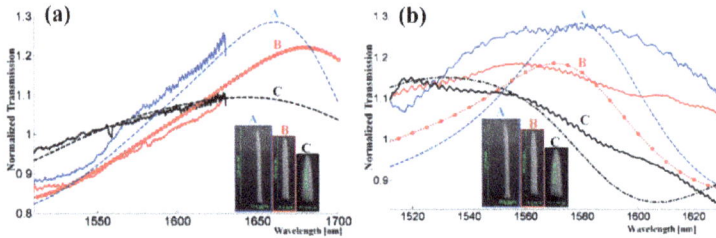

Fig. 36. Spectral transmission of the sample with different wire antenna embedded in (a) air and (b) polyamide, normalized by the single ring antenna transmission (adapted from Ref. 41).

2.6.3. *Composite tip integrated with spiral plasmonic lens*

Although the axially symmetric probes presented above offers extremely high field enhancement, the singularity center of the radially polarized illumination need to be aligned to the center of the structure in the experiment. It necessitates a scanning mechanism that limits the speed and realistic scanning area achievable of these probes in near-field optical imaging, sensing and lithographic patterning applications. The spiral plasmonic lens with circularly polarized excitation demonstrated in Section 2.5 provides efficient plasmonic focusing while eliminating the stringent alignment requirement. Base on this observation, a composite plasmonic near-field probe design that consists of a spiral plasmonic lens and a sharp conical tip has been proposed (Fig. 37).[43] Instead of a subwavelength detector, a tip is integrated in the center of the spiral to efficiently couple and channel the plasmonic wave adiabatically towards the tip apex.

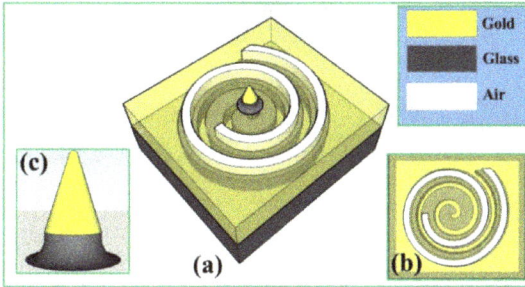

Fig. 37. (a) Diagram of the proposed near-field plasmonic probe consists of two spiral plasmonic lens and a composite conical tip at the center. Radially polarized beam illuminate the structure from the substrate side. (b) Top-view of the spiral plasmonic lens. (c) Zoom-in of the tip combines a dielectric base and a metallic tip (adapted from Ref. 43).

A trade-off between the SPP coupling efficiency and far-field radiation leakage of the illumination needs to be made. Spiral plasmonic lens with wider slot gives rise to a higher coupling efficiency from the excitation to the SPP mode. However, wider slot opening leads to stronger far-field radiation that should be avoided in near-field optical applications. In order to solve this dilemma, a double-layer spiral plasmonic lens has been adopted (see Fig. 37). The lower spiral with wider opening provides high SPP collection efficiency, while the upper spiral maintains narrow slot width to reduce the far-field radiation.

To obtain high field enhancement at the tip apex, a composite dielectric/metal tip is designed as opposed to the conventional full-metal tip. The dielectric base of the composite tip can be regarded as an optical nano-fiber. Hence, by adjusting the radius of the dielectric base, the coupling efficiency from the SPP surface mode in the vicinity of the focus to the guided mode of the nano-fiber could be optimized when these two modes are matched similar to the butt-coupling.[44]

Because of the spin dependence of the spiral plasmonic lens, the field intensity near the tip apex changes dramatically if the polarization of illumination is switched between RHC and LHC. Figures 38(a) and (b) illustrates the simulation results of intensity distributions on the near-field probe with LHS and RHS spiral structures under the same RHC polarized illumination. Numerical simulations show that this optimized probe achieve an electric field enhancement of 366 and circular polarization extinction ratio higher than 80.

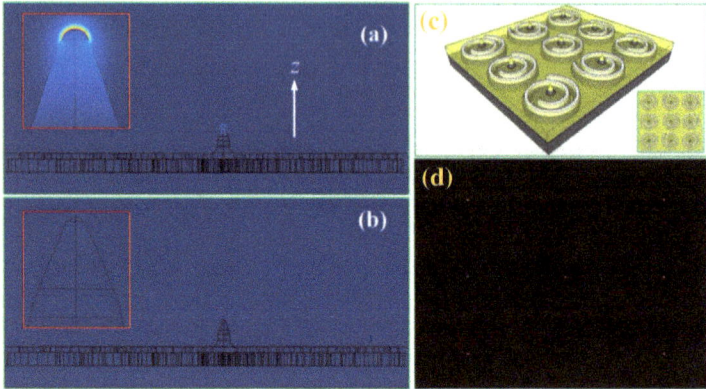

Fig. 38. Simulated intensity distribution on the probe with (a) LHS and (b) RHS plasmonic lens under the same RHC polarized illumination. (c) Diagram of the proposed array of the near-field probe. (d) Simulated intensity distribution at 10 nm above the tip apex for RHC polarized illumination (adapted from Ref. 43).

The spin-dependent behavior of the plasmonic field enables the hot spot at the tip apex to be switched on and off by modulating the polarization handedness of the illumination. Figure 38(c) shows a 3×3 array containing both LHS and RHS spiral plasmonic lenses. Under RHC polarized illumination, the LHS elements corresponding to the letter "N" pattern produce the hot spots (simulation result shown in Fig. 38(d)). As demonstrated in this example, by eliminating alignment requirement while maintaining very high field enhancement, this type of probe can be designed in an array format that is highly suitable for large area parallel near-field imaging, sensing and photolithography applications.

3. Conclusions

In this chapter, recent progresses on plasmonics focusing and localization with the use of vectorial optical fields as excitation are reviewed. Specifically, it has been discovered that radial polarization is the ideal source for SPP excitation with axially symmetric plasmonic structure. Various plasmonic structures with axial symmetric are designed to achieve extremely strong field enhancement with spot size far beyond the diffraction limit. By matching the azimuthally polarized illumination to spatially arranged triangular apertures, a tightly focusing plasmonic field could also be achieved. Recently, an experimental technique has also been reported to offer insight into the optical properties of subwavelength nanosturctures by adapting the polarization pattern of the excitation field to the nanostructure.[45] These examples allow us to introduce a generalized polarization mode matching concept within the context of the antenna radiation theory to explain the optimal plasmonic focusing effect with these vectorial optical fields. Plasmonic focusing devices can be regarded as receiving antenna for the incoming optical signal. To optimize the field enhancement effect, the illumination polarization pattern needs to match the corresponding polarization mode pattern that would radiate from the same plasmonic structure as if it were used as a transmitter. This concept shows a very important perspective in the design and study of metamaterials/plasmonic nanostructures that can be used as guidelines for the design and optimization of future functional metamaterials/plasmonic devices. These optimized nanostructures under their matching vectorial optical excitations may find important applications in areas such as near-field optical imaging, sensing, optical manipulations of nanoparticles, lithography and polarimetric imaging.

References

1. B. Hecht, H. Bielefeldt, L. Novotny, Y. Inouye and D. W. Pohl, *Phys. Rev. Lett.* **77**, 1889 (1996).
2. J. Pendry, *Science* **285**, 1687 (1999).
3. K. Kneipp, Y. Wang, H. Kneipp, L. T. Perelman, I. Itzkan, R. R. Dasari and M. S. Feld, *Phys. Rev. Lett.* **78**, 1667 (1997).
4. S. M. Nie and S. R. Emery, *Science* **275**, 1102 (1997).
5. J. Homola, SS, Yee and G. Gauglitz, *Sens. Actuators B*, **54**, 3 (1999).
6. N. Fang, H. Lee, C. Sun and X. Zhang, *Science* **308**, 534 (2005).
7. X. Luo and T. Ishihara, *Appl. Phys. Lett.* **84**, 4780 (2004).
8. H. Raether, *Surface Plasmons on Smooth and Rough Surfaces and on Gratings* (Springer, Berlin, Heidelberg 1988).
9. W. Chen and Q. Zhan, *Proc. SPIE* **6450**, 64500D (2007).
10. H. Kano, S. Mizuguchi and S. Kawata, *J. Opt. Soc. Am. B* **15**, 1381 (1998).
11. Q. Zhan, *Opt. Lett.* **31**, 1726 (2006).
12. W. Chen and Q. Zhan, *Opt. Lett.* **34**, 6 (2009).
13. A. Bouhelier, F. Ignatovich, A. Bruyant, C. Huang, G. Colas Des Francs, J.-C. Weeber, A. Dereux, G. P. Wiederrecht and L. Novotny, *Opt. Lett.* **32**, 2535 (2007).

14. L. Novotny and B. Hecht, *Principles of Nano-Optics* (Cambridge U. Press, 2006).
15. S. Ruschin and A. Leizer, *J. Opt. Soc. Am. A* **15**, 1139 (1998).
16. G. Rui, Y. Lu, P. Wang, H. Ming and Q. Zhan, *J. Appl. Phys.* **108**, 074304 (2010).
17. G. Rui, Y. Lu, P. Wang, H. Ming and Q. Zhan, *Opt. Commun.* **283**, 2272 (2010).
18. A. Drezet, A. L. Stepanov, H. Ditlbacher, A. Hohenau, B. Steinberger, F. R. Aussenegg, A. Leitner and J. R. Krenn, *Appl. Phys. Lett.* **86**, 074104 (2005).
19. Z. Liu, J. M. Steele, W. Srituravanich, Y. Pikus, C. Sun and X. Zhang, *Nano Lett.* **5**, 1726 (2005).
20. Z. Liu, J. M. Steele, H. Lee and X. Zhang, *Appl. Phys. Lett.* **88**, 171108 (2006).
21. G. M. Lerman, A. Yanai and U. Levy, *Nano Lett.* **9**, 2139 (2009).
22. J. M. Steele, Z. Liu, Y. Wang and X. Zhang, *Opt. Express* **14**, 5664 (2006).
23. W. Chen, D. C. Abeysinghe, R. L. Nelson and Q. Zhan, *Nano Lett.* **9**, 4320 (2009).
24. A. Yanai and U. Levy, *Opt. Express* **17**, 924 (2009).
25. G. M. Lerman, M. Grajower, A. Yanai and U. Levy, *Opt. Lett.* **36**, 3972 (2011).
26. W. Chen, W. Han, D. C. Abeysinghe, R. L. Nelson and Q. Zhan, *J. Opt.* **13**, 015003 (2011).
27. T. W. Ebbesen, H. J. Lezec, H. F. Ghaemi, T. Thio and P. A. Wolff, *Nature* **391**, 667 (1998).
28. D. Crouse and P. Keshavareddy, *Opt. Express* **15**, 1415 (2006).
29. P. Banzer, J. Kindler, S. Quabis, U. Peschel and G. Leuchs, *Optics Express* **18**, 10896 (2010).
30. G. Rui, Q. Zhan and H. Ming, *Plasmonics* **6**, 521 (2011).
31. Q. Zhan, *Opt. Lett.* **31**, 867 (2006).
32. S. Yang, W. Chen, R. L. Nelson and Q. Zhan, *Opt. Lett.* **34**, 3047(2009).
33. W. Chen, D. C. Abeysinghe, R. L. Nelson and Q. Zhan, *Nano Lett.* **10**, 2075 (2010).
34. Z. Wu, W. Chen, D. C. Abeysinghe, R. L. Nelson and Q. Zhan, *Opt. Lett.* **35**, 1755 (2010).
35. Z. Wu, P. E. Powers, A. M. Sarangan and Q. Zhan, *Opt. Lett.* **33**, 1653 (2008).
36. W. Chen, R. L. Nelson and Q. Zhan, *Opt. Lett.* **15**, 581 (2012).
37. W. Chen, R. L. Nelson and Q. Zhan, *Opt. Lett.* **37**, 1442 (2012).
38. A. Bouhelier, J. Renger, M. R. Beversluis and L. Novotny, *J. Microsc.* **210**, 220 (2003).
39. W. Chen and Q. Zhan, *Opt. Express* **15**, 4106 (2007).
40. G. Rui, W. Chen, Y. Lu, P. Wang, H. Ming and Q. Zhan, *J. Opt.* **12**, 035004 (2010).
41. P. Ginzburg, A. Nevet, N. Berkovitch, A. Normatov, G. M. Lerman, A. Yanai, U. Levy and M. Orenstein, *Nano Lett.* **11**, 220 (2011).
42. A. Normatov, P. Ginzburg, N. Berkovitch, G. M. Lerman, A. Yanai, U. Levy and M. Orenstein, *Opt. Express* **18**, 14079 (2010).
43. G. Rui, W. Chen and Q. Zhan, *Opt. Express* **19**, 5187 (2011).
44. X. Chen, V. Sandoghdar and M. Agio, *Nano Lett.* **9**, 3756 (2009).
45. P. Banzer, U. Peschel, S. Quabis and G. Leuchs, *Optics Express* **18**, 10905 (2010).

CHAPTER 6

OPTICAL MEASUREMENT TECHNIQUES UTILIZING VECTORIAL OPTICAL FIELDS

Qiwen Zhan

Electro-Optics Program, University of Dayton
300 College Park, Dayton, Ohio 45469, USA
E-mail: qzhan1@udayton.edu

Interaction and propagation of optical polarization have been widely explored in optical measurement techniques. The polarization diversity within the beam cross-section of vectorial optical fields offers interesting new opportunities in precise and rapid optical measurements. In this chapter, several optical measurement techniques that utilize the spatial variant state of polarization of the vector fields will be briefly introduced as examples. Rotational polarization symmetry has been exploited in microellipsometer designs and instrumentations to improve signal-to-noise ratio (SNR) while maintaining high spatial resolution. An extreme sensitive radial polarization interferometer is devised base on the rotational polarization symmetry of cylindrical vector beams. Polarization diversities of vectorial optical fields for rapid and parallel optical measurements are also demonstrated in a rapid Mueller matrix polarimetry for anisotropic material characterization and an atomic spin analyzer.

1. Introduction

Optical inspection and metrology are widely used in materials diagnostics and characterization. From the simple qualitative information obtained from imaging and phase contrast optics to the highly precise measurements provided by interferometry, polarimetry, and ellipsometry, optical techniques have been instrumental in advancing the state-of-the-art in materials science, microelectronics, biology, and many other disciplines. The non-destructive nature of most optical measurements and their inherent simplicity have made them valuable to modern research and fabrication.

Optical polarization as the vector nature of optical field plays an important role in many optical measurement techniques. The polarization diversity within the beam cross-section of vectorial optical fields offers interesting opportunities in optical measurement. In this chapter, the polarization symmetry is firstly exploited in a radially symmetric microellipsometer design to boost the SNR in a spatially resolved ellipsometer and achieve sub-micron level spatial resolution. A nulling microellipsometer is also illustrated as an improvement of the radially symmetric microellipsometer. A radial polarization interferometer is demonstrated with improved phase sensitivity compared with a traditional Michelson interferometer, where the phase difference between the interferometer's arms is manifested as spatially varying intensity distribution through interfering radially and azimuthally polarized beams. A rapid Mueller matrix polarimetry that can extract twelve Mueller matrix elements from a single intensity image enabled by the parallel polarization measurement with the use of vector beams is shown and applied to the characterization of anisotropic samples including metamaterials. Recently, polarization diversity of cylindrical vector (CV) beams have also been used as atomic spin analyzers and applied in rapid, single-shot atomic spin visualizations in a vapor cell.

2. Manipulation techniques for vectorial optical fields

In order to make use of vectorial optical fields in precise optical measurement applications, devices that can perform basic manipulations such as reflection, polarization rotation and retardation modulation are necessary. The key for these operations is to preserve the polarization distribution with the beam cross-section. Taking CV beams as examples, the challenge is to maintain the polarization symmetry through these operations. When CV beans are reflected and steered, polarization symmetry could be broken due to the non-equal reflection coefficients for s- and p-polarizations. Even if the magnitudes of these reflection coefficients are close, the phase difference can still destroy the polarization symmetry. In principle, metallic mirror should preserve the polarization symmetry better. However, many metallic mirrors have protective coatings that could give rise to different reflection coefficients for s- and p-polarizations. Combination of two identical beam-splitters (ideally picked up from the same coating run) with twisted orientation has been devised to maintain polarization symmetry while provides the steering function for CV beams (Fig. 1).[1] Similar arrangement with two identically coated metallic mirrors can be used to achieve higher throughput.

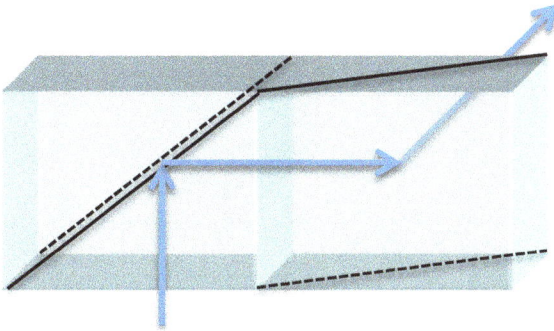

Fig. 1. Schematic drawing of a beam-splitter pair made of two identical beam-splitters that can preserve the spatial polarization distribution of the vectorial optical fields while redirecting it.

Polarization rotation can be achieved with active material (such as quartz) or Faraday rotators. However, these types of rotators lack tunability in terms of the rotation angle. And it is difficult to manufacture Faraday rotator with large clear aperture that is necessary for some applications. Polarization rotator using two cascaded $\lambda/2$-plates (shown in Fig. 2) has been designed and demonstrated to solve this problem.[1,2] If the angle between the fast axes of the two cascaded $\lambda/2$-plates (oriented at θ_1 and θ_2, respectively) is $\Delta\theta = \theta_2 - \theta_1$, the Jones matrix of this device can be shown to be:

$$T = \begin{pmatrix} \cos(2\Delta\theta) & -\sin(2\Delta\theta) \\ \sin(2\Delta\theta) & \cos(2\Delta\theta) \end{pmatrix} = R(2\Delta\theta), \qquad (1)$$

which is a pure polarization rotation function that is independent of the incident polarization. The polarization rotation angle can be tuned by adjusting the angle between the fast axes of the two plates. This type of devices can be used to rotate the polarization pattern of a certain CV beam into other desired generalized CV beam polarization patterns (refer to Fig. 1 in Chapter 1 for examples).

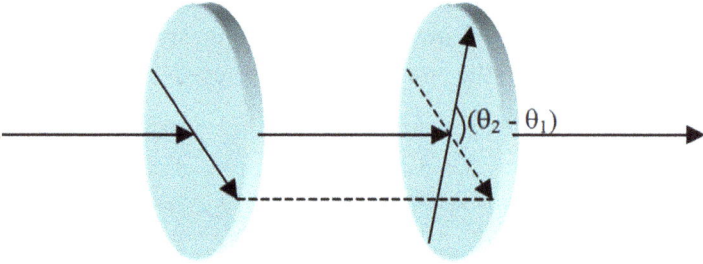

Fig. 2. An optical polarization rotator made of two cascaded λ/2-plates.

The polarization rotator using two cascaded λ/2-plates requires mechanical rotation of one of the λ/2-plates to realize the polarization rotation function. For applications that require high speed dynamic polarization rotation, this mechanism may cause vibrational noise and pose a limit on the ultimately achievable speed. Non-mechanical polarization rotator using liquid crystal (LC) or electro-optics (EO) retarders sandwiched between two orthogonally oriented λ/4-plates (Fig. 3) has been designed and demonstrated[1]. The Jones matrix of these devices can be shown to be:

$$T = R(-\frac{\pi}{2})\begin{pmatrix} 1 & 0 \\ 0 & -j \end{pmatrix}R(\frac{\pi}{2})R(-\frac{\pi}{4})\begin{pmatrix} 1 & 0 \\ 0 & e^{-j\delta} \end{pmatrix}R(\frac{\pi}{4})\begin{pmatrix} 1 & 0 \\ 0 & -j \end{pmatrix}$$

$$= -je^{-j\frac{\delta}{2}}R(\frac{\delta}{2}), \qquad (2)$$

where δ is the retardation of the LC or EO retarder. These devices can be made with very large clear aperture and the amount of rotation can be adjusted and modulated.

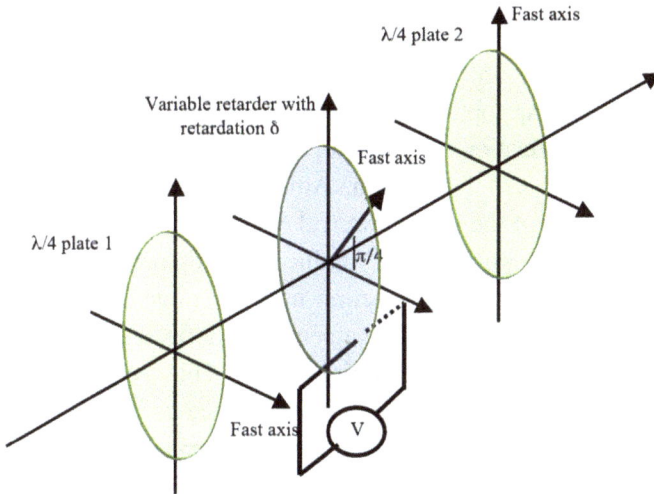

Fig. 3. Diagram of a non-mechanical polarization rotator using liquid crystal or electro-optic variable retarder sandwiched between two orthogonally oriented λ/4-plates.

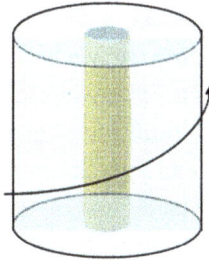

Fig. 4. Schematic of polarization rotator made of optical fiber with axial torsion.

In addition to these free space optics, for fiber optic devices, the axial torsion in optical fiber induces a birefringence in the fiber. It can be shown that the uniformly twisted fiber behaves like a medium exhibiting rotatory power (Fig. 4).[3] This provides a convenient means of polarization control. However, careful control is necessary due to the sensitivity. Besides these manipulation techniques, a fast EO radial polarization retarder reported in Ref. 4 can be used to provide retardation between radial and azimuthal polarization components. Beams with cylindrical symmetrically distributed elliptical polarization states can be generated with the use of this device.

3. Microellipsometer with rotational symmetry

Thin film structures are commonly involved in the materials research, biological sciences, optics, semiconductor, data storage industries. Accurate characterization of the properties of these thin films plays a critical role in modern fabrication. Among different optical techniques, ellipsometry is one of the most powerful optical metrological techniques of measuring thin films properties.[5, 6] It allows simultaneous determination of film thickness and index of refraction in an accurate and non-destructive manner. Optical ellipsometry can be generally defined as the measurement of the state of polarization of the light wave. This technique is based on exploiting the polarization modification that occurs as polarized light is transmitted through or reflected from the interfaces of thin films. By measuring this modification, one can extract the information about this interface, such as the thickness of the film, the refraction index of the material, the roughness of the interface, etc.

The rise of micro- and nano-structured microelectronic materials and devices, along with their ever-decreasing feature sizes, has prompted researchers to develop new ellipsometric capabilities in order to meet the challenges of fast and accurate characterization of these materials and devices. The applications of traditional ellipsometers with collimated beams are limited due to the poor spatial resolution arising from the large beam footprint on the surface under examination. This type of ellipsometers cannot be used to resolve small features that may be present on the surface. Improving the lateral resolution and the SNR of ellipsometric techniques remains of great interests to the community.

3.1. *Microellipsometer with rotational symmetry*

In order to improve the SNR while still maintain a high ellipsometric sensitivity with high spatial resolution, a spatially resolved micro-ellipsometer design utilizing rotational polarization symmetry has been developed and demonstrated.[1] The idea is best illustrated in Fig. 5. A standard ellipsometer is illustrated in Fig. 5(a). In Fig. 5(b), a beam normally illuminates the sample. A high numerical aperture (NA) objective lens is inserted in the optical path to focus the beam to a tight

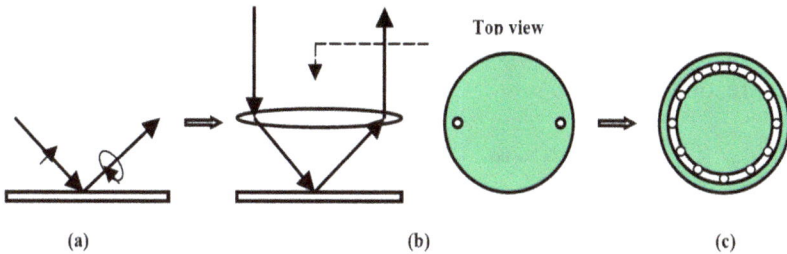

Fig. 5. Schematic illustration of the concept of rotationally symmetric microellipsometer. (a) A conventional ellipsometer; (b) One channel of a microelipsometer using high NA objective lens. (c) Repeating the single channel in a rotational symmetric manner to increase SNR.

spot on the sample. Each ray path in Fig. 5(b) acts just like a standard ellipsometer. If we use the rotating analyzer technique to detect this channel, we will observe a sinusoidal signal on the detector and obtain the ellipsometric information carried by this ray.[5] Considering one ray a standard rotating analyzer ellipsometric "channel", we can repeat this channel in a rotationally symmetric manner such that the micro-ellipsometer acts like a multi-channel conventional ellipsometer (Fig. 5(c)). Every individual channel located at a different angular location inside a common annular region will be designed to give the identical sinusoidal signal on the detector. The high numerical aperture illumination allows a high spatial resolution and an ellipsometric measurement of the sample. Due to the notational symmetry, each channel contributes to the signal collaboratively and a single large-area detector can be used to capture the total power from all the channels for subsequent analysis. Consequently, the SNR of the system is improved.

To carry out this idea, the rotational symmetryhas to be maintained throughout the optical system. A rotationally symmetric polarization signal is achieved using the combination of circularly polarized illumination, a polarization rotator, and a radial analyzer (Fig. 6). An annular aperture (Fig. 5(c)) is used to perform back focal plane spatial filtering by selecting the signal generated from a narrow annular cone of illumination for high ellipsometric sensitivity.[1] Enabled by the polarization signal's symmetry, the SNR is improved by collecting the ellipsometric signal within an entire annular region in the back focal plane, instead of a single azimuthal position. The combination of high NA focusing and back focal plane polarization analysis yields a high spatial resolution.

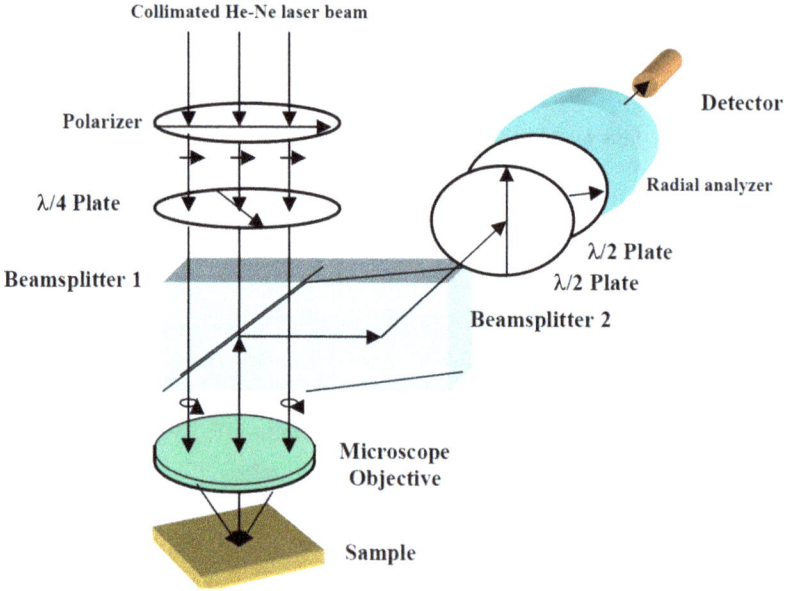

Fig. 6. Schematic of the realization of a rotationally symmetric microellipsometer design.

Suppose that for each channel, the ellipse of polarization has ellipticity angle ε and elevation angle θ_0. After a tedious derivation using Jones calculus, the signal on the detector can be shown to be[1]

$$P = K\left\{1 + \cos 2\varepsilon \cos\left[2(F - \theta_0)\right]\right\}, \tag{3}$$

where K is a constant, and F is the amount of polarization rotation introduced by the cascaded two $\lambda/2$-plates. The ellipsometric information is in the amplitude and phase of the signal, which can be detected by use of standard synchronous detection techniques. The (ε, θ_0) can be related to the more commonly used ellipsometric pairs via:

$$\tan^2 \Psi = \frac{\sin^2 \theta_0 \cos^2 \varepsilon + \cos^2 \theta_0 \sin^2 \varepsilon}{\cos^2 \theta_0 \cos^2 \varepsilon + \sin^2 \theta_0 \sin^2 \varepsilon}, \tag{4}$$

$$\Delta = \tan^{-1}\left(\frac{\tan\varepsilon}{\tan\theta_0}\right) + \tan^{-1}(\tan\theta_0\tan\varepsilon) - \frac{\pi}{2}. \qquad (5)$$

Either the (ε, θ_0) pair or the (Ψ, Δ) pair can be used in a standard regression procedure to infer the thickness and index of refraction of thin films within an extremely small illuminated area.

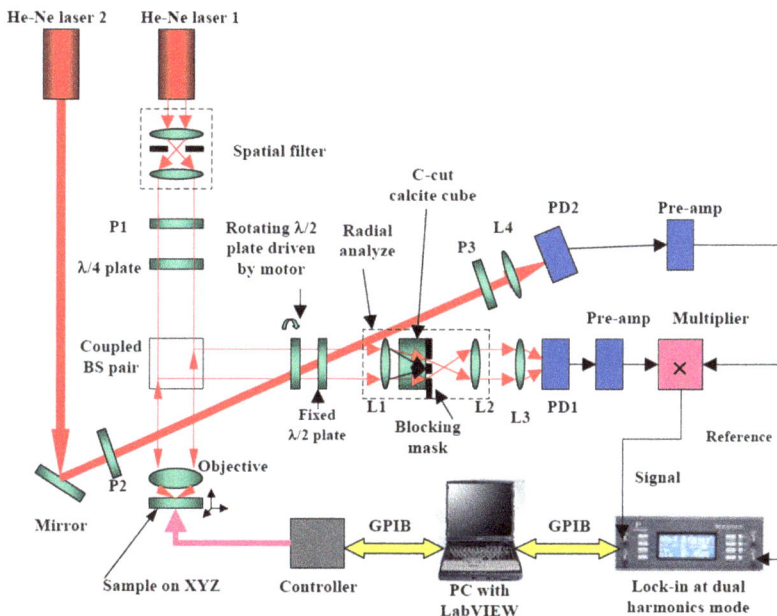

Fig. 7. Experimental setup of a rotationally symmetric microellipsometer.

The experimental setup is illustrated in Fig. 7. In the signal path, a high NA lens (NA = 0.8) focuses the circularly polarized incident light onto the sample and the collects the reflected light. The reflected light passes through the coupled beam-splitter pair, variable pure polarization rotator, radial analyzer and is brought to focus on the photodiode PD1. A motor is used to drive the first half-wave plate at a constant speed in the variable circular retarder. The radial analyzer module consists of two lenses and a c-cut calcite cube.[1] A pre-amp is used to amplify the signal. Besides the signal path, another laser was used to generate the reference signal for synchronous detection. This He-Ne laser beam passes through a polarizer, the circularly variable retarder, another polarizer and is brought to focus on the photodiode PD2. An identical pre-amp is used to amplify the reference signal.

In order to detect the modulation in Eq. (3), a multiplier is used to mix the signal and the reference such that the dc term is raised to the first harmonic and the ac term is raised to the second harmonic. The output of the multiplier is connected to the signal input of a lock-in amplifier running at dual-harmonics mode. The reference signal is also connected to the reference input of the lock-in amplifier. The lock-in returns both the phase and amplitude of the first and the second harmonics simultaneously. With the above measured results, one can calculate the (ε, θ_0) pair and further derive the refractive index and thickness of the thin film through regression models used in standard ellipsometry. The sample is mounted on a translation stage with electro-strictive actuators. Ellipsometric images of the sample can be obtained through scanning. The capability of this rotationally symmetric microellipsometer is

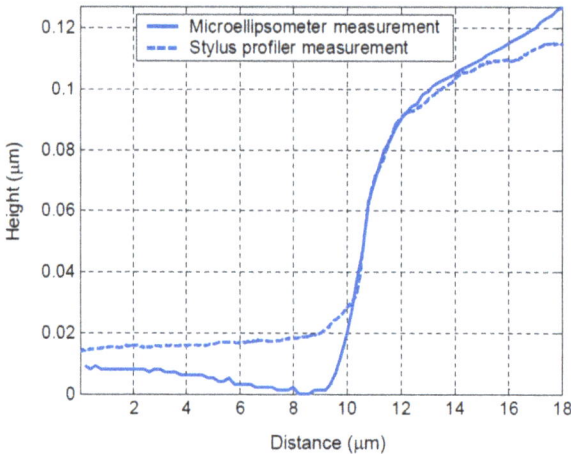

Fig. 8. Experimental surface profile measurement results of a micro-prism and comparison with scanning stylus profiler results.[1]

confirmed and demonstrated by comparing the surface profiles of a micro-prism measured by the microellipsometer and a standard mechanical stylus profiler and good agreements have been obtained (Fig. 8). The results also clearly demonstrated the advantages of microellipsometer in terms of the achievable spatial resolution and its non-contact nature.[1]

3.2. *Nulling microellipsometer with rotational symmetry*

In the microellipsometer implementation described in Section 3.1, a mechanical motor was used to drive the polarization rotator, causing vibrational noise and limiting the modulation frequency of the polarization signal. In a continuation work, the mechanically driven polarization rotator is replaced with an EO polarization rotator (as shown in Fig. 3).[7] The use of this EO polarization rotator eliminates the vibrational noise and enables higher modulation frequency for the polarization signal. Furthermore, it makes it possible to implement a nulling detection scheme that was proposed in Ref. 1.

The schematic experimental setup of this nulling microellipsometer is shown in Fig. 9. The beam goes through the intermediate optical elements for spatial filtering and collimation, and ends up focused by the objective lens (Nikon LU Plan, 100x, NA = 0.9) on the sample. The sample is placed on a computer-controlled XYZ translation stage, and stage is controlled through a GPIB interface. The reflected polarization is collected by the objective lens and redirected toward the analyzer arm by a beam-splitter pair, before entering the polarization rotator. The electro-optics variable retarder (EOVR) in the polarization rotator is biased by a range of amplified voltages supplied by a function generator and amplified by a high voltage amplifier. A small ac voltage is supplied by the same function generator for signal modulation. The beam leaving the polarization rotator goes through a radial analyzer, and then through an annular aperture for spatial filtering of the central portion of the beam attributed to low angles of illumination. After spatial filtering, the beam is focused onto the detector. The output of the detector is split into two parts: one part is sent to a lock-in amplifier for the detection of the first harmonic signal. The other part of the detector's output is sent to a data acquisition (DAQ) card. The temporal signal acquired with the DAQ card is pre-processed via low-pass filtering to get rid of the high frequency noise, and Fourier transformed to retrieve the dc signal.

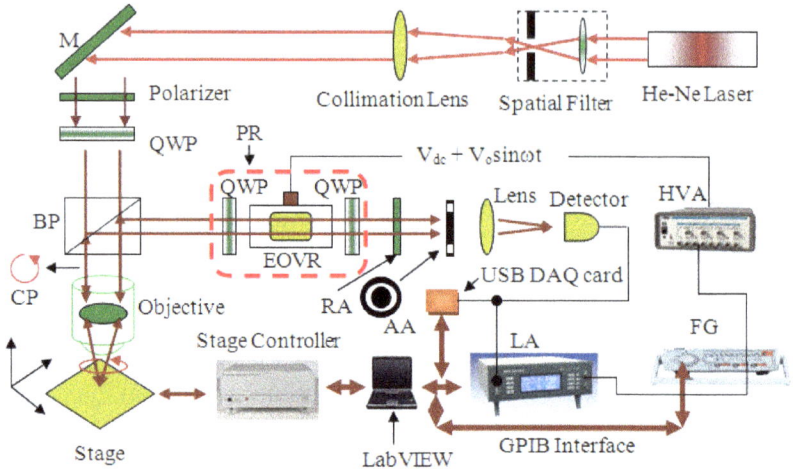

Fig. 9. Experimental setup of a nulling microellipsometer with rotational symmetry.[7]

When a dc bias and an ac modulation voltage are applied across the EO modulator the ellipse of polarization in each channel is rotated by

$$\Phi = \Phi_{\text{bias}} + \alpha_m \cos \omega t, \qquad (6)$$

where Φ_{bias} is the part of the rotation angle attributed to the dc bias voltage, α_m is the amplitude of the modulation angle attributed to the ac voltage, and ω is the modulation frequency. The general form of the signal is similar to that of a rotating analyzer ellipsometer. In our case the general signal can be written as

$$P = K\left\{1 + \cos 2\varepsilon \cos\left[2\left(\Phi_{\text{bias}} + \alpha_m \cos \omega t + \theta_0\right)\right]\right\}, \qquad (7)$$

where K is a constant. Assuming a small angular modulation, i.e., $\alpha_m \ll 1$ and using the Taylor's expansion, the cosine term in Eq. (7) can be expanded as

$$\cos\left[2\left(\Phi_{\text{bias}} + \alpha_m \cos \omega t + \theta_0\right)\right]$$
$$\cong \cos 2\left(\Phi_{\text{bias}} + \theta_0\right)\left(1 - \alpha_m^2 + \alpha_m^2 \cos 2\omega t\right) - 2\alpha_m \sin 2\left(\Phi_{bias} + \theta_0\right)\cos \omega t. \qquad (8)$$

Substitution of Eq. (8) into Eq. (7) reveals three most relevant harmonic components in the general signal:

$$P(0) = K\left[1 + \cos 2\varepsilon\left(1 - \alpha_m^2\right)\cos 2\left(\Phi_{\text{bias}} + \theta_0\right)\right], \qquad (9)$$

$$P(\omega) = 2K\alpha_m \cos 2\varepsilon \sin 2\left(\Phi_{bias} + \theta_0\right)\cos\left(\omega t + \pi\right) , \quad (10)$$

$$P(2\omega) = K \cos 2\varepsilon\alpha_m^2 \cos 2\left(\Phi_{bias} + \theta_0\right)\cos 2\omega t . \quad (11)$$

Equations (9)–(11) are the dc, first, and second harmonics of the signal. They all directly depend on the ellipticity angle ε, the modulation amplitude α_m, and the ellipse orientation θ_0. It can be deduced from Eq. (8) that when $\Phi_{bias} = -\theta_0$ or $\Phi_{bias} = -\theta_0 \pm \pi/2$ the first harmonic becomes *null*, i.e. $P(\omega) = 0$, hence, the nulling detection scheme. The ratio of the dc signal at two consecutive null positions, i.e., $\Phi_{bias} = -\theta_0$ and $\Phi_{bias} = -\theta_0 \pm \pi/2$, gives $\tan^2 \varepsilon$. This means that by measuring two consecutive nulls, one can determine both the elevation angle $\Phi_{bias} = -\theta_0$ and the ellipticity angle ε.

In order to check the nulling detection scheme, a SiO_2 film was deposited on a silicon substrate. Its thickness and index of refraction of were characterized using a conventional spectroscopic ellipsometer. The sample was then analyzed with the microellipsometer, scanning the dc bias voltage range to obtain the first harmonic and dc signal (shown in Fig. 10 (a) and (b)). The two minima in the first harmonic can be clearly seen with a voltage difference between the two minima corresponds to an angular rotation of 90°, confirming the nulling detection principle. Figure 10(c) shows two profile measurements, corresponding to the first line scan and the second (repeated) line scan, respectively. The second line scan (dashed curve) is used as a test to check for measurements

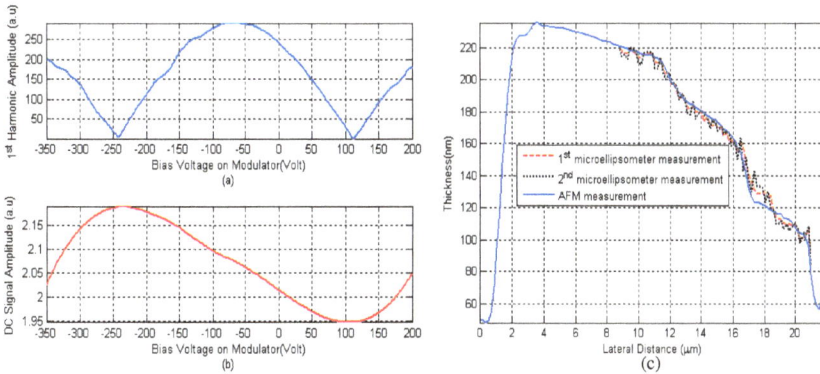

Fig. 10. Confirmation of the nulling detection scheme with (a) First harmonic and (b) DC signal vs. bias voltage. (c) Surface profile of a micro-prism measured by the nulling microellipsometer and comparison with AFM scanning results.[7]

repeatability. Both profiles are found to be in good agreement with the atomic force microscope (AFM) result.

4. Radial polarization interferometer

Interferometry is another sensitive and widely used optical measurement techniques. Typically interferometry is performed with two beams with the same polarization states. Interference with orthogonally spatially homogeneously polarized beams was developed for translational and velocity measurements. Recently, a radial polarization interferometer (RPI) was demonstrated that combines the concepts of spatially variant polarization with orthogonal polarization interferometry.[8] Essentially, the RPI generates a spatially varying intensity distribution that is dependent on the phase differences between the two arms of the interferometer. The spatial variance in the intensity pattern helps to overcome the achievable accuracy limit imposed by bit quantization of a CCD camera used to capture the interferogram.

The setup of a RPI is illustrated in Fig. 11. Linearly polarized laser beam is rotated to 45° with respect to the x-axis by the λ/2-plate and spit into two orthogonally polarization by the polarization beam-splitter to two interferometer arms. Each of the beam passes through a λ/4-plate oriented at 45° from the x-axis and reflected by the mirror. One mirror is fixed and used as reference. The other mirror is movable, creating phase differences

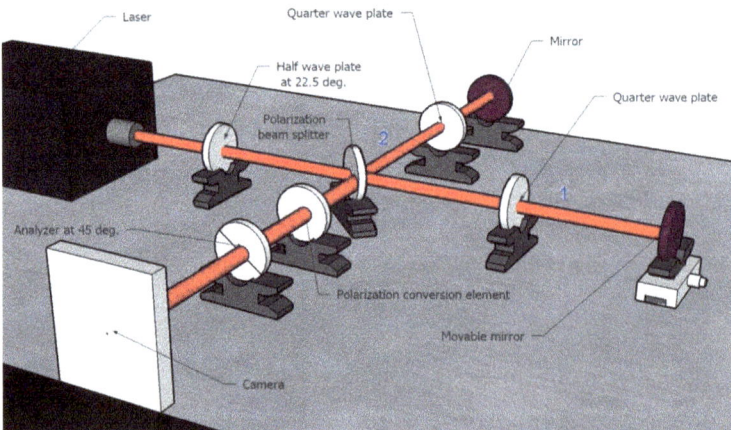

Fig. 11. Experimental setup of a radial polarization interferometer (RPI).[8]

between the two arms. The two reflected beams are spatially overlapped and linearly polarized orthogonally. The combined beams passes through a polarization conversion element that converts the two orthogonal linear polarization into radial and azimuthal polarization respectively. Finally, the superposition of the radial and azimuthal polarizations passes through a linear analyzer at 45° and the interferogram is capture on a CCD camera for further analysis.

Assuming the phase difference between the two arms is ϕ and the azimuthal angle in the polar coordinates is θ. It can be shown that the interferogram on the CCD camera should take the form of:

$$I(\phi, \theta) = (1 + cos2\theta \cdot cos\phi)/2. \tag{12}$$

Clearly, the phase difference is encoded into the intensity variations along the azimuthal direction. By measuring the contrast of the interferogram:

$$C = [\max(I(\theta)) - \min(I(\theta))]/[\max(I(\theta)) - \min(I(\theta))], \tag{13}$$

one can determine the phase difference. The contrast changes from 1 to 0 and back to 1 upon a phase change of π. In addition, the orientation of the interferogram rotates $\pi/2$ during the same π phase change, offering another valuable characteristics for the phase determination. Examples of the resulted interferograms are shown in Fig. 12.

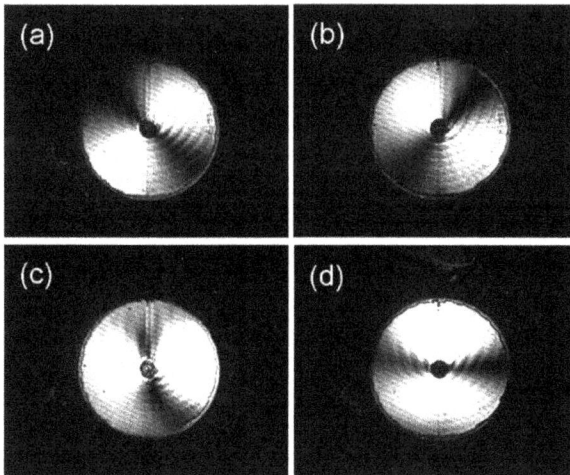

Fig. 12. Intensity profiles of the radially (a), and azimuthally (b) polarized arms passing through the linear analyzer; (c) and (d) are the interferograms of the two arms with 0 and π phase differences, respectively.[8]

Based on the above principle, the feasibility and capability of a RPI has been experimentally verified and demonstrated. It is shown that under the digital quantization limit of a CCD camera, the RPI is capable of measuring much smaller phase changes compared with a conventional Michelson interferometer (CMI). On average, the minimum detectable phase changes by the RPI could be 3-4 orders of magnitude smaller compared with the CMI. Several other forms of the RPI are also demonstrated in this work. Other than moving one of the arm to create variable phase differences, tunable laser source are used instead to create phase difference by adjusting the laser wavelength. In addition to the translational distance measurement, it is also shown that the RPI can be adapted to retrieve 2D phase profile of a test object inserted into one of the interferometer's arms.

5. Rapid Mueller matrix polarimetry

Similar to ellipsometry, polarimetry is an optical measurement technique that utilizes the detection of polarization changes upon light-matter interactions for optical characterization, imaging and sensing applications. The main difference relies in the fact that ellipsometry mostly deals with totally polarized and coherent light while partially polarized and partially coherent optical fields are typically measured by polarimetry. In order to handle partially polarized and partially coherent light fields, Stokes parameters are used in polarimetry to describe the light polarization. Consequently, the polarization responses of samples and testing materials is represented by a 4 by 4 Mueller matrix and the characterization task is turned into the determination of the 16 elements of this matrix. Determination of the elements of Mueller matrix in a Mueller matrix polarimetry (MMP) is done by analyzing the reflected or transmitted optical polarization states with several known input polarization states created by a polarization state generator (PSG). Traditionally, these known polarization states are created in a sequentially manner with the use of rotating or modulating optical devices, which ultimately limits the speed of a MMP and its applications in real-time characterization of dynamic events.

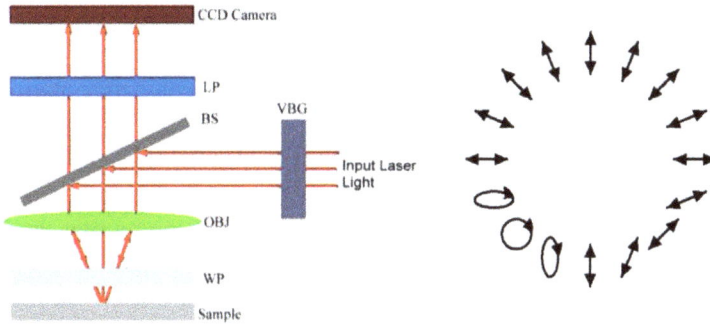

Fig. 13. (left) Schematic drawing of the working principle of a rapid MMP; and (right) the polarization pattern of the incident vectorial optical field (adapted from [9].

The developments in vectorial optical fields offer unique opportunity to improve the speed of a MMP through parallelizing the PSG. One of such examples is illustrated in Fig. 13.[9] An input vectorial optical field with spatially designed known polarization states are used as the illumination at the input pupil of an objective lens. The objective lens focuses the input vectorial optical field onto the sample through a $\lambda/4$-plate and the reflected signal passes the same $\lambda/4$-plate, objective lens and is projected onto a CCD camera after passing through a linear polarizer. The resulted intensity patterns are used for the determination of the Mueller matrix of the sample under test. Through the used of an vectorial optical field as the input, the PSG function is parallelized and a rapid MMP measurement is realized with a single shot. The performance of this rapid MMP is numerically investigated with the use of anisotropic samples such as anisotropic thin film and stratified metal-dielectric metamaterials.

6. Atomic spin analyzer

In an idea similar to the rapid MMP described above, modified CV beams have been applied for rapid detecting and preparing the atomic spin of a warm rubidium vapor in a spatially dependent manner.[10] The modified CV beam is created by passing a radial polarization through a $\lambda/4$-plate (Fig. 14(a)). It is shown that this modified vector beam as a probe can serve as an atomic spin analyzer for an optically pumped medium to provide rapid, single-shot atomic spin visualizations in a vapor cell (Fig. 14(b)). The parallelism offered by the spatially variant polarization across the beam cross-section drastically lowers measurement time and improves reproducibility.

Additionally, higher-power modified CV beams can be used as pumps to spatially modulate the spin of the atoms in the vapor. These beams have azimuthally varying elliptical polarization so that the space-variant spin density in the laser beam can be transferred to the atomic medium. They are thus beneficial for providing a means of preparing samples with position-dependent spin. For an atomic sample that is uniformly-prepared, such as a vapor cell or large ensemble of cold atoms, the effect of incident state of polarization on the spin transfer can be visualized in a single shot measurement. Such a parallelism enables the recording of temporal dynamics of the spin polarization with a pulsed Gaussian probe beam.

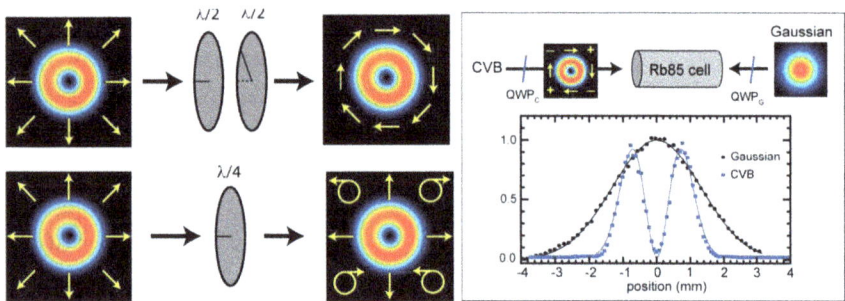

Fig. 14. (left) The modified CV beam; and (right) its use as probe to serve as rapid atomic spin analyzer (adapted from Ref. 10).

7. Summary

In this chapter, several optical measurement techniques that utilize the spatial variant state of polarization of vectorial optical fields are briefly reviewed. Rotational polarization symmetry has been exploited in microellipsometer and interferometer designs to improve measurement signal-to-noise-ratio and accuracy. Polarization diversities of vectorial optical fields are utilized as a parallelization scheme for rapid optical measurements in a Mueller matrix polarimetry design for anisotropic material characterization and an atomic spin analyzer. These examples clearly demonstrate the potential benefits offered by the polarization diversity within the beam cross-section of vectorial optical fields. With further progresses of vectorial optical fields, particularly the maturity of various generation, manipulation and characterization techniques, more studies and applications of vectorial optical fields in the development of advanced optical measurement tools are highly anticipated.

References

1. Q. Zhan and J. R. Leger, *Appl. Opt.* **41**, 4630-4637 (2002).
2. Q. Zhan and J. R. Leger, *Opt. Express*, **10**, 324, (2002).
3. S. Huard, *Polarization of Light*, Masson, Paris, (1996).
4. B. C. Lim, P. B. Phua, W. J. Lai and M. H. Hong, *Opt. Lett.* **33**, 950-952 (2008).
5. R. M. A. Azzam and N. M. Bashara, *Ellipsometry and polarized light*, North Holland Publishing Company, New York (1977).
6. H. G. Tompkins, *A users' guide to ellipsometry*, Academic Press, Boston (1993).
7. A. Tschimwang and Q. Zhan, *Appl. Opt.* **49**, 1574-1580, (2010).
8. G. M. Lerman and U. Levy, *Opt. Express* **17**, 23234-23246 (2009).
9. S. Tripathi and K. C. Toussaint, *Opt. Express* **17**, 21396-21407 (2009).
10. F. K. Fatemi, *Opt. Express* **19**, 25143-25150 (2011).

CHAPTER 7

PARTIALLY COHERENT VECTOR BEAMS: FROM THEORY TO EXPERIMENT

Yangjian Cai*, Fei Wang, Chengliang Zhao, Shijun Zhu, Gaofeng Wu, and Yiming Dong

School of Physical Science and Technology,
Soochow University, Suzhou 215006, China
*E-mail: yangjiancai@suda.edu.cn

Partially coherent vector beams have displayed many unique properties, which are much different from coherent vector beams and are useful in many applications, such as free-space optical communications, remote sensing, optical imaging, material thermal processing and particle trapping. A review of the recent developments on partially coherent vector beams with uniform or non-uniform polarization is provided. Characterizations of partially coherent vector beams are introduced. A tensor method is introduced for treating the paraxial propagation of partially coherent vector beam with uniform polarization (i.e., stochastic electromagnetic beam). Effect of coherence and polarization of a stochastic electromagnetic beam on the coincidence Fractional Fourier transform and degree of paraxiality are discussed. Methods for generating partially coherent vector beams with controllable spatial coherence have been developed. Effect of spatial coherence of a partially coherent vector beam with non-uniform polarization (i.e., cylindrical vector partially coherent beam) on its paraxial propagation properties and its tight focusing properties is also discussed.

1. Introduction

Coherence and polarization are two important properties of a light beam, and were studied separately in the past decades.[1-3] Coherence is regarded as a consequence of correlations between some components of the fluctuating electric field at two (or more) points, while polarization is a manifestation of correlations involving components of the fluctuating electric field at a single point.[2] It is usually assumed that the state of polarization of a light beam is invariant as the beam propagates in free space. The state of polarization and the degree of coherence of a light beam have been regarded to be independent of each other. In 1994, James showed that the degree of polarization of a light beam generated by a partially coherent source may change on propagation in free space.[4] In 2003, Wolf proposed a unified theory of coherence and polarization for partially coherent vector beams,[5] which can be used conveniently to study the changes of the spectral density, the spectral degree of polarization and the spectral degree of coherence as the beam propagates, and it was shown that the coherence and polarization of a partially coherent beam are interrelated.[6] Since then, numerous efforts have been paid to characterization, generation and propagation of partially coherent vector beams due to their important applications in free-space optical communication, remote sensing, optical imaging, particle trapping, material thermal processing and holography.[7]

Vector beams can be classified as beams with spatially uniform state of polarization (e.g., elliptically polarized beam and circularly polarized beam) and beams with spatially non-uniform state of polarization (e.g., radially polarized beam, azimuthally polarized beam and cylindrical vector beam). Partially coherent vector beam with spatially uniform state of polarization usually is called partially coherent electromagnetic beam or partially coherent and partially polarized beam or stochastic electromagnetic beam.[7-8] Partially coherent beam with spatially non-uniform state of polarization is called cylindrical vector partially coherent beam.[9] Partially coherent radially or azimuthally polarized beam can be regarded as a special case of cylindrical vector partially coherent beam.[10-11] In this chapter, we will introduce recent developments on partially coherent vector beams with uniform or non-uniform state of polarization both theoretically and experimentally.

2. Characterizations of partially coherent vector beams

Consider a planar, secondary source, located in the plane $z = 0$, that generates a partially coherent vector beam which propagates into the half-space $z > 0$, in a direction close to the positive z-axis. The source is assumed to be statistically stationary and obey Gaussian statistics. Based on the unified theory of coherence and polarization proposed by Wolf,[5] the second-order correlation properties of the partially coherent vector beam in spatial-frequency domain can be characterized by the so-called 2×2 cross-spectral density matrix,

$$\overset{\leftrightarrow}{W}(\mathbf{r}_1,\mathbf{r}_2) = \begin{pmatrix} W_{xx}(\mathbf{r}_1,\mathbf{r}_2) & W_{xy}(\mathbf{r}_1,\mathbf{r}_2) \\ W_{yx}(\mathbf{r}_1,\mathbf{r}_2) & W_{yy}(\mathbf{r}_1,\mathbf{r}_2) \end{pmatrix}, \tag{1}$$

with elements

$$W_{\alpha\beta}(\mathbf{r}_1,\mathbf{r}_2) = \left\langle E_\alpha^*(\mathbf{r}_1)E_\beta(\mathbf{r}_2) \right\rangle, \quad (\alpha = x, y; \beta = x, y), \tag{2}$$

where E_x and E_y denote the components of the random electric vector, with respect to two mutually orthogonal, x and y directions, perpendicular to the z-axis. $\mathbf{r}_1$ and $\mathbf{r}_2$ are transverse position vectors. The asterisk denotes the complex conjugate and the angular brackets denote ensemble average.

The spectral density of the partially coherent vector beam at point $\mathbf{r}$ is given by the formula[7]

$$\left\langle I(\mathbf{r}) \right\rangle = \mathrm{tr}\overset{\leftrightarrow}{W}(\mathbf{r},\mathbf{r}) = W_{xx}(\mathbf{r},\mathbf{r}) + W_{yy}(\mathbf{r},\mathbf{r}), \tag{3}$$

Where tr denotes the trace of the matrix.

The spectral degree of coherence of the partially coherent vector beam at a pair of transverse points with position vectors $\mathbf{r}_1$ and $\mathbf{r}_2$ is defined by the formula[7]

$$\mu(\mathbf{r}_1,\mathbf{r}_2) = \frac{\mathrm{tr}\overset{\leftrightarrow}{W}(\mathbf{r}_1,\mathbf{r}_2)}{\sqrt{\mathrm{tr}\overset{\leftrightarrow}{W}(\mathbf{r}_1,\mathbf{r}_1)\,\mathrm{tr}\overset{\leftrightarrow}{W}(\mathbf{r}_2,\mathbf{r}_2)}}. \tag{4}$$

The spectral degree of polarization of the partially coherent vector beam at point $\mathbf{r}$ is defined by the expression[7]

$$P(\mathbf{r}) = \sqrt{1 - \frac{4\det \ddot{\mathbf{W}}(\mathbf{r},\mathbf{r})}{[\mathrm{tr}\ddot{\mathbf{W}}(\mathbf{r},\mathbf{r})]^2}}, \tag{5}$$

where det stands for the determinant of the matrix.

The spectral degree of cross-polarization of the partially coherent vector beam at a pair of transverse points with position vectors $\mathbf{r}_1$ and $\mathbf{r}_2$ is defined as[12]

$$P(\mathbf{r}_1,\mathbf{r}_2) = \sqrt{\frac{2\mathrm{tr}[\ddot{\mathbf{W}}^\dagger(\mathbf{r}_1,\mathbf{r}_2)\cdot\ddot{\mathbf{W}}(\mathbf{r}_1,\mathbf{r}_2)]}{\left|\mathrm{tr}\ddot{\mathbf{W}}(\mathbf{r}_1,\mathbf{r}_2)\right|^2}} - 1, \tag{6}$$

where the dot denotes tensor product and the dagger denotes the Hermitian adjoint.

The intensity fluctuations of the beam at point $\mathbf{r}$ is expressed as

$$\Delta I(\mathbf{r}) = I(\mathbf{r}) - \langle I(\mathbf{r})\rangle, \tag{7}$$

where $\langle I(\mathbf{r})\rangle$ is given by Eq. (3), and $I(\mathbf{r})$ is expressed as

$$I(\mathbf{r}) = E_x^*(\mathbf{r})E_x(\mathbf{r}) + E_y^*(\mathbf{r})E_y(\mathbf{r}). \tag{8}$$

The correlation between the intensity fluctuations at a pair of transverse points with position vectors $\mathbf{r}_1$ and $\mathbf{r}_2$ is expressed as[12-14]

$$\langle\Delta I(\mathbf{r}_1)\Delta I(\mathbf{r}_2)\rangle = \frac{1}{2}\left\{1 + \left[P(\mathbf{r}_1,\mathbf{r}_2)\right]^2\right\}\left|\mu(\mathbf{r}_1,\mathbf{r}_2)\right|^2\langle I(\mathbf{r}_1)\rangle\langle I(\mathbf{r}_2)\rangle, \tag{9}$$

where $\mu(\mathbf{r}_1,\mathbf{r}_2)$ and $P(\mathbf{r}_1,\mathbf{r}_2)$ are given by Eqs. (4) and (6), respectively.

The cross-spectral density matrix of a partially coherent vector beam at point $\mathbf{r}$ can be locally represented as a sum of completely polarized beam and a completely unpolarized beam[15]

$$\ddot{\mathbf{W}}(\mathbf{r},\mathbf{r}) = \ddot{\mathbf{W}}^{(u)}(\mathbf{r},\mathbf{r}) + \ddot{\mathbf{W}}^{(p)}(\mathbf{r},\mathbf{r}), \tag{10}$$

where

$$\ddot{\mathbf{W}}^{(u)}(\mathbf{r},\mathbf{r}) = \begin{pmatrix} A(\mathbf{r},\mathbf{r}) & 0 \\ 0 & A(\mathbf{r},\mathbf{r}) \end{pmatrix},$$

$$\ddot{\mathbf{W}}^{(p)}(\mathbf{r},\mathbf{r}) = \begin{pmatrix} B(\mathbf{r},\mathbf{r}) & D(\mathbf{r},\mathbf{r}) \\ D^*(\mathbf{r},\mathbf{r}) & C(\mathbf{r},\mathbf{r}) \end{pmatrix},$$

(11)

with

$$A(\mathbf{r},\mathbf{r}) = \frac{1}{2}\left[W_{xx}(\mathbf{r},\mathbf{r}) + W_{yy}(\mathbf{r},\mathbf{r}) - \sqrt{\left[W_{xx}(\mathbf{r},\mathbf{r}) - W_{yy}(\mathbf{r},\mathbf{r}) \right]^2 + 4\left| W_{xy}(\mathbf{r},\mathbf{r}) \right|^2} \right],$$

$$B(\mathbf{r},\mathbf{r}) = \frac{1}{2}\left[W_{xx}(\mathbf{r},\mathbf{r}) - W_{yy}(\mathbf{r},\mathbf{r}) + \sqrt{\left[W_{xx}(\mathbf{r},\mathbf{r}) - W_{yy}(\mathbf{r},\mathbf{r}) \right]^2 + 4\left| W_{xy}(\mathbf{r},\mathbf{r}) \right|^2} \right],$$

$$C(\mathbf{r},\mathbf{r}) = \frac{1}{2}\left[W_{yy}(\mathbf{r},\mathbf{r}) - W_{xx}(\mathbf{r},\mathbf{r}) + \sqrt{\left[W_{xx}(\mathbf{r},\mathbf{r}) - W_{yy}(\mathbf{r},\mathbf{r}) \right]^2 + 4\left| W_{xy}(\mathbf{r},\mathbf{r}) \right|^2} \right],$$

$$D(\mathbf{r},\mathbf{r}) = W_{xy}(r,r).$$

(12)

The elements of the matrix $\ddot{\mathbf{W}}^{(p)}$ may be written as products of "equivalent monochromatic field" components, say ε_x and ε_y.[15] The quadratic form associated with such a matrix can be shown to represent the ellipse, known as a spectral polarization ellipse, of the form

$$C(\mathbf{r},\mathbf{r})\varepsilon_x^{(r)2}(\mathbf{r},\mathbf{r}) - 2\operatorname{Re} D(\mathbf{r},\mathbf{r})\varepsilon_x^{(r)}(\mathbf{r},\mathbf{r})\varepsilon_y^{(r)}(\mathbf{r},\mathbf{r}) + B(\mathbf{r},\mathbf{r})\varepsilon_y^{(r)2}(\mathbf{r},\mathbf{r})$$
$$= [\operatorname{Im} D(\mathbf{r},\mathbf{r})]^2,$$

(13)

where Re and Im stand for real and imaginary parts of complex numbers, $\varepsilon_x^{(r)}(\mathbf{r},\mathbf{r}) = \sqrt{B(\mathbf{r},\mathbf{r})}\cos(\omega t + \delta_x)$, $\varepsilon_y^{(r)}(\mathbf{r},\mathbf{r}) = \sqrt{C(\mathbf{r},\mathbf{r})}\cos(\omega t + \delta_y)$, and $\delta_y - \delta_x = \arg[D(\mathbf{r},\mathbf{r})]$. The major and minor semi-axes of the ellipse, A_1 and A_2, as well as its degree of ellipticity, ε, and its orientation angle, θ, can be related directly to the elements of the cross-spectral density matrix with the help of the expressions

$$A_{1,2}(\mathbf{r},\mathbf{r}) = \frac{1}{\sqrt{2}}\left[\sqrt{\left(W_{xx}(\mathbf{r},\mathbf{r})-W_{yy}(\mathbf{r},\mathbf{r})\right)^2 + 4\left|W_{xy}(\mathbf{r},\mathbf{r})\right|^2} \right.$$

$$\left. \pm\sqrt{\left(W_{xx}(\mathbf{r},\mathbf{r})-W_{yy}(\mathbf{r},\mathbf{r})\right)^2 + 4[\mathrm{Re}\,W_{xy}(\mathbf{r},\mathbf{r})]^2} \right]^{1/2}, \quad (14)$$

$$\varepsilon(\mathbf{r},\mathbf{r}) = \frac{A_2(\mathbf{r},\mathbf{r})}{A_1(\mathbf{r},\mathbf{r})}, \tag{15}$$

$$\theta(\mathbf{r},\mathbf{r}) = \frac{1}{2}\arctan\left(\frac{2\,\mathrm{Re}\,W_{xy}(\mathbf{r},\mathbf{r})}{W_{xx}(\mathbf{r},\mathbf{r})-W_{yy}(\mathbf{r},\mathbf{r})}\right). \tag{16}$$

In Eq. (14) signs "+" and "−" between the two square roots correspond to A_1 and A_2, respectively.

The polarization properties of a coherent vector beam at a point in space can be determined either by use of the Stokes parameters or the coherence matrix (sometimes called the polarization matrix). Generalized Stokes parameters of a partially coherent vector beam which depend on two spatial variables are introduced in Ref. 16, and defined as

$$S_0(\mathbf{r}_1,\mathbf{r}_2) = W_{xx}(\mathbf{r}_1,\mathbf{r}_2) + W_{yy}(\mathbf{r}_1,\mathbf{r}_2), \tag{17}$$

$$S_1(\mathbf{r}_1,\mathbf{r}_2) = W_{xx}(\mathbf{r}_1,\mathbf{r}_2) - W_{yy}(\mathbf{r}_1,\mathbf{r}_2), \tag{18}$$

$$S_2(\mathbf{r}_1,\mathbf{r}_2) = W_{xy}(\mathbf{r}_1,\mathbf{r}_2) + W_{yx}(\mathbf{r}_1,\mathbf{r}_2), \tag{19}$$

$$S_3(\mathbf{r}_1,\mathbf{r}_2) = i[W_{yx}(\mathbf{r}_1,\mathbf{r}_2) - W_{xy}(\mathbf{r}_1,\mathbf{r}_2)]. \tag{20}$$

The generalized Stokes parameters contain information about both the polarization and the coherence properties of the beam. The spectral degree of coherence (Eq. (4)) can be expressed in terms of the generalized Stokes parameters as follows

$$\mu(\mathbf{r}_1,\mathbf{r}_2) = \frac{S_0(\mathbf{r}_1,\mathbf{r}_2)}{\sqrt{S_0(\mathbf{r}_1,\mathbf{r}_1)}\sqrt{S_0(\mathbf{r}_2,\mathbf{r}_2)}}. \tag{21}$$

The spectral degree of polarization (Eq. (5)) can be expressed in terms of the generalized Stokes parameters as follows

$$P(\mathbf{r}) = \frac{\sqrt{S_1^2(\mathbf{r},\mathbf{r}) + S_2^2(\mathbf{r},\mathbf{r}) + S_3^2(\mathbf{r},\mathbf{r})}}{S_0(\mathbf{r},\mathbf{r})}. \tag{22}$$

Above results are obtained under the paraxial condition. When the beam width and the wavelength of the partially coherent vector beam are comparable, such as the optical near fields and the tightly focused beams, the paraxial condition isn't valid anymore. In this case, the 2×2 cross-spectral density matrix is not adequate for describing the polarization and coherence properties of the partially coherent vector beam. The 3×3 cross-spectral density matrix is introduced to describe the second-order correlation properties the partially coherent vector beam. Different definitions of the spectral degree of coherence, the spectral degree of polarization and generalized Stokes parameters have been introduced to describe the coherence and polarization properties of the three-dimensional partially coherent vector beam.[17-28]

3. Partially coherent vector beams with uniform state of polarization: theory

3.1. *Partially coherent electromagnetic Gaussian Schell-model beam*

Partially coherent electromagnetic Gaussian Schell-model (EGSM) beam is a typical partially coherent vector beam with uniform state of polarization. The elements of the cross-spectral density matrix of the EGSM source have the form[7, 29]

$$W_{\alpha\beta}(\mathbf{r}_1,\mathbf{r}_2) = \sqrt{\langle I_\alpha(\mathbf{r}_1)\rangle}\sqrt{\langle I_\beta(\mathbf{r}_1)\rangle}\mu_{\alpha\beta}(\mathbf{r}_1 - \mathbf{r}_2), \quad (\alpha = x, y; \beta = x, y), \tag{23}$$

where $\langle I_\alpha(\mathbf{r}_1)\rangle$ and $\langle I_\beta(\mathbf{r}_1)\rangle$ denote the spectral densities of the α and β components of the electric field, and $\mu_{\alpha\beta}$ denote the degree of correlation between the two components. These quantities are expressed as[29-30]

$$\langle I_\alpha(\mathbf{r})\rangle = A_\alpha^2 \exp\left(-\frac{\mathbf{r}^2}{2\sigma_\alpha^2}\right), \quad (\alpha = x, y), \tag{24}$$

$$\mu_{\alpha\beta}\left(\mathbf{r}_1,\mathbf{r}_2\right) = B_{\alpha\beta}\exp\left[-\frac{\left(\mathbf{r}_1-\mathbf{r}_2\right)^2}{2\delta_{\alpha\beta}}\right], \quad (\alpha=x,y;\beta=x,y), \qquad (25)$$

where A_α is the square root of the spectral density of electric field component E_α, $B_{\alpha\beta}=|B_{\alpha\beta}|\exp(i\phi)$ is the correlation coefficient between the E_x and E_y field components and satisfy the relation $B_{\alpha\beta}=B^*_{\beta\alpha}$, $|B_{\alpha\beta}|=1$ when $\alpha=\beta$, and $|B_{\alpha\beta}|\leq 1$ when $\alpha\neq\beta$, σ_α and $\delta_{\alpha\beta}$ denote the widths of the spectral density and the spectral degrees of correlation, respectively, and $\delta_{xy}=\delta_{yx}$. We stress here that all the parameters A_α, $B_{\alpha\beta}$, σ_α and $\delta_{\alpha\beta}$ are independent of position but, in general, depend on the frequency. A more general EGSM source can be anisotropic (i.e., astigmatic) or possess a twist phase, more information about anisotropic EGSM source or EGSM source with twist phase can be found in Ref. 31 and Ref. 32, respectively.

The beam conditions on the elements of the cross-spectral density matrix of an EGSM source in order that the source generate an EGSM beam are expressed as follows[30]

$$\frac{1}{4\sigma_x^2}+\frac{1}{\delta_{xx}^2}\leq\frac{2\pi^2}{\lambda^2}, \quad \frac{1}{4\sigma_y^2}+\frac{1}{\delta_{yy}^2}\leq\frac{2\pi^2}{\lambda^2}, \qquad (26)$$

where λ is the wavelength of the beam.

The sufficient condition on the choice of the parameters needed to describe a physically realizable EGSM source is expressed as[29, 33]

$$\max\left\{\delta_{xx},\delta_{yy}\right\}\leq\delta_{xy}\leq\min\left\{\frac{\delta_{xx}}{\sqrt{|B_{xy}|}},\frac{\delta_{yy}}{\sqrt{|B_{xy}|}}\right\}. \qquad (27)$$

The necessary condition on the choice of the parameters needed to describe a physically realizable EGSM source is expressed as[33]

$$\frac{A_x^2\sigma_x^4\delta_{xx}^2}{\delta_{xx}^2+4\sigma_x^2}-\frac{2A_xA_y|B_{xy}|\sigma_x^2\sigma_y^2\delta_{xy}^2}{\delta_{xy}^2+2\sigma_x^2+2\sigma_y^2}+\frac{A_y^2\sigma_y^4\delta_{yy}^2}{\delta_{yy}^2+4\sigma_y^2}\geq 0, \qquad (28)$$

$$\frac{2\sigma_x^2\delta_{xx}^2}{\delta_{xx}^2+4\sigma_x^2}-\frac{\left(\sigma_x^2+\sigma_y^2\right)\delta_{xy}^2}{\delta_{xy}^2+2\sigma_x^2+2\sigma_y^2}+\frac{2\sigma_y^2\delta_{yy}^2}{\delta_{yy}^2+4\sigma_y^2}\leq 0. \qquad (29)$$

The necessary and sufficient condition for nonnegativeness for the class of electromagnetic Schell-model sources can be found in Ref. 34.

3.2. *Tensor method for treating the paraxial propagation of partially coherent electromagnetic Gaussian Schell-model beam*

Tensor method was first introduced by Arnaud to treat the paraxial propagation of a coherent astigmatic beam about 50 years ago.[35, 36] Cai *et al.* extended the tensor method to treat the paraxial propagation and imaging of scalar GSM beam with twist phase.[37-43] Recently, it was adopted to treat the paraxial propagation and imaging of an EGSM beam. [31, 32, 44-52] In this section, we will introduce briefly the tensor method.

Applying Eqs. (23)-(25), the elements of the cross-spectral density matrix of the EGSM beam is expressed as

$$W_{\alpha\beta}(\mathbf{r}_1, \mathbf{r}_2) = A_\alpha A_\beta B_{\alpha\beta} \exp\left[-\frac{\mathbf{r}_1^2}{4\sigma_\alpha^2} - \frac{\mathbf{r}_2^2}{4\sigma_\beta^2} - \frac{(\mathbf{r}_1 - \mathbf{r}_2)^2}{2\delta_{\alpha\beta}^2}\right], \quad (\alpha = x, y; \beta = x, y).$$

(30)

Eq. (30) can alternatively be expressed in the following tensor form[37, 44, 46]

$$W_{\alpha\beta}(\tilde{\mathbf{r}}) = A_\alpha A_\beta B_{\alpha\beta} \exp\left[-\frac{ik}{2}\tilde{\mathbf{r}}^T \mathbf{M}_{0\alpha\beta}^{-1}\tilde{\mathbf{r}}\right], \quad (\alpha = x, y; \beta = x, y),$$

(31)

where $\tilde{\mathbf{r}}^T = \begin{pmatrix} \mathbf{r}_1^T & \mathbf{r}_2^T \end{pmatrix}$ and $\mathbf{M}_{0\alpha\beta}^{-1}$ has the form

$$\mathbf{M}_{0\alpha\beta}^{-1} = \begin{pmatrix} \dfrac{1}{ik}\left(\dfrac{1}{2\sigma_\alpha^2} + \dfrac{1}{\delta_{\alpha\beta}^2}\right)\mathbf{I} & \dfrac{i}{k\delta_{\alpha\beta}^2}\mathbf{I} \\[2ex] \dfrac{i}{k\delta_{\alpha\beta}^2}\mathbf{I} & \dfrac{1}{ik}\left(\dfrac{1}{2\sigma_\beta^2} + \dfrac{1}{\delta_{\alpha\beta}^2}\right)\mathbf{I} \end{pmatrix},$$

(32)

with $\mathbf{I}$ being a 2×2 unit matrix. Eq. (31) also can be used to characterize the elements of the cross-spectral density matrix of an astigmatic EGSM beam or an EGSM beam with twist phase just by replacing $\mathbf{M}_{0\alpha\beta}^{-1}$ of the EGSM beam with $\mathbf{M}_{0\alpha\beta}^{-1}$ of those beams.[31-32]

After propagating through a general astigmatic (i.e., nonsymmetric) ABCD optical system,[37] the elements of the cross-spectral density matrix can be expressed in the following tensor form

$$W_{\alpha\beta}(\tilde{\rho}) = A_{\alpha}A_{\beta}B_{\alpha\beta}\left[\det\left(\bar{\mathbf{A}} + \bar{\mathbf{B}}\mathbf{M}_{0\alpha\beta}^{-1}\right)\right]^{-1/2}\exp\left[-\frac{ik}{2}\tilde{\rho}^{T}\mathbf{M}_{1\alpha\beta}^{-1}\tilde{\rho}\right],$$

$$(\alpha = x, y; \beta = x, y), \tag{33}$$

where $\tilde{\rho}^{T} = \begin{pmatrix} \rho_{1}^{T} & \rho_{2}^{T} \end{pmatrix}$ with ρ_{1} and ρ_{2} being the transverse position vectors in the output plane, while $\mathbf{M}_{1\alpha\beta}^{-1}$ and $\mathbf{M}_{0\alpha\beta}^{-1}$ are related by the following known tensor ABCD law

$$\mathbf{M}_{1\alpha\beta}^{-1} = \left(\bar{\mathbf{C}} + \bar{\mathbf{D}}\mathbf{M}_{0\alpha\beta}^{-1}\right)\left(\bar{\mathbf{A}} + \bar{\mathbf{B}}\mathbf{M}_{0\alpha\beta}^{-1}\right)^{-1}. \tag{34}$$

Here $\bar{\mathbf{A}}$, $\bar{\mathbf{B}}$, $\bar{\mathbf{C}}$ and $\bar{\mathbf{D}}$ are 4×4 matrices of the form:

$$\bar{\mathbf{A}} = \begin{pmatrix} \mathbf{A} & 0\mathbf{I} \\ 0\mathbf{I} & \mathbf{A}^{*} \end{pmatrix}, \quad \bar{\mathbf{B}} = \begin{pmatrix} \mathbf{B} & 0\mathbf{I} \\ 0\mathbf{I} & -\mathbf{B}^{*} \end{pmatrix}, \quad \bar{\mathbf{C}} = \begin{pmatrix} \mathbf{C} & 0\mathbf{I} \\ 0\mathbf{I} & -\mathbf{C}^{*} \end{pmatrix}, \quad \bar{\mathbf{D}} = \begin{pmatrix} \mathbf{D} & 0\mathbf{I} \\ 0\mathbf{I} & \mathbf{D}^{*} \end{pmatrix}, \tag{35}$$

where $\mathbf{A}, \mathbf{B}, \mathbf{C}$ and $\mathbf{D}$ are the 2×2 sub-matrices of the general astigmatic ABCD optical system, and "*" denotes the complex conjugate.

If we set $\mathbf{A}, \mathbf{B}, \mathbf{C}$ and $\mathbf{D}$ as follows

$$\mathbf{A} = \begin{pmatrix} 1 & 0 \\ 0 & 1 \end{pmatrix}, \mathbf{B} = \begin{pmatrix} z & 0 \\ 0 & z \end{pmatrix}, \quad \mathbf{C} = \begin{pmatrix} 0 & 0 \\ 0 & 0 \end{pmatrix}, \quad \mathbf{D} = \begin{pmatrix} 1 & 0 \\ 0 & 1 \end{pmatrix}, \tag{36}$$

Equation (33) reduces to the following expression for the elements of the cross-spectral density matrix of the EGSM beam in free space

$$W_{\alpha\beta}(\tilde{\rho}) = A_{\alpha}A_{\beta}B_{\alpha\beta}\left[\det\left(\bar{\mathbf{I}} + \bar{\mathbf{B}}\mathbf{M}_{0\alpha\beta}^{-1}\right)\right]^{-1/2}\exp\left[-\frac{ik}{2}\tilde{\rho}^{T}\left(\mathbf{M}_{0\alpha\beta} + \bar{\mathbf{B}}\right)^{-1}\tilde{\rho}\right],$$

$$(\alpha = x, y; \beta = x, y), \tag{37}$$

with $\bar{\mathbf{I}}$ being a 4×4 unit matrix. Applying Eqs (3)-(22), (36) and (37), one can study the statistical properties of an EGSM beam in free space conveniently.

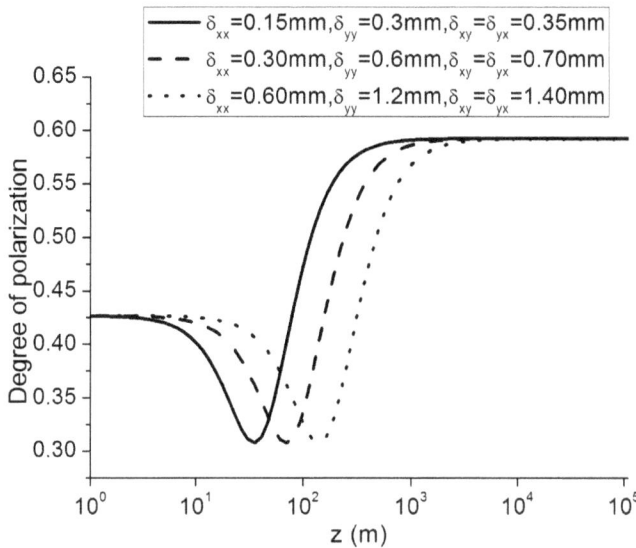

Fig. 1. On-axis spectral degree of polarization of an EGAM beam as a function of the propagation distance z in free space for different values of the correlation coefficients with $A_x = 1.5$, $A_y = 1$, $B_{xy} = B_{yx} = 0.2$, $\sigma_x = \sigma_y = 20$mm.

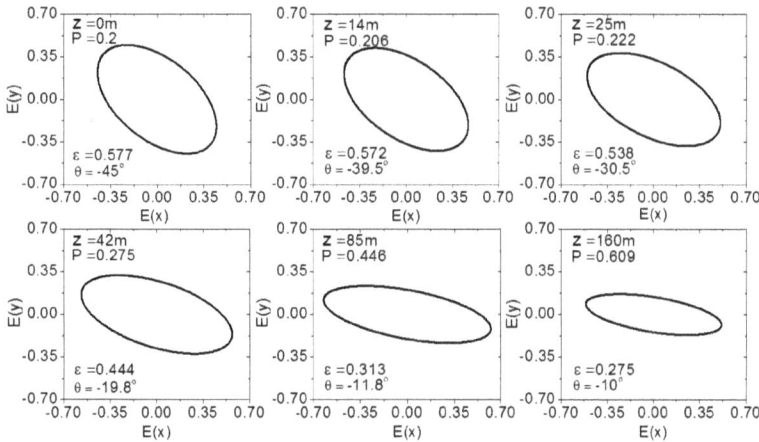

Fig. 2. On-axis polarization ellipse of an EGSM beam as a function of the propagation distance z in free space with $A_x = 1$, $A_y = 1$, $|B_{xy}| = 0.2$, $\phi = \pi/3$, $\sigma_x = \sigma_y = 20$mm, $\delta_{xx} = 0.3$mm, $\delta_{yy} = 0.6$mm, $\delta_{xy} = 0.7$mm.

Figures 1 and 2 show the on-axis spectral degree of polarization and the on-axis polarization ellipse of an EGSM beam as a function of the propagation distance z in free space. Both the spectral degree of polarization and the polarization ellipse don't remain invariant on propagation but vary on propagation, and this variation depends closely on the correlation coefficients $\delta_{xx}, \delta_{xy}, \delta_{yy}$ (i.e., spectral degree of coherence). More information about the statistical properties of an EGSM beam in free space can be found in Refs. 4-16, 31, 32, 53-57.

3.3. *Statistics properties of a partially coherent electromagnetic Gaussian Schell-model beam in a Gaussian cavity*

The conventional theory of laser resonators was originally confined to completely spatially coherent light fields until Wolf, Agarwal, and Gori generalized it for light fields with any state of coherence.[58-61] Palma and co-workers have then studied the behavior of the degree of coherence and of the spectral properties of scalar partially coherent beams in a Gaussian cavity.[62-63] However, almost all previous works on the subject were restricted to the scalar theory. In Ref. 64, the theory of resonator modes was developed for partially coherent electromagnetic fields (see also Ref. 65 for an alternative theory), where the transverse electromagnetic modes have been related to the classic Fox–Li modes of the cavity. Recently, with the help of the tensor method, the statistics properties of an EGSM beam in a Gaussian cavity were studied.[46-47, 66-69] In this section, we will introduce the statistical properties in a Gaussian cavity.

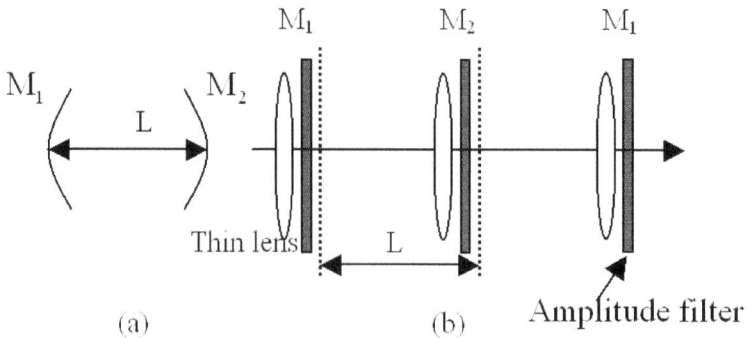

Fig. 3. Schematic diagram of a Gaussian cavity and its equivalent (unfolded) version.

A bare Gaussian cavity is composed of two reflecting surfaces (mirrors) with the same sizes and curvatures (see Fig. 3(a)). Originally, the EGSM beam is sent to one of the mirrors in such a way that it then travels back and forth in the cavity. For simplicity we will assume that both mirrors are spherical, with radius of curvature R, mirror spot size η, and distance between the centers of the mirrors L. The interaction of beam with such a resonator is equivalent to its propagation through a sequence of thin spherical lenses of focal lengths $f = R/2$ combined with filters with a Gaussian amplitude transmission function.[62-63] Figure 3(b) shows the equivalent "unfolded" version of the resonator. Depending on the value of the stability parameter $g = 1 - L/R$, the resonators are classified as stable $0 \le g < 1$ or unstable $g \ge 1$.[70]

If the EGSM beam travels once between two mirrors (see Fig. 3(b)), the matrices $\mathbf{A}, \mathbf{B}, \mathbf{C}$ and $\mathbf{D}$ take the form

$$
\begin{pmatrix} \mathbf{A} & \mathbf{B} \\ \mathbf{C} & \mathbf{D} \end{pmatrix} = \begin{pmatrix} \mathbf{I} & L \cdot \mathbf{I} \\ \left(-\dfrac{2}{R} - i\dfrac{\lambda}{\pi \eta^2} \right) \mathbf{I} & \left(1 - \dfrac{2L}{R} - i\dfrac{\lambda L}{\pi \eta^2} \right) \mathbf{I} \end{pmatrix}. \tag{38}
$$

Here η can be regarded as the "soft" mirror spot size. After the EGSM beam travels between the two mirrors for N times, $\mathbf{A}, \mathbf{B}, \mathbf{C}$ and $\mathbf{D}$ for the equivalent optical system become

$$
\begin{pmatrix} \mathbf{A} & \mathbf{B} \\ \mathbf{C} & \mathbf{D} \end{pmatrix} = \begin{pmatrix} \mathbf{I} & L \cdot \mathbf{I} \\ \left(-\dfrac{2}{R} - i\dfrac{\lambda}{\pi \eta^2} \right) \mathbf{I} & \left(1 - \dfrac{2L}{R} - i\dfrac{\lambda L}{\pi \eta^2} \right) \mathbf{I} \end{pmatrix}^N. \tag{39}
$$

Applying Eqs (3)-(22), (33)-(35), (38) and (39), we can study the behavior of the statistical properties of the EGSM beam in a Gaussian cavity.

Figure 4 shows the on-axis spectral degree of polarization versus N for different values of cavity parameter g and the initial correlation coefficients with $\eta = 0.8$mm, $A_x = A_y = 0.707$, $B_{xy} = B_{yx} = 0.2$ and $\sigma_x = \sigma_y = 1$mm. One sees that the spectral degree of polarization increases as N increase, and its value approaches different constant values for different resonators when N is enough large ($N > 30$). The spectral degree of polarization exhibits growth with oscillations in stable resonators

$(0 \leq g < 1)$ while growth is monotonic for unstable resonators $(g \geq 1)$. Also, for unstable resonators the spectral degree of polarization decreases for higher values of g. The spectral degree of polarization decreases as the correlation coefficients in the input plane take larger values both in stable and unstable resonators.

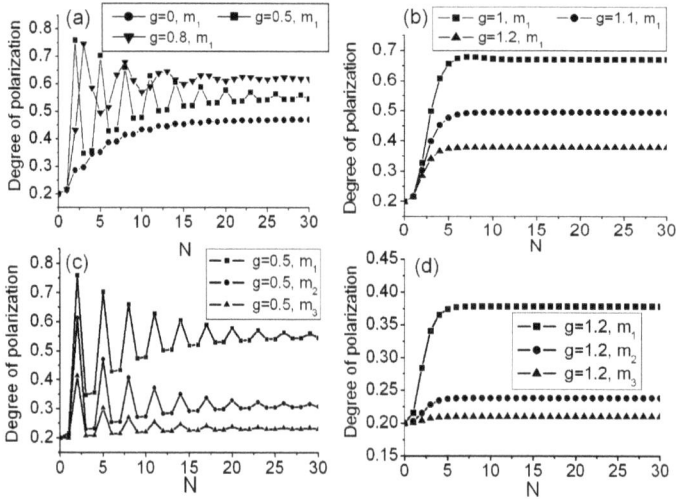

Fig. 4. On-axis spectral degree of polarization versus N for different values of cavity parameter g and the source correlation coefficients. $m_1 : \delta_{xx} = \delta_{yy} = 0.1\text{mm}$, $\delta_{xy} = \delta_{yx} = 0.2\text{mm}$. $m_2 : \delta_{xx} = \delta_{yy} = 0.25\text{mm}$, $: \delta_{xy} = \delta_{yx} = 0.5\text{mm}$. $m_3 : \delta_{xx} = \delta_{yy} = 0.5\text{mm}$, $\delta_{xy} = \delta_{yx} = 1\text{mm}$.

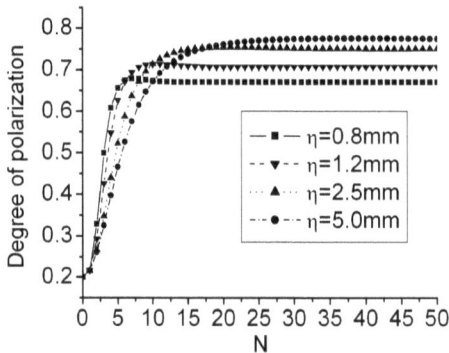

Fig. 5. On-axis spectral degree of polarization versus N for different values of mirror spot size ε in a Gaussian plane-parallel cavity $(g = 1)$ with $\delta_{xx} = \delta_{yy} = 0.1\text{mm}$ and $\delta_{xy} = \delta_{yx} = 0.2\text{mm}$.

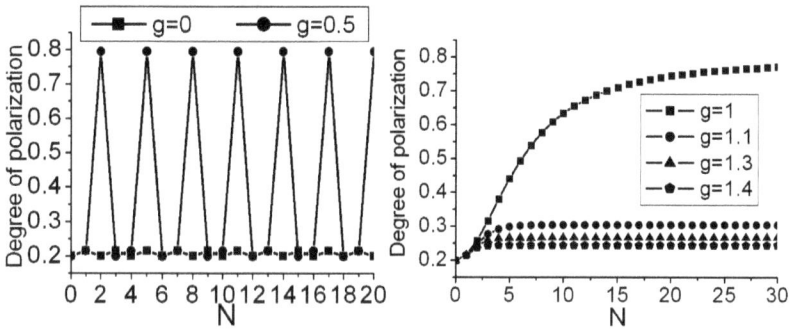

Fig. 6. On-axis spectral degree of polarization versus N for different values of g in a lossless cavity $(\varepsilon \to \infty)$ with $\delta_{xx} = \delta_{yy} = 0.1$mm and $\delta_{xy} = \delta_{yx} = 0.2$mm.

Figure 5 shows the on-axis spectral degree of polarization versus N for different values of η in a Gaussian plane-parallel cavity $(g = 1)$. We can observe that the growth of the spectral degree of polarization is more pronounced in response to higher values of η (i.e., the losses per trip are smaller) when N is sufficiently large. In the ideal case of a lossless cavity $(\eta \to \infty)$, we see from Fig. 6 that in stable lossless cavities, the spectral degree of polarization has an oscillatory behavior, and its value does not saturate even for large values of N. In unstable cavities, the spectral degree of polarization grows rapidly and reaches a saturation value in a few trips, and this value decreases as g increases.

More information about the evolution properties of the statistical properties, such as the state of polarization, correlation properties, spectral shift and propagation factor, of an EGSM beam can be found in Refs. 47, 66-68. Thermal lens effect induced changes of polarization, coherence and spectrum of an EGSM beam in a Gaussian cavity can be found in Ref. 69.

3.4. *Propagation of a partially coherent electromagnetic Gaussian Schell-model beam in turbulent atmosphere*

Propagation characteristics of different types of beams propagating in the turbulent atmosphere are of interest for optical communications, imaging and remote sensing applications.[40-41, 48, 50, 71-80] Statistical properties of EGSM beams propagating in the atmosphere have been studied.[48, 81-90] More importantly, it was found that under suitable conditions the EGSM beams may have reduced levels of intensity fluctuations (scintillations) compared to the scalar GSM beams,[85] which makes them attractive for free-space optical communications. In this section, we introduce the propagation of an EGSM beam in turbulent atmosphere with the help of the tensor method. Figure 7 shows the propagation geometry.

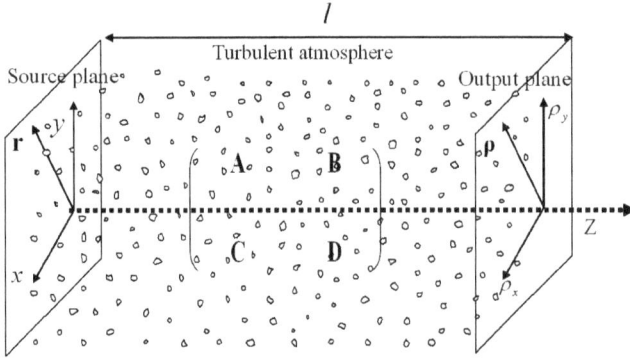

Fig. 7. Propagation geometry of an EGSM beam in turbulent atmosphere.

Within the validity of the paraxial approximation, the propagation of the elements of the cross-spectral density matrix of a partially coherent vector beam through a general astigmatic ABCD optical system situated in the turbulent atmosphere can be studied with the help of the following generalized Huygens-Fresnel integral[72, 91-92]

$$
\begin{aligned}
W_{\alpha\beta}(\boldsymbol{\rho}_1,\boldsymbol{\rho}_2) =& \frac{1}{\lambda^2[\det(\mathbf{B})]^{1/2}[\det(\mathbf{B}^*)]^{1/2}} \int_{-\infty}^{\infty}\int_{-\infty}^{\infty}\int_{-\infty}^{\infty}\int_{-\infty}^{\infty} W_{\alpha\beta}(\mathbf{r}_1,\mathbf{r}_2) \\
& \times \left\langle \exp\left[\Psi(\mathbf{r}_1,\boldsymbol{\rho}_1) + \Psi^*(\mathbf{r}_2,\boldsymbol{\rho}_2)\right]\right\rangle \\
& \times \exp\left[-\frac{ik}{2}\left(\mathbf{r}_1^T\mathbf{B}^{-1}\mathbf{A}\mathbf{r}_1 - 2\mathbf{r}_1^T\mathbf{B}^{-1}\boldsymbol{\rho}_1 + \boldsymbol{\rho}_1^T\mathbf{D}\mathbf{B}^{-1}\boldsymbol{\rho}_1\right)\right] \\
& \times \exp\left[\frac{ik}{2}\left(\mathbf{r}_2^T\left(\mathbf{B}^*\right)^{-1}\mathbf{A}^*\mathbf{r}_2 - 2\mathbf{r}_2^T\left(\mathbf{B}^*\right)^{-1}\boldsymbol{\rho}_2 + \boldsymbol{\rho}_2^T\mathbf{D}^*\left(\mathbf{B}^*\right)^{-1}\boldsymbol{\rho}_2\right)\right] \\
& \times d\mathbf{r}_1 d\mathbf{r}_2,
\end{aligned}
$$

(40)

where "*" denotes the complex conjugate. The expression in the angular brackets in Eq. (40) can be expressed as

$$
\begin{aligned}
&\left\langle \exp\left[\Psi(\mathbf{r}_1,\boldsymbol{\rho}_1) + \Psi^*(\mathbf{r}_2,\boldsymbol{\rho}_2) \right] \right\rangle \\
&= \exp\left[-\frac{(\mathbf{r}_1-\mathbf{r}_2)^2}{\rho_0^2} - \frac{(\mathbf{r}_1-\mathbf{r}_2)(\boldsymbol{\rho}_1-\boldsymbol{\rho}_2)}{\rho_0^2} - \frac{(\boldsymbol{\rho}_1-\boldsymbol{\rho}_2)^2}{\rho_0^2} \right],
\end{aligned}
\tag{41}
$$

ρ_0 being the coherence length of a spherical wave propagating in the turbulent medium given by the expression

$$
\rho_0 = \det[\mathbf{B}]^{1/2}\left(1.46k^2 C_n^2 \int_0^l \det[\mathbf{B}(z)]^{5/6}\,dz \right)^{-3/5}.
\tag{42}
$$

Here $\mathbf{B}(z)$ is the sub-matrix for back-propagation from output plane to propagation distance z, C_n^2 is the structure constant of turbulent atmosphere, l is the distance between the source plane and output plane. Here we have applied the Kolmogorov turbulence spectrum and a quadratic approximation for wave structure function.

Equation (40) can be expressed in the following alternative form[48]

$$
\begin{aligned}
W_{\alpha\beta}(\tilde{\boldsymbol{\rho}}) =\; & \frac{k^2}{4\pi^2\left[\mathrm{Det}\left(\bar{\mathbf{B}}\right)\right]^{1/2}} \int_{-\infty}^{\infty}\int_{-\infty}^{\infty}\int_{-\infty}^{\infty}\int_{-\infty}^{\infty} W_{\alpha\beta}(\tilde{\mathbf{r}}) \\
& \times \exp\left[-\frac{ik}{2}\left(\tilde{\mathbf{r}}^T \bar{\mathbf{B}}^{-1}\bar{\mathbf{A}}\tilde{\mathbf{r}} - 2\tilde{\mathbf{r}}^T\bar{\mathbf{B}}^{-1}\tilde{\boldsymbol{\rho}} + \tilde{\boldsymbol{\rho}}^T \bar{\mathbf{D}}\bar{\mathbf{B}}^{-1}\tilde{\boldsymbol{\rho}} \right) \right] \\
& \times \exp\left[-\frac{ik}{2}\tilde{\mathbf{r}}^T\bar{\mathbf{P}}\,\tilde{\mathbf{r}} - \frac{ik}{2}\tilde{\mathbf{r}}^T\bar{\mathbf{P}}\,\tilde{\boldsymbol{\rho}} - \frac{ik}{2}\tilde{\boldsymbol{\rho}}^T\bar{\mathbf{P}}\,\tilde{\boldsymbol{\rho}} \right] d\tilde{\mathbf{r}},
\end{aligned}
\tag{43}
$$

where $\bar{\mathbf{A}}$, $\bar{\mathbf{B}}$, $\bar{\mathbf{C}}$ and $\bar{\mathbf{D}}$ are given by Eq. (35), and

$$
\bar{\mathbf{P}} = \frac{2}{ik\rho_0^2}\begin{pmatrix} \mathbf{I} & -\mathbf{I} \\ -\mathbf{I} & \mathbf{I} \end{pmatrix}.
\tag{44}
$$

Substituting Eq. (31) into Eq. (43), after some vector integration and tensor operations, we obtain the following expression for the elements of the cross-spectral density matrix of an EGSM beam after propagating through an astigmatic ABCD optical system in a turbulent atmosphere[48]

$$W_{\alpha\beta}(\tilde{\boldsymbol{\rho}}) = \frac{A_\alpha A_\beta B_{\alpha\beta}}{\left[\det\left(\bar{\mathbf{A}} + \bar{\mathbf{B}}\mathbf{M}_{0\alpha\beta}^{-1} + \bar{\mathbf{B}}\bar{\mathbf{P}}\right)\right]^{1/2}} \exp\left[-\frac{ik}{2}\tilde{\boldsymbol{\rho}}^T \mathbf{M}_{1\alpha\beta}^{-1}\tilde{\boldsymbol{\rho}}\right]$$

$$\exp\left[-\frac{ik}{2}\tilde{\boldsymbol{\rho}}^T \bar{\mathbf{P}}\tilde{\boldsymbol{\rho}} - \frac{ik}{2}\tilde{\boldsymbol{\rho}}^T\left(\bar{\mathbf{B}}^{-1T} - \frac{1}{4}\bar{\mathbf{P}}^T\right)\left(\mathbf{M}_{0\alpha\beta}^{-1} + \bar{\mathbf{B}}^{-1}\bar{\mathbf{A}} + \bar{\mathbf{P}}\right)^{-1}\bar{\mathbf{P}}\tilde{\boldsymbol{\rho}}\right],$$

(45)

with

$$\mathbf{M}_{1\alpha\beta}^{-1} = \left(\bar{\mathbf{C}} + \bar{\mathbf{D}}\mathbf{M}_{0\alpha\beta}^{-1} + \bar{\mathbf{D}}\bar{\mathbf{P}}\right)\left(\bar{\mathbf{A}} + \bar{\mathbf{B}}\mathbf{M}_{0\alpha\beta}^{-1} + \bar{\mathbf{B}}\bar{\mathbf{P}}\right)^{-1}.$$

(46)

In the absence of an optical system but with presence of atmospheric turbulence, the transformation matrix between the source plane and the output plane is given by

$$\begin{pmatrix} \mathbf{A} & \mathbf{B} \\ \mathbf{C} & \mathbf{D} \end{pmatrix} = \begin{pmatrix} \mathbf{I} & l\mathbf{I} \\ 0 & \mathbf{I} \end{pmatrix},$$

(47)

then Eq. (45) reduces to

$$W_{\alpha\beta}(\tilde{\boldsymbol{\rho}}) = \frac{A_\alpha A_\beta B_{\alpha\beta}}{\left[\det\left(\bar{\mathbf{I}} + \bar{\mathbf{B}}\mathbf{M}_{0\alpha\beta}^{-1} + \bar{\mathbf{B}}\bar{\mathbf{P}}\right)\right]^{1/2}} \exp\left[-\frac{ik}{2}\tilde{\boldsymbol{\rho}}^T\left(\bar{\mathbf{P}} + \bar{\mathbf{B}}\right)\tilde{\boldsymbol{\rho}}\right]$$

$$\times \exp\left[\frac{ik}{2}\tilde{\boldsymbol{\rho}}^T\left(\bar{\mathbf{B}}^{-1} - \frac{1}{2}\bar{\mathbf{P}}\right)^T\left(\mathbf{M}_1^{-1} + \bar{\mathbf{B}}^{-1} + \bar{\mathbf{P}}\right)^{-1}\left(\bar{\mathbf{B}}^{-1} - \frac{1}{2}\bar{\mathbf{P}}\right)\tilde{\boldsymbol{\rho}}\right],$$

(48)

with $\rho_0 = \left(0.545 C_n^2 k^2 l\right)^{-3/5}$. Applying Eqs (3)-(22), (44)-(48), we can study the statistical properties of an EGSM beam in turbulent atmosphere.

The on-axis spectral degree of polarization and the on-axis polarization ellipse of an EGSM beam as a function of the propagation distance z in turbulent atmosphere are shown in Figs. 8 and 9. The evolution properties in turbulent atmosphere are much different from those in free space. One finds that after propagating over sufficiently long distance z in turbulent atmosphere, the spectral degree of polarization of an EGSM beam regains it's initial value, and the polarization ellipse acquires the same shape as in the source plane. These properties may be used in polarization modulation schemes for transmitting data through random media.[83] More information about the statistical properties of an EGSM beam in turbulent atmosphere can be found in Refs. 48, 81-90 . Equations (44)-(48) also can be applied for studying stochastic electromagnetic beams for LIDAR systems operating through turbulent atmosphere,[50] and for sensing of semi-rough targets embedded in atmospheric turbulence by means of stochastic electromagnetic beams.[50, 93]

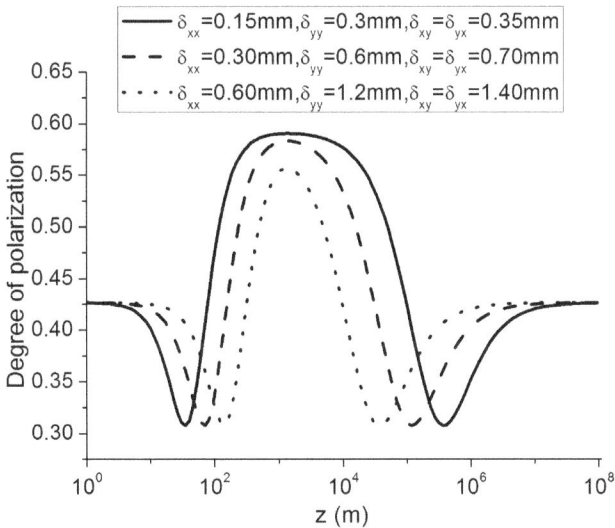

Fig. 8. On-axis spectral degree of polarization of an EGAM beam as a function of the propagation distance z in turbulent atmosphere for different values of the correlation coefficients with $A_x = 1.5$, $A_y = 1$, $B_{xy} = B_{yx} = 0.2$, $\sigma_x = \sigma_y = 20$mm, $C_n^2 = 5 \times 10^{-14}$ m$^{-2/3}$.

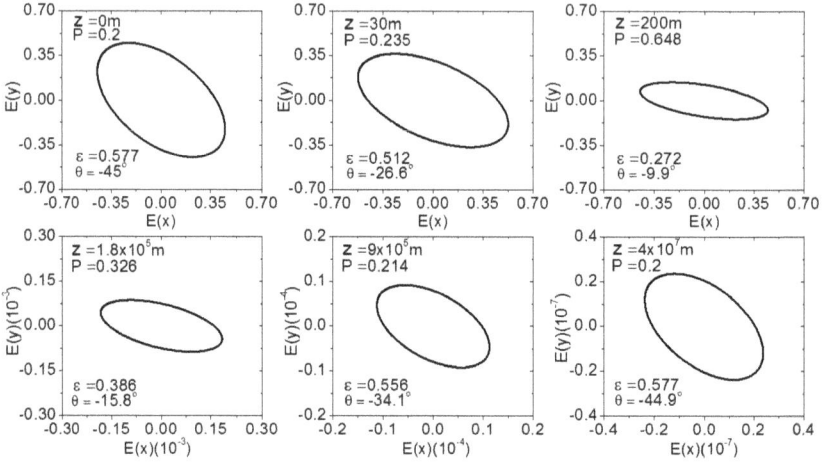

Fig. 9. On-axis polarization ellipse of an EGSM beam as a function of the propagation distance z in turbulent atmosphere with $A_x = 1$, $A_y = 1$, $|B_{xy}| = 0.2$, $\phi = \pi/3$, $\sigma_x = \sigma_y = 20\text{mm}$, $\delta_{xx} = 0.3\text{mm}$, $\delta_{yy} = 0.6\text{mm}$, $\delta_{xy} = 0.7\text{mm}$, $C_n^2 = 5 \times 10^{-14}\,\text{m}^{-2/3}$.

3.5. *Coincidence fractional Fourier transform with a partially coherent electromagnetic Gaussian Schell-model beam*

The concept of fractional Fourier transform (FRT) was proposed by Namias in 1980 as the generalization of conventional Fourier transform.[94] In 1993, Ozaktas, Mendlovic and Lohmanm introduced FRT into optics and designed optical systems to achieve FRT.[95, 96] Since then, the FRT has been used widely in signal processing, optical image encryption and beam analysis.[97] In 2005, Cai *et al.* introduced the concept of coincidence FRT and designed the optical system for implementing coincidence FRT with incoherent light, partially coherent light and entangled photon pairs[98, 99] as an extension of coincidence Fourier transform (also named ghost imaging and interference).[100-107] Coincidence FRT is a method to obtain the FRT pattern of an object by measuring the coincidence counting rate (i.e., the fourth order correlation) of two detected signals going through two different light paths, and the object is located in one light path.[98, 99] Coincidence subwavelength FRT was studied in Ref. 108. Lensless optical implementation of coincidence FRT was analyzed in Ref. 109. Wang *et al.* successfully observed the coincidence FRT pattern of an object implemented with a partially coherent light in experiment.[110] More recently, coincidence FRT with an EGSM beam was explored in Ref. 111. In this section, we will introduce the theory of coincidence FRT with an EGSM beam.

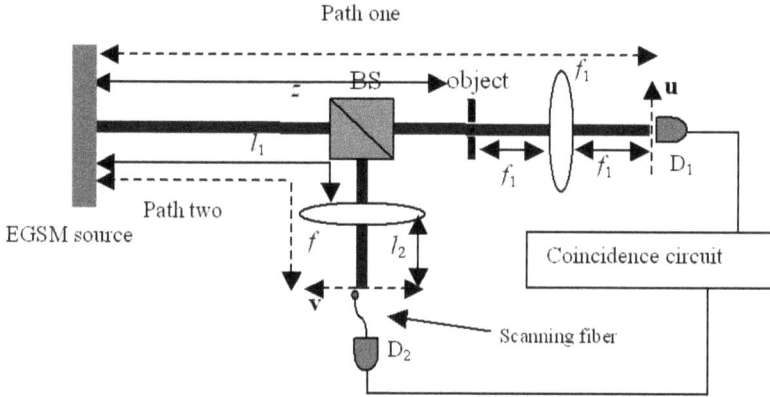

Fig. 10. Optical system for implementing coincidence FRT with an EGSM beam.

The optical system for implementing coincidence FRT with an EGSM beam is shown in Fig. 10. The EGSM beam emitted from the source is firstly split into two beams by a 50:50 beam splitter (BS). The transmitted beam propagates through path one to a single-photon detector D_1 which is located at $\mathbf{u} = 0$. In path one, there is an object with transmission function $H(\xi)$ between the BS and D_1, and there is a thin lens with focal length f_1 between the object and D_1. The distance from the light source to the object, the distance from the object to the lens and the distance from the lens to D_1 are z, f_1, respectively. The reflected beam propagates through path two to single-photon detector D_2. In path two, there is a thin lens with focal length f between the BS and D_2, and the distance from the light source to the thin lens and the distance from the thin lens to D_2 are l_1 and l_2, respectively. D_2 is connected with a single mode optical fiber whose tip is scanning on the transverse plane $\mathbf{v}$. The output signals from D_1 and D_2 are then sent to an electronic coincidence circuit to measure the coincidence counting rate (i.e., the fourth-order correlation, $G^{(2)}(\mathbf{u}, \mathbf{v})$) of two detected signals.

Based on the unified theory of coherence and polarization and by use of the moment theorem, the fourth-order correlation function $G^{(2)}$ $(\mathbf{u}, \mathbf{v})$ between two detectors D_1, D_2 can be expressed in terms of the second-order correlation function as follows[111]

$$G^{(2)}(\mathbf{u},\mathbf{v}) = \langle I_1(\mathbf{u}) I_2(\mathbf{v}) \rangle$$
$$= \langle E_x^*(\mathbf{u}) E_x(\mathbf{u}) E_x^*(\mathbf{v}) E_x(\mathbf{v}) \rangle + \langle E_x^*(\mathbf{u}) E_x(\mathbf{u}) E_y^*(\mathbf{v}) E_y(\mathbf{v}) \rangle$$
$$+ \langle E_y^*(\mathbf{u}) E_y(\mathbf{u}) E_x^*(\mathbf{v}) E_x(\mathbf{v}) \rangle + \langle E_y^*(\mathbf{u}) E_y(\mathbf{u}) E_y^*(\mathbf{v}) E_y(\mathbf{v}) \rangle \quad (49)$$
$$= \langle I_x(\mathbf{u}) \rangle \langle I_x(\mathbf{v}) \rangle + \langle I_x(\mathbf{u}) \rangle \langle I_y(\mathbf{v}) \rangle + \langle I_y(\mathbf{u}) \rangle \langle I_x(\mathbf{v}) \rangle$$
$$+ \langle I_y(\mathbf{u}) \rangle \langle I_y(\mathbf{v}) \rangle + F^{(2)}(\mathbf{u},\mathbf{v}),$$

where $\mathbf{u} \equiv (u_1, u_2)$, $\mathbf{v}_2 \equiv (v_1, v_2)$, $\langle I_i(\mathbf{u}) \rangle$ and $\langle I_i(\mathbf{v}) \rangle$ $(i = x$ or $y)$ are the average intensities of the i component of the electric field at D_1 and D_2, respectively. $F^{(2)}(\mathbf{u},\mathbf{v})$ is expressed as

$$F^{(2)}(\mathbf{u},\mathbf{v}) = |W_{yy}(\mathbf{u},\mathbf{v})|^2 + |W_{xx}(\mathbf{u},\mathbf{v})|^2 + |W_{xy}(\mathbf{u},\mathbf{v})|^2 + |W_{yx}(\mathbf{u},\mathbf{v})|^2. \quad (50)$$

where the term W_{ij} represents the second-order correlation between two detectors.

W_{ij} and $\langle I_i \rangle$ at two detectors are given by

$$\langle I_i(\eta) \rangle = \int_{-\infty}^{\infty}\int_{-\infty}^{\infty}\int_{-\infty}^{\infty}\int_{-\infty}^{\infty} h_l^*(\mathbf{r}_1,\eta) h_l(\mathbf{r}_2,\eta) W_{ii}(\mathbf{r}_1,\mathbf{r}_2) d^2\mathbf{r}_1 d^2\mathbf{r}_2, \quad (51)$$
$$\times (\eta = \mathbf{u}, \mathbf{v};\ i = x, y;\ l = 1,2),$$

$$W_{ij}(\mathbf{u},\mathbf{v}) = \int_{-\infty}^{\infty}\int_{-\infty}^{\infty}\int_{-\infty}^{\infty}\int_{-\infty}^{\infty} h_1^*(\mathbf{r}_1,\mathbf{u}) h_2(\mathbf{r}_2,\mathbf{v}) W_{ij}(\mathbf{r}_1,\mathbf{r}_2) d^2\mathbf{r}_1 d^2\mathbf{r}_2, \quad (52)$$
$$\times (i, j = x, y),$$

where $h_1(\mathbf{r},\mathbf{u})$ and $h_2(\mathbf{r},\mathbf{v})$ are the response functions of path one and path two, respectively and are expressed as

$$h_1(\mathbf{r},\mathbf{u}) = -\frac{1}{\lambda^2 z f_1} \int_{-\infty}^{\infty}\int_{-\infty}^{\infty} H(\xi) \exp\left[-\frac{i\pi}{\lambda z}\left(\mathbf{r}^2 - 2\mathbf{r}\cdot\xi + \xi^2\right) + \frac{2i\pi}{\lambda f_1}\xi\cdot\mathbf{u}\right] \quad (53)$$
$$\times d^2\xi,$$

$$h_2(\mathbf{r}, \mathbf{v}) = -\frac{i}{\lambda b_2} \exp\left[-\frac{i\pi}{\lambda b_2}\left(a_2\mathbf{r}^2 - 2\mathbf{r}\cdot\mathbf{v} + d_2\mathbf{v}^2 \right) \right]. \tag{54}$$

Here $H(\xi)$ denotes the transmission function of the object, a_2, b_2, c_2, d_2 are the elements of the transfer matrix of path 2 given by

$$\begin{pmatrix} a_2 & b_2 \\ c_2 & d_2 \end{pmatrix} = \begin{pmatrix} 1 - l_2/f & l_1 + l_2 - \dfrac{l_1 l_2}{f} \\ -1/f & 1 - l_1/f \end{pmatrix}. \tag{55}$$

If the parameters satisfy the following relations

$$l_1 = z + f_e \tan\frac{\phi}{2},\ l_2 = f_e \tan\frac{\phi}{2},\ f = f_e / \sin\phi, \tag{56}$$

where $\phi = p\pi/2$, the optical system in Fig. 10 can be used to implement coincidence FRT with an EGSM beam.[111] Here p is called the fractional order of FRT, f_e is called "standard focal length".

We assume the object in Fig. 10 to be a double slit, whose transmission function is written as

$$H(\xi_1, \xi_2) = \begin{cases} 1, & \dfrac{d}{2} - \dfrac{a}{2} \le |\xi_1| \le \dfrac{d}{2} + \dfrac{a}{2}, \\ 0, & \text{others} \end{cases} \tag{57}$$

where a and d are the slits width and the distance between two slits, respectively.

Substituting Eq. (23) and (57) into Eqs. (49)-(52), we can calculate the influence of the beam parameters of the EGSM beam on the coincidence FRT pattern of a double slit. One finds that the quality of the coincidence FRT pattern becomes better as the correlation coefficients δ_{xx}, δ_{yy} decreases (i.e., the degree of coherence of the source beam decreases) or as the width of the spectral density σ_0 increases (see Fig. 11). Furthermore, the coincidence FRT pattern of the object is also affected by the degree of polarization of the source beam (see Fig. 12), particularly for the case of $A_x > A_y$. For the case of $A_x > A_y$, with the decrease of the degree of polarization of the source beam, the side lobes of the normalized coincidence FRT pattern decreases gradually, which means the quality the coincidence FRT pattern is reduced. For the case of $A_x < A_y$, no distinguishable difference appears as the degree of polarization of the source beam varies. This phenomenon is caused by the fact that the contribution of the element $W_{xx}(\mathbf{r}_1,\mathbf{r}_2)$ to the coincidence FRT pattern of the object dominates that of the element $W_{yy}(\mathbf{r}_1,\mathbf{r}_2)$ for the case of $A_x > A_y$, and the contribution of the $W_{yy}(\mathbf{r}_1,\mathbf{r}_2)$ plays a dominant role otherwise. More information about the influence of the beam parameters on the quality and visibility of the coincidence FRT pattern can be found in Ref. 111

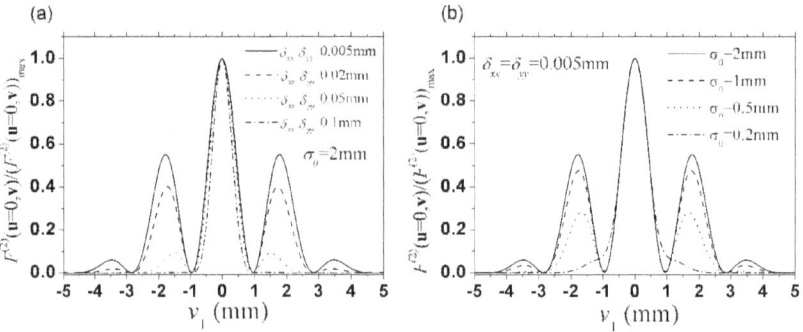

Fig. 11. Cross line ($v_2 = 0$) of the normalized coincidence FRT pattern of a double slit for different values of the correlation coefficients δ_{xx}, δ_{yy} and the width of the spectral density σ_0 implemented with an EGSM beam with $a = 80\mu m$, $d=200\mu m$, $f_e = 600mm$, $f_1 = 250mm$, $z = 150mm$, $p = 1$, $A_x = 1.5$, $A_y = 1$, $\sigma_x = \sigma_y = \sigma_0$, $B_{xy} = 0$.

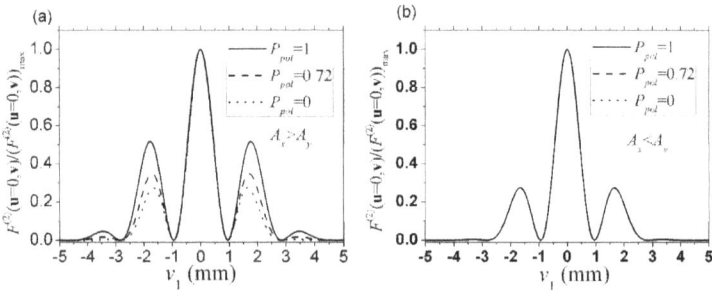

Fig. 12. Cross line ($v_2 = 0$) of the normalized coincidence FRT pattern of a double slit for different values of the degree of polarization of the EGSM beam with $a = 80\mu m$, $d=200\mu m$, $f_e = 600mm$, $f_1 = 250mm$, $z = 150mm$, $p = 1$, $\sigma_x = \sigma_y = \sigma_0 = 2mm$, $A_y = 1$, $\delta_{xx} = 0.01mm$, $\delta_{yy} = 0.03mm$, $B_{xy} = 0$.

3.6. *Degree of paraxiality of a partially coherent electromagnetic Gaussian Schell-model beam*

The description of an optical beam in the nonparaxial regime is becoming more and more important with the advent of new optical structures, e.g., microcavities and photonic bandgap crystals, characterized by linear dimensions or spatial scales of variation comparable to or even smaller than wavelength. One interesting aspect in dealing with nonparaxial beams is quantification of how paraxial a light field is. Gawhary and Severini introduced the concept of the degree of paraxiality, which provides an effective way for characterizing the paraxiality of a monochromatic light beam.[112-114] Wang *et al.* extended the concept of degree of paraxiality to partially coherent fields.[115] More recently, the degree of paraxiality of an EGSM beam was explored.[116] In this section, we introduced the degree of paraxiality of an EGSM beam.

The degree of paraxiality of the monochromatic light beam at certain propagation distance z was defined in Ref. 114 by the expression

$$P = \frac{\iint_{p^2+q^2<1/\lambda^2} |A_0(p,q)|^2 \sqrt{1-\lambda^2(p^2+q^2)}dpdq}{\iint_{p^2+q^2<1/\lambda^2} |A_0(p,q)|^2 dpdq}, \tag{58}$$

where $A_0(p,q)$ is the angular spectrum of the light beam at the source plane with p and q being the coordinates in spatial-frequency domain. The light beam is regarded as a completely paraxial beam if $P=1$, and it is regarded as a completely nonparaxial beam if $P=0$. The light beam is regarded as partially paraxial beam if $0<P<1$. The physical meaning of P can be considered as the average value of the consine of the angle of propagation that each homogeneous plane wave, constituting the real beam, makes with the optical axis.

The degree of paraxiality of a stochastic electromagnetic field is extended as[116]

$$P = \frac{\iint_{p^2+q^2<1/\lambda^2} \sum_\alpha A_{\alpha\alpha}(p,q,p,q)\sqrt{1-\lambda^2(p^2+q^2)}dpdq}{\iint_{p^2+q^2<1/\lambda^2} \sum_\alpha A_{\alpha\alpha}(p,q,p,q)dpdq}, \quad (\alpha = x, y),$$

(59)

where $A_{\alpha\alpha}$ is the angular correlation function given by the formula

$$A_{\alpha\alpha}(p_1,q_1,p_2,q_2) = \int_{-\infty}^{\infty}\int_{-\infty}^{\infty}\int_{-\infty}^{\infty}\int_{-\infty}^{\infty} W_{\alpha\alpha}(x_1,y_1,x_2,y_2)$$
$$\times \exp\left[-2\pi i\left(p_2x_2 - p_1x_1 + q_2y_2 - q_1y_1\right)\right]dx_1dy_1dx_2dy_2.$$

(60)

Substituting Eq. (23) into Eqs. (59) and (60), we obtain the following expression for the degree of paraxiality of an EGSM beam

$$P = \frac{A_x^2\beta_x + A_y^2\beta_y}{A_x^2C_x + A_y^2C_y},$$

(61)

where

$$\beta_x = 1 - \sqrt{\pi}\exp(-\gamma_x)Erfi[\sqrt{\gamma_x}]/\sqrt{4\gamma_x},$$
$$\beta_y = 1 - \sqrt{\pi}\exp(-\gamma_y)Erfi[\sqrt{\gamma_y}]/\sqrt{4\gamma_y},$$

$$\gamma_x = 2k^2\sigma_0^2\delta_{xx}^2/(\delta_{xx}^2 + 4\sigma_0^2), \quad \gamma_y = 2k^2\sigma_0^2\delta_{yy}^2/(\delta_{yy}^2 + 4\sigma_0^2),$$
$$Erfi[s] = Erf[is]/i,$$

$$C_x = 1 - \exp(-\gamma_x), \quad C_y = 1 - \exp(-\gamma_y), \quad \sigma_x = \sigma_y = \sigma_0. \quad (62)$$

The degree of paraxiality of an EGSM beam is closely determined by the width of the spectral density and the widths of the spectral degree of correlations (see Fig. 13). The degree of paraxiality decreases as the width of the spectral density decreases. Under the condition of $\sigma_0 \to 0$, the field reduces to a point source, but the value of the degree of paraxiality tends to 2/3, which represents the residual paraxiality as described in Ref. 114. Under the condition of $\sigma_0 \to \infty$, the degree of paraxiality of an EGSM beam tends to a constant. The lower δ_{xx} and δ_{yy}

are, the smaller is the constant. This phenomenon can be explained by the fact that the EGSM beam becomes highly divergent as δ_{xx} and δ_{yy} decrease (i.e., degree of coherence decreases), thus leading to the decrease in the degree of paraxiality. Under the condition of $\delta_{xx} \to 0$ and $\delta_{yy} \to 0$ (i.e., the EGSM beam becomes completely incoherent), the degree of paraxiality tends to 2/3, which means a completely incoherent EGSM beam also has a residual paraxiality, the same as a point source (i.e., $\sigma_0 \to 0$) as suggested in Ref. 114. More information about the influence of the spectral degree of polarization and the linear polarizer on the degree of paraxiality of an EGSM beam can be found in Ref. 116.

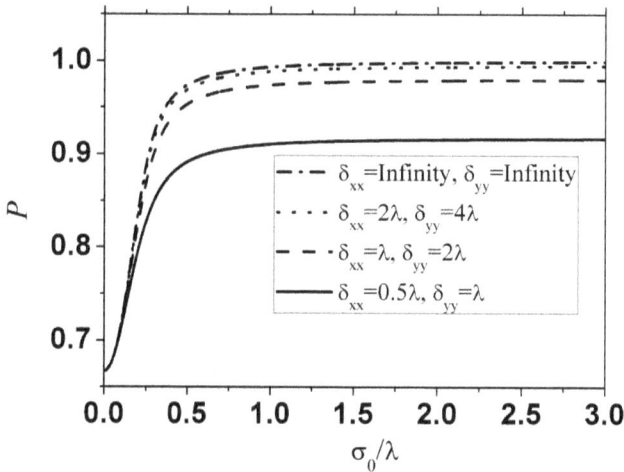

Fig. 13. Dependence of the degree of paraxiality of an EGSM beam on the normalized width (i.e., σ_0 / λ) of the spectral density for different values of the widths (i.e., δ_{xx} and δ_{yy}) of the spectral degree of correlations with $A_x = 2A_y$.

4. Partially coherent vector beams with uniform state of polarization: experiment

4.1. *Experimental generation and measurement of a partially coherent electromagnetic beam*

Up to now, although partially coherent electromagnetic beams have been studied widely in theory, only few papers were published on generation and measurement of such beams.[117-126] Different methods have been proposed to generate a partially coherent electromagnetic beam.

Experimental measurement of the cross-spectral density matrix and the generalized Stokes parameters was reported in Refs. 121-124. More recently, Coherence-induced polarization changes in a stochastic electromagnetic beam were demonstrated experimentally.[125] Wang *et al.* reported experimental measurement of the beam parameters of an EGSM beam.[126] In this section, we will introduce experimental generation and measurement of an EGSM beam.

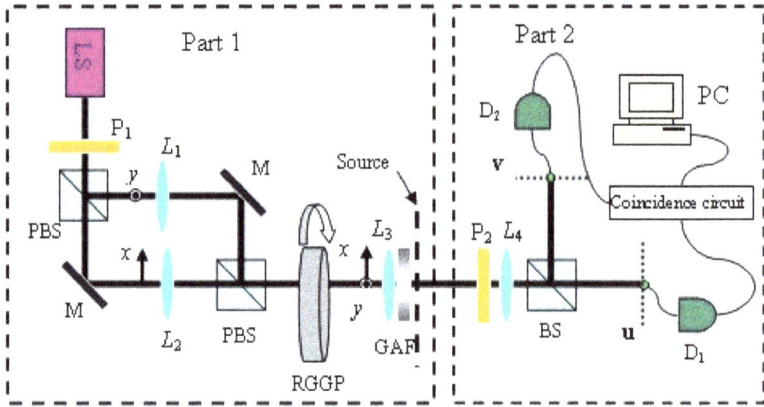

Fig. 14. Experimental setup for generating an EGSM beam and measuring its beam parameters. LS, He-Ne laser; P_1, P_2, linear polarizer; BS, 50:50 beam splitter; PBS, polarization beam splitter; L_1, L_2, L_3, L_4, thin lens; M, reflecting mirror; RGGP, rotating ground-glass plate; GAF, Gaussian amplitude filter; D_1, D_2, single photon detector.

Part 1 of Fig. 14 shows the experimental setup for generating an EGSM beam. After passing through linear polarizer P_1, the initial linearly polarized He-Ne laser beam passes through a Mach-Zehnder interferometer (MZI). Two orthogonally polarized beams are superimposed together at the output of the MZI, then the combined beam illuminates a rotating ground-glass plate (RGGP) producing a partially coherent electromagnetic beam. After passing through a collimating thin lens L_3 and a Gaussian amplitude filter (GAF), an EGSM beam is generated. P_1 whose transmission axis forms an angle θ with the x-axis is used to adjust the ratio A_x / A_y of the beam by varying θ. L_1 and L_2 in Fig. 14 are used to control the focused beam spot sizes of the x- and y-linearly polarized beams on the RGGP. The parameters σ_x and σ_y of the generated EGSM beam are only determined by the transmission function of the GAF. The parameters δ_{xx}, δ_{yy} and δ_{xy} of the generated EGSM beam are mainly determined by the focused beam spot sizes on the RGGP. L_1 and L_2 have the same focal length ($f = 10$cm), and the eikonals along two arms of the MZI are the same. Then the transmitted light just behind the GAF is regarded as an EGSM beam. In the experiment, ϕ_{xy} is assumed to be zero due to the same eikonals along two arms of the MZI.

Part 2 of Fig. 14 shows the experimental setup for measuring the parameters of the generated EGSM beam. The EGSM beam first passes through a linear polarizer P_2 and a thin lens L_4 with focal length $f = 15$cm, then is split into two beams by a beam splitter. The transmitted and reflected beams going through two separated optical paths will arrive at single photon detectors D_1 and D_2, which scan the transverse planes of u and v, respectively. Both the distances from the GAF to L_4 and from L_4 to D_1 and D_2 are $2f$ (i.e., $2f$-imaging system). The output signals from D_1 and D_2 are sent to an electronic coincidence circuit to measure the fourth-order correlation function between the two detectors (i.e., intensity correlation function).[127]

First, we adjust P_2 to set its transmission axis along x-axis. In this case, only the element W_{xx} exists behind P_2. By measuring the intensity distribution and its maximum intensity at plane u or v with the help of the single photon detector D_1 or D_2, we can obtain the values of the parameters σ_x and A_x. The fourth-order correlation function between D_1 and D_2 is given by the following expression

$$G_{xx}^{(2)}(u_1 - v_1, \tau) = \frac{\langle I_x(u_1, t) I_x(v_1, t + \tau) \rangle}{\langle I_x(u_1, t) \rangle \langle I_x(v_1, t + \tau) \rangle},\tag{63}$$

where τ denotes the delay time of the photon flux of two optical paths. Applying the Gausian moment theorem, $G_{xx}^{(2)}(u_1 - v_1, \tau)$ with $\tau = 0$ can be simplified as

$$G_{xx}^{(2)}(u_1 - v_1, \tau = 0) = 1 + \exp\left[-\frac{(u_1 - v_1)^2}{\delta_{xx}^2}\right].\tag{64}$$

We fix D_2 at $v=0$, and D_1 scans along the plane u. The coincidence circuit records the fourth-order correlation between D_1 and D_2. Then we can obtain the distribution of the normalized fourth-order correlation function $G_{xx}^{(2)}(u_1, \tau = 0)$ with $\tau = 0$. From the curve of the Gaussian fit for the experimental results, we can obtain the value of δ_{xx}. If we adjust P_2 to set its transmission axis along y-axis, only the element W_{yy} exists behind P_2. Then through a similar operation for obtaining σ_x, A_x and δ_{xx}, we can measure the values of the parameters σ_y, A_y and δ_{yy}.

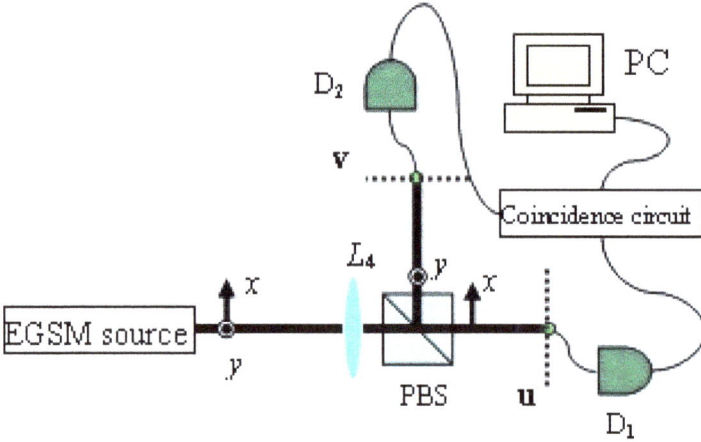

Fig. 15. Experimental scheme for measuring the parameters $\left| B_{xy} \right|$ and δ_{xy}.

Figure 15 shows the experimental scheme for measuring the parameters $\left| B_{xy} \right|$ and δ_{xy}. Different from the part 2 of Fig. 14, in this case, P_2 is removed, and the BS is replaced with a polarization beam splitter (PBS). After passing through the PBS, the x component and y component of the field will arrive at D_1 and D_2, respectively. The normalized fourth-order correlation function between D_1 and D_2 with $\tau = 0$ can be expressed as

$$G_{xy}^{(2)}(u_1 - v_1, \tau = 0) = \frac{\langle I_x(u_1,t) I_y(v_1,t) \rangle}{\langle I_x(u_1,t) \rangle \langle I_y(v_1,t) \rangle} = 1 + \left| B_{xy} \right|^2 \exp\left[-\frac{(u_1 - v_1)^2}{\delta_{xy}^2} \right]. \quad (65)$$

Following the same procedure for obtaining δ_{xx}, D_2 is fixed at $v=0$, and D_1 scans along the plane u, and the coincidence circuit record the distribution of the normalized fourth-order correlation function $G_{xy}^{(2)}(u_1, \tau = 0)$. From the curve of the Gaussian fit for the experimental results, we can obtain the value of δ_{xy}. From Eq. (65), $\left|B_{xy}\right|$ can be obtained by the relation $\left|B_{xy}\right| = \sqrt{G_{xy}^2(u_1, \tau = 0) - 1}$ under the condition of $u_1 = 0$.

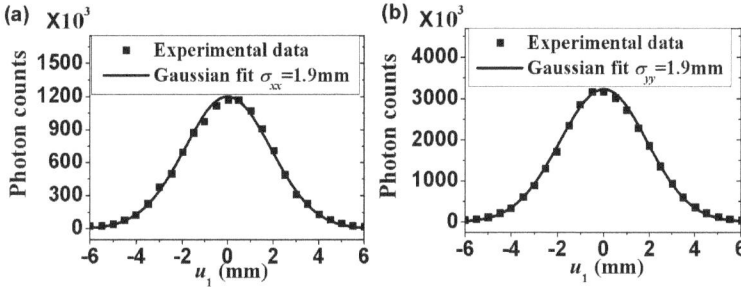

Fig. 16. Experimental results (dotted curves) and corresponding Gaussian fit (solid curves) of the intensity distributions of the x-component and y-component of the EGSM beam.

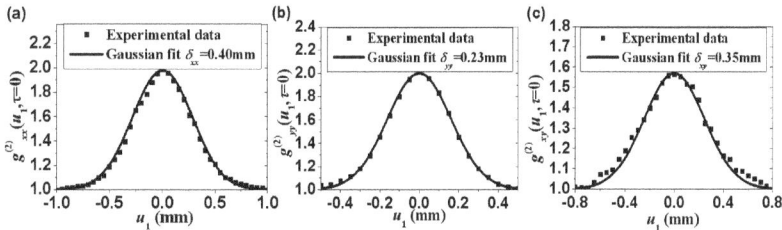

Fig. 17. Experimental results (dotted curves) and corresponding Gaussian fit (solid curves) of the normalized fourth-order correlation functions.

Figure 16 shows the experimental results (dotted curves) and corresponding Gaussian fit (solid curves) of the intensity distributions of the x-component and y-component of the EGSM source. Figure 17 shows the experimental results (dotted curves) and corresponding Gaussian fit (solid curves) of the normalized fourth-order correlation functions. From Fig. 16, we obtain $\sigma_x = \sigma_y = 1.9$ mm, $A_x / A_y = 0.61$. From Fig. 17, we obtain $\delta_{xx} = 0.40$ mm, $\delta_{yy} = 0.23$ mm, $\delta_{xy} = 0.35$ mm, $\left|B_{xy}\right| = 0.74$.

With the measured beam parameters of the EGSM beam, we can simulate the propagation properties of the generated EGSM beam quantitatively, and compare the results with experimental results. More information can be found in Ref. 126.

4.2. *Experimental coupling of a partially coherent electromagnetic Gaussian Schell-model beam into a single-mode optical fiber*

The coupling of light into optical fibers has been investigated extensively due to its wide applications in optical communications, biomedical optics, LIDARs, stellar interferometry, and wavefront sensing since Cohen carried out an experimental study of the coupling of the GaAs injection laser beams into optical fibers in 1972.[128] The majority of the previously published papers were devoted to coupling of coherent beams into optical fibers. In some practical applications, such as free-space optical communications and LIDARs, coupling of a stochastic beam into an optical fiber is inevitable encountered because the coherence of laser beam was degraded by the atmospheric turbulence during propagation.[129, 130] Thus, it is of great importance to study the coupling of a stochastic beam into an optical fiber. Recently, Salem *et al.* theoretically studied the effects of coherence and polarization on the coupling of an EGSM beam into optical fibers based on the theory of coherence and polarization and found that the coupling efficiency is closely related to the states of coherence and polarization.[131-133] More recently, Zhao *et al.* reported experimental coupling of an EGSM beam into a single-mode optical fiber.[134] In this section, we will introduce the procedure of coupling of an EGSM beam into a single-mode optical fiber.

According to Ref. 133, the coupling efficiency of an EGSM beam into a single-mode fiber may be approximately expressed as

$$\eta = \left(\frac{\pi}{\lambda f W} \right) \frac{\sum_{\alpha=x,y} A_\alpha^2 w_\alpha^2 \left[C_{\alpha\alpha}^2 - (1/4\delta_{\alpha\alpha}^4) \right]^{-1}}{\sum_{\alpha=x,y} A_\alpha^2 \sigma_\alpha^2 \left(W^2 + 2\sigma_\alpha^2 \right)^{-1}}, \tag{66}$$

where f is the focal length of the coupling lens, $W = D/2\sqrt{2}$, with D being the aperture diameter of the coupling lens, w_α is the width of the mode polarized along direction α in the optical fiber, $C_{\alpha\alpha}$ is given by expression

$$C_{\alpha\alpha} = 1/4\sigma_\alpha^2 + 1/2\delta_{\alpha\alpha}^2 + \pi^2 w_\alpha^2 / \lambda^2 f^2 + 1/W^2. \tag{67}$$

From Eq. (66), one sees that the coupling efficiency is independent of the anti-diagonal elements of the cross-spectral density matrix, but is closely related with the parameters δ_{xx}, δ_{yy} and the spectral degree of polarization when the parameters of the fiber and the coupling lens are fixed.

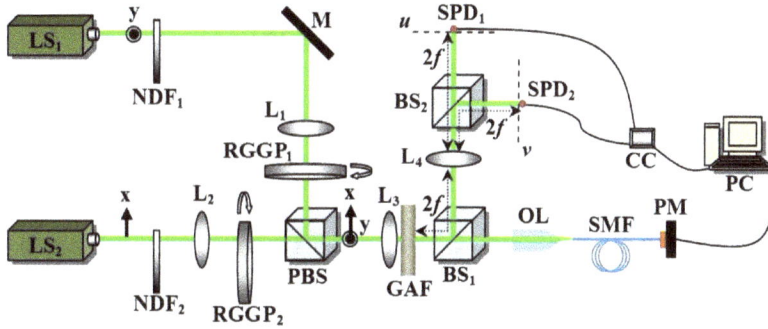

Fig. 18. Experimental setup for generating an EGSM beam without anti-diagonal elements, measuring the beam parameters and coupling the EGSM beam into a single-mode fiber. LS_1, LS_2, Diode-pumped solid-state lasers; NDF_1, NDF_2, neutral density filters; M, reflecting mirror; $RGGP_1$, $RGGP_2$, rotating ground-glass disks; PBS, polarization beam splitter; GAF, Gaussian amplitude filter; BS_1, BS_2, 50:50 beam splitters; L_1, L_2, L_3, L_4, thin lenses; OL, objective lens; SMF, single-mode fiber; PM, power meter; PC, personal computer; SPD_1, SPD_2, single photon detectors; CC, coincidence circuit.

Figure 18 shows the experimental setup for generating an EGSM beam without anti-diagonal elements, measuring the beam parameters and coupling of the EGSM beam into a single-mode fiber. The y (or x) linearly polarized beam $(\lambda = 532\text{nm})$ generated from LS_1 (or LS_2) first passes through the NDF_1 (or NDF_2) and the L_1 (or L_2), then illuminates on the $RGGP_1$ (or $RGGP_2$), producing y (or x) linearly polarized partially coherent beam. Then the polarization beam splitter combines two orthogonally polarized partially coherent beams, producing a partially coherent electromagnetic beam. After passing through the L_3 and the GAF, the partially coherent electromagnetic beam becomes an EGSM beam. NDF_1 and NDF_2 are used for modulating the parameters A_x and A_y. The correlation coefficients δ_{xx} and δ_{yy} are mainly determined by focused beam spot sizes on the $RGGP_1$ and $RGGP_2$. σ_x and σ_y are determined by the transmission function of the GAF. The parameters of the EGSM beam can be measured through the similar procedure described in Section 4.1. In the experiment, we have $\sigma_x = \sigma_y = 1.9$ mm.

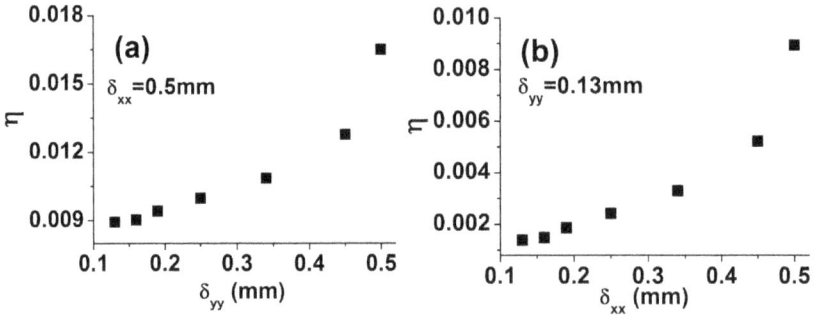

Fig. 19. Experimental results of the coupling efficiency versus the correlation coefficients δ_{xx} and δ_{yy} of the EGSM beam.

After the beam parameters of the generated EGSM beam were measured, we can obtain the value of the spectral degree of polarization of the generated EGSM beam and study the coupling of the beam into a single-mode fiber. In the experiment, the generated EGSM beam (i.e., the transmitted beam from BS_1) is coupled into the single-mode fiber with the objective lens. The numerical aperture (NA) of the objective lens equals to 0.1. The single-mode fiber with NA = 0.13 is made of fused silica (S460 HP produced by the THORLAB) and its operating wavelength ranges from 450 nm to 600 nm. The power meter is used to measure the power of the beam just before the objective lens and the power at the output of the single-mode fiber. The experimental results of the coupling efficiency versus the correlation coefficients δ_{xx} and δ_{yy} of the generated EGSM beam is shown in Fig. 19. The experimental results of the coupling efficiency versus the spectral degree of polarization of the generated EGSM beam is shown in Fig. 20. One finds that the coupling efficiency increases with the increase of δ_{xx} or δ_{yy} (i.e., spectral degree of coherence), decreases with the increase of spectral degree of polarization for the case $A_x > A_y$ and increases with the increase of spectral degree of polarization for the case $A_x < A_y$. The experimental results agree well with the theoretical predictions

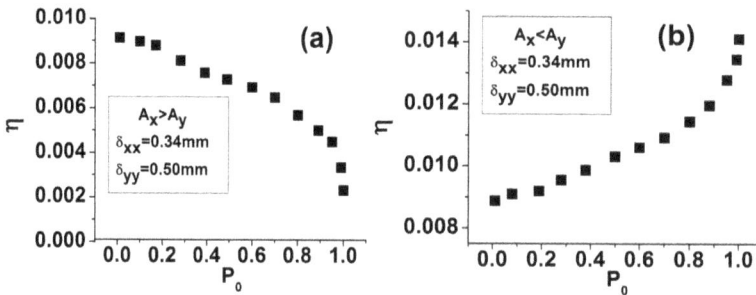

Fig. 20. Experimental results of the coupling efficiency versus the spectral degree of polarization of the EGSM beam.

5. Partially coherent vector beams with non-uniform state of polarization: theory

In the past several years, laser beams with cylindrical symmetry in polarization (i.e., cylindrical vector beams) have been widely investigated and applied in optical trapping, laser making, dark field imaging, free-space optical communications, singular optics, data storage, 3D tailoring of the focus shape, optical inspection and metrology.[3] Different methods have been developed to generate cylindrical vector beam. The tight focusing properties, propagation properties and second-harmonic generation of cylindrical vector beams have been explored in detail. More recently, the cylindrical vector beam to the partially coherent case based on the unified theory of coherence and polarization.[9] Nonparaxial propagation properties of a cylindrical vector partially coherent field were explored in Ref. 27. Experimental generation of a radially or azimuthally polarized beam with controllable spatial coherence was reported recently.[10-11] Now we will introduce recent theoretical development on cylindrical vector partially coherent beam

5.1. *Cylindrical vector partially coherent beam and its paraxial propagation*

In cylindrical coordinates, the electric field of coherent cylindrically polarized Laguerre-Gaussian (LG) beam at $z = 0$ is expressed as follows Ref. 135

$$\vec{E}(r,\phi) = \exp\left(-\frac{r^2}{w_0^2}\right)\left(\frac{2r^2}{w_0^2}\right)^{(n\pm1)/2} L_p^{n\pm1}\left(\frac{2r^2}{w_0^2}\right)\begin{Bmatrix} \cos(n\phi)\vec{e}_\phi \mp \sin(n\phi)\vec{e}_r \\ \pm\sin(n\phi)\vec{e}_\phi + \cos(n\phi)\vec{e}_r \end{Bmatrix}, \tag{68}$$

where r and ϕ are the radial and azimuthal (angle) coordinates, $L_p^{n\pm1}$ denotes the Laguerre polynomial with mode orders p and $n\pm1$, w_0 is the beam width of the fundamental Gaussian mode. When $p=0$ and $n=0$, Eq. (68) degenerates to the electric field of a radially or azimuthally polarized Gaussian beam.

Equation (68) can be expressed in the alternative form in Cartesian coordinates[9]

$$\vec{E}(x,y) = \begin{cases} E_x(x,y)\vec{e}_x + E_y(x,y)\vec{e}_y \\ E_y(x,y)\vec{e}_x + E_x(x,y)\vec{e}_y \end{cases}$$

$$= \exp\left(-\frac{x^2+y^2}{w_0^2}\right)\frac{(-1)^p}{2^{2p+n\pm1}p!}\left\{\begin{array}{l} \dfrac{1}{2i}\displaystyle\sum_{m=0}^{p}\sum_{s=0}^{n\pm1} i^s\left[1-(-1)^s\right]\binom{p}{m}\binom{n\pm1}{s} \\[4pt] \times H_{2m+n\pm1-s}\left(\dfrac{\sqrt{2}x}{w_0}\right)H_{2p-2m+s}\left(\dfrac{\sqrt{2}y}{w_0}\right)\vec{e}_x \\[10pt] \dfrac{1}{2}\displaystyle\sum_{m=0}^{p}\sum_{s=0}^{n\pm1} i^s\left[1+(-1)^s\right]\binom{p}{m}\binom{n\pm1}{s} \\[4pt] \times H_{2m+n\pm1-s}\left(\dfrac{\sqrt{2}x}{w_0}\right)H_{2p-2m+s}\left(\dfrac{\sqrt{2}y}{w_0}\right)\vec{e}_x \end{array}\right.$$

$$+\frac{1}{2}\sum_{m=0}^{p}\sum_{s=0}^{n\pm1} i^s\left[1+(-1)^s\right]\binom{p}{m}\binom{n\pm1}{s}H_{2m+n\pm1-s}\left(\frac{\sqrt{2}x}{w_0}\right)$$

$$\times H_{2p-2m+s}\left(\frac{\sqrt{2}y}{w_0}\right)\vec{e}_y$$

$$\left.+\frac{1}{2i}\sum_{m=0}^{p}\sum_{s=0}^{n\pm1} i^s\left[1-(-1)^s\right]\binom{p}{m}\binom{n\pm1}{s}H_{2m+n\pm1-s}\left(\frac{\sqrt{2}x}{w_0}\right)\right\}$$

$$\times H_{2p-2m+s}\left(\frac{\sqrt{2}y}{w_0}\right)\vec{e}_y$$

$$\tag{69}$$

Based on the unified theory of coherence and polarization, for the case of $\vec{E}(x,y) = E_x(x,y)\vec{e}_x + E_y(x,y)\vec{e}_y$, we can express the elements of the cross-spectral density matrix of a cylindrical vector partially coherent LG beam as follows

$$W_{xx}(x_1,y_1,x_2,y_2) = \frac{1}{4}\exp\left[-\frac{x_1^2 + x_2^2 + y_1^2 + y_2^2}{w_0^2} - \frac{(x_1-x_2)^2 + (y_1-y_2)^2}{2\delta_{xx}^2}\right]$$

$$\times\frac{1}{2^{4p+2n\pm2}(p!)^2}\sum_{m=0}^{p}\sum_{s=0}^{n\pm1}\sum_{l=0}^{p}\sum_{h=0}^{n\pm1}i^s(-i)^h\left[1-(-1)^s\right]$$

$$\times\left[1-(-1)^h\right]\binom{p}{m}\binom{p}{l}\binom{n\pm1}{s}\binom{n\pm1}{h}H_{2m+n\pm1-s}\left(\frac{\sqrt{2}x_1}{w_0}\right)$$

$$\times H_{2l+n\pm1-h}\left(\frac{\sqrt{2}x_2}{w_0}\right)H_{2p-2m+s}\left(\frac{\sqrt{2}y_1}{w_0}\right)H_{2p-2l+h}\left(\frac{\sqrt{2}y_2}{w_0}\right),$$

$$(70)$$

$$W_{xy}(x_1,y_1,x_2,y_2) = \frac{B_{xy}}{4i}\exp\left[-\frac{x_1^2 + x_2^2 + y_1^2 + y_2^2}{w_0^2} - \frac{(x_1-x_2)^2 + (y_1-y_2)^2}{2\delta_{xy}^2}\right]$$

$$\times\frac{1}{2^{4p+2n\pm2}(p!)^2}\sum_{m=0}^{p}\sum_{s=0}^{n\pm1}\sum_{l=0}^{p}\sum_{h=0}^{n\pm1}i^s(-i)^h\left[1-(-1)^s\right]$$

$$\times\left[1+(-1)^h\right]\binom{p}{m}\binom{p}{l}\binom{n\pm1}{s}\binom{n\pm1}{h}H_{2m+n\pm1-s}(\frac{\sqrt{2}x_1}{w_0})$$

$$\times H_{2l+n\pm1-h}(\frac{\sqrt{2}x_2}{w_0})H_{2p-2m+s}(\frac{\sqrt{2}y_1}{w_0})H_{2p-2l+h}(\frac{\sqrt{2}y_2}{w_0}),$$

$$(71)$$

$$W_{yx}(x_1,y_1,x_2,y_2) = \left[W_{xy}(x_2,y_2,x_1,y_1)\right]^*, \qquad (72)$$

$$W_{yy}\left(x_1,y_1,x_2,y_2\right)=\frac{1}{4}\exp\left[-\frac{x_1^2+x_2^2+y_1^2+y_2^2}{w_0^2}-\frac{\left(x_1-x_2\right)^2+\left(y_1-y_2\right)^2}{2\delta_{yy}^2}\right]$$

$$\times\frac{1}{2^{4p+2n\pm2}\left(p!\right)^2}\sum_{m=0}^{p}\sum_{s=0}^{n\pm1}\sum_{l=0}^{p}\sum_{h=0}^{n\pm1}i^s(-i)^h\left[1+(-1)^s\right]$$

$$\times\left[1+(-1)^h\right]\binom{p}{m}\binom{p}{l}\binom{n\pm1}{s}\binom{n\pm1}{h}H_{2m+n\pm1-s}\left(\frac{\sqrt{2}x_1}{w_0}\right)$$

$$\times H_{2l+n\pm1-h}\left(\frac{\sqrt{2}x_2}{w_0}\right)H_{2p-2m+s}\left(\frac{\sqrt{2}y_1}{w_0}\right)H_{2p-2l+h}\left(\frac{\sqrt{2}y_2}{w_0}\right),$$

$$(73)$$

where $B_{\alpha\beta}$ is the correlation coefficient between the E_x and E_y field components, $\delta_{\alpha\beta}$ denotes the width of the spectral degree of correlation.

In a similar way, for the case of $\vec{E}(x,y)=E_y(x,y)\vec{e}_x+E_x(x,y)\vec{e}_y$, we can express the elements of the cross-spectral density matrix of a cylindrical vector partially coherent LG beam as follows

$$W_{1xx}\left(x_1,y_1,x_2,y_2\right)=W_{yy}\left(x_1,y_1,x_2,y_2\right),\ W_{1yy}\left(x_1,y_1,x_2,y_2\right)=W_{xx}\left(x_1,y_1,x_2,y_2\right),$$
$$W_{1xy}\left(x_1,y_1,x_2,y_2\right)=W_{yx}\left(x_1,y_1,x_2,y_2\right),\ W_{1yx}\left(x_1,y_1,x_2,y_2\right)=W_{xy}\left(x_1,y_1,x_2,y_2\right),$$

$$(74)$$

where W_{xx}, W_{xy}, W_{yx}, W_{yy} are given by Eqs. (70)-(73).

Substituting Eqs. (70)-(73) into the generalized Collins formula for treating the paraxial propagation of a partially coherent beam, one can obtain analytical propagation formulas for the elements of the cross-spectral density matrix of a cylindrical vector partially coherent LG beam and study the effect of coherence on the propagation properties of a cylindrical vector partially coherent LG beam conveniently.[9]

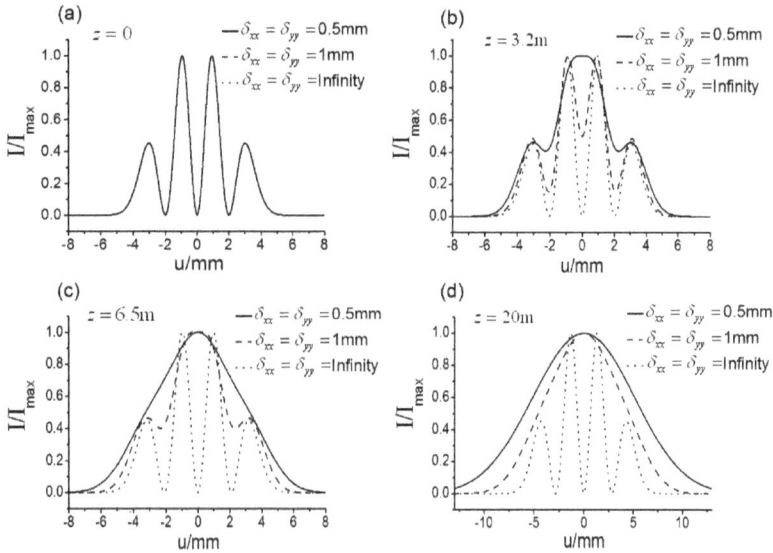

Fig. 21. Normalized intensity distribution (cross line) of a cylindrical vector partially coherent LG beam for different values of the correlation coefficients δ_{xx}, δ_{yy} at several propagation distances in free space with $p = 1$, $n \pm 1 = 1$, $w_0 = 2$mm.

One finds from Fig. 21 that the intensity distribution properties of a cylindrical vector partially coherent LG beam are very different from those of a cylindrical vector coherent LG beam, and is closely determined by the correlation coefficients δ_{xx}, δ_{yy} . For a cylindrical vector coherent LG beam, its initial source beam profile remains invariant on propagation although its beam spot increases. For a cylindrical vector partially coherent LG beam, its initial source beam profile doesn't remain invariant on propagation, but gradually disappears on propagation and eventually takes a Gaussian shape. As the initial correlation coefficients δ_{xx}, δ_{yy} decrease (i.e., the initial degree of coherence decreases), the transition from a cylindrical vector LG beam into a Gaussian beam occurs more quickly and the beam spreads more rapidly.

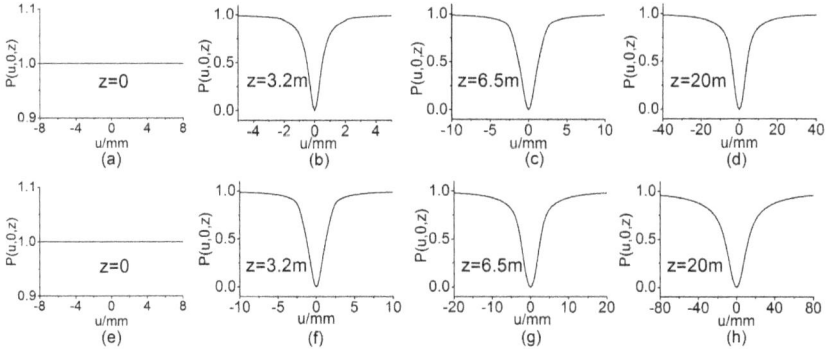

Fig. 22. Degree of polarization (cross line, $v = 0$) of a cylindrical vector partially coherent LG beam for different values of the initial correlation coefficients δ_{xx}, δ_{xy}, δ_{yy} at several propagation distances in free space with $w_0 = 2$mm, $B_{xy} = 1$, $p = 1$, $n \pm 1 = 1$, $w_0 = 2$mm.

One finds from Fig. 22 that the evolution properties of degree of polarization of a cylindrical vector partially coherent LG beam are also much different that of a cylindrical vector coherent LG beam. The degree of polarization of a cylindrical vector coherent LG beam in the source plane equals 1 for all the points across the entire transverse plane, and it's value remains invariant during propagation. For a cylindrical vector partially coherent LG beam, although the degree of polarization equals 1 for all the points across the entire transverse plane and is independent of the initial correlation coefficients δ_{xx}, δ_{xy}, δ_{yy} in the source plane, it doesn't remain invariant during propagation, on the contrary, the degree of polarization varies on propagation and is closely determined by the initial correlation coefficients (i.e., the initial degree of coherence). Thus, one comes to the conclusion that the polarization structure of a cylindrical vector partially coherent LG beam is destroyed during propagation in free space (i.e., a cylindrical vector partially coherent LG beam is depolarized during propagation).

More information about the statistical properties of a cylindrical vector partially coherent LG beam in free space can be found in Ref. 9. Statistical properties of a nonparaxial cylindrical vector partially coherent LG beam can be found in Ref. 27.

5.2. Tight focusing properties of a partially coherent azimuthally polarized beam

In many practical applications, the cylindrical vector beam usually is focused by a high numerical aperture (NA) objective lens and the focused beam spot size is comparable to the wavelength of the beam. In this case, the Richards-Wolf vectorial diffraction integral is adopted to study the focusing properties of the cylindrical vector beam. In this section, we introduce the tight focusing properties of a partially coherent azimuthally polarized beam focused by a high NA objective lens. Figure 23 shows the scheme of tightly focusing system.

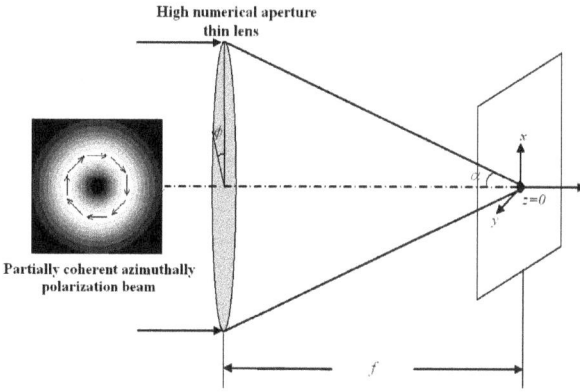

Fig. 23. Scheme of tight focusing system.

Based on the Richards-Wolf diffraction integral, in the cylindrical coordinate system, the vectorial electric field of a tightly focused cylindrical vector beam at the focal plane is expressed as[135]

$$
\vec{E}(r,\varphi,z) = -\frac{if\sqrt{n_1}}{\lambda} \int_0^{\theta_{max}} \int_0^{2\pi} \sqrt{\cos\theta}\sin\theta \exp\left[ik_1(z\cos\theta + r\sin\theta\cos(\phi-\varphi))\right]
$$

$$
\times \begin{bmatrix} l_x(\theta,\phi)(\cos\theta + \sin^2\phi(1-\cos\theta)) + l_y(\theta,\phi)\cos\phi\sin\phi(\cos\theta-1) \\ l_x(\theta,\phi)\cos\phi\sin\phi(\cos\theta-1) + l_y(\theta,\phi)(\cos\theta + \cos^2\phi(1-\cos\theta)) \\ -l_x(\theta,\phi)\cos\phi\sin\theta - l_y(\theta,\phi)\sin\phi\sin\theta \end{bmatrix}
$$

$$
\times d\theta d\phi,
$$

(75)

where r, φ, z are the cylindrical coordinates of an observation point, φ is the azimuthal angle of the incident beam, f is the focal length of the lens, $k_1 = kn_1 = 2\pi n_1 / \lambda$ is the wave number in the surrounding medium with n_1 being the refractive index of the surrounding medium, θ is the NA angle, and θ_{max} is the maximal NA angle which is related with the NA by the formula $\theta_{max} = \arcsin NA / n_1$, $l_x(\theta, \phi)$ and $l_y(\theta, \phi)$ are the pupil apodization functions at the aperture surface and are derived by setting $r = f \sin \theta$ in $E_x(x, y)$ and $E_y(x, y)$, respectively.

Based on the unified theory of coherence and polarization, the 3×3 cross-spectral density matrix of the tightly focused cylindrical vector partially coherent beam in the focal region is expressed as[11]

$$\ddot{W}(r_1,\varphi_1,r_2,\varphi_2,z) = \begin{pmatrix} W_{xx}(r_1,\varphi_1,r_2,\varphi_2,z) & W_{xy}(r_1,\varphi_1,r_2,\varphi_2,z) & W_{xz}(r_1,\varphi_1,r_2,\varphi_2,z) \\ W_{yx}(r_1,\varphi_1,r_2,\varphi_2,z) & W_{yy}(r_1,\varphi_1,r_2,\varphi_2,z) & W_{yz}(r_1,\varphi_1,r_2,\varphi_2,z) \\ W_{zx}(r_1,\varphi_1,r_2,\varphi_2,z) & W_{zy}(r_1,\varphi_1,r_2,\varphi_2,z) & W_{zz}(r_1,\varphi_1,r_2,\varphi_2,z) \end{pmatrix},$$

$$(76)$$

where

$$W_{xx}(r_1,\varphi_1,r_2,\varphi_2,z) = \frac{f^2 n_1}{\lambda^2} \left\{ \int_0^{\theta_{max}} \int_0^{\theta_{max}} \int_0^{2\pi} \int_0^{2\pi} \sqrt{\cos\theta_1 \cos\theta_2} \sin\theta_1 \sin\theta_2 \right.$$

$$\times \exp\left[-ik_1(z\cos\theta_2 + r_2\sin\theta_2\cos(\phi_2 - \varphi_2))\right]$$

$$\times \exp\left[ik_1(z\cos\theta_1 + r_1\sin\theta_1\cos(\phi_1 - \varphi_1))\right]$$

$$\times \left\{ \begin{array}{l} W_{xx}(\theta_1,\phi_1,\theta_2,\phi_2)\left[\cos\theta_1 + \sin^2\phi_1(1-\cos\theta_1)\right] \\ \times\left[\cos\theta_2 + \sin^2\phi_2(1-\cos\theta_2)\right] \end{array} \right.$$

$$+ W_{xy}(\theta_1,\phi_1,\theta_2,\phi_2)\left[\cos\theta_1 + \sin^2\phi_1(1-\cos\theta_1)\right]$$

$$\times \cos\phi_2\sin\phi_2(\cos\theta_2 - 1) + W_{yx}(\theta_1,\phi_1,\theta_2,\phi_2)$$

$$\times \cos\phi_1\sin\phi_1(\cos\theta_1 - 1)\left[\cos\theta_2 + \sin^2\phi_2(1-\cos\theta_2)\right]$$

$$\left. + W_{yy}(\theta_1,\phi_1,\theta_2,\phi_2)\cos\phi_1\sin\phi_1(\cos\theta_1 - 1) \right.$$

$$\left. \times\cos\phi_2\sin\phi_2(\cos\theta_2 - 1) \right\} d\theta_1 d\theta_2 d\phi_1 d\phi_2 \right\},$$

$$(77)$$

$$W_{xy}(r_1,\varphi_1,r_2,\varphi_2,z) = \frac{f^2 n_1}{\lambda^2}\left\{\int_0^{\theta_{max}}\int_0^{\theta_{max}}\int_0^{2\pi}\int_0^{2\pi}\sqrt{\cos\theta_1\cos\theta_2}\,\sin\theta_1\sin\theta_2\right.$$

$$\times\exp\left[-ik_1(z\cos\theta_2 + r_2\sin\theta_2\cos(\phi_2-\varphi_2))\right]$$

$$\times\exp\left[ik_1(z\cos\theta_1 + r_1\sin\theta_1\cos(\phi_1-\varphi_1))\right]$$

$$\times\left\{W_{xx}(\theta_1,\phi_1,\theta_2,\phi_2)\left[\cos\theta_1+\sin^2\phi_1(1-\cos\theta_1)\right]\right.$$

$$\times\cos\phi_2\sin\phi_2(\cos\theta_2-1)+W_{xy}(\theta_1,\phi_1,\theta_2,\phi_2)$$

$$\times\left[\cos\theta_1+\sin^2\phi_1(1-\cos\theta_1)\right]$$

$$\times\left[\cos\theta_2+\cos^2\phi_2(1-\cos\theta_2)\right]+W_{yx}(\theta_1,\phi_1,\theta_2,\phi_2)$$

$$\times\cos\phi_1\sin\phi_1(\cos\theta_1-1)\cos\phi_2\sin\phi_2(\cos\theta_2-1)$$

$$+W_{yy}(\theta_1,\phi_1,\theta_2,\phi_2)\cos\phi_1\sin\phi_1(\cos\theta_1-1)$$

$$\left.\left.\times\left[\cos\theta_2+\cos^2\phi_2(1-\cos\theta_2)\right]\right\}d\theta_1 d\theta_2 d\phi_1 d\phi_2\right\},$$

(78)

$$W_{xz}(r_1,\varphi_1,r_2,\varphi_2,z) = -\frac{f^2 n_1}{\lambda^2}\left\{\int_0^{\theta_{max}}\int_0^{\theta_{max}}\int_0^{2\pi}\int_0^{2\pi}\sqrt{\cos\theta_1\cos\theta_2}\,\sin\theta_1\sin\theta_2\right.$$

$$\times\exp\left[-ik_1(z\cos\theta_2 + r_2\sin\theta_2\cos(\phi_2-\varphi_2))\right]$$

$$\times\exp\left[ik_1(z\cos\theta_1 + r_1\sin\theta_1\cos(\phi_1-\varphi_1))\right]$$

$$\times\left\{W_{xx}(\theta_1,\phi_1,\theta_2,\phi_2)\left[\cos\theta_1+\sin^2\phi_1(1-\cos\theta_1)\right]\cos\phi_2\sin\theta_2\right.$$

$$+W_{xy}(\theta_1,\phi_1,\theta_2,\phi_2)\left[\cos\theta_1+\sin^2\phi_1(1-\cos\theta_1)\right]\sin\phi_2\sin\theta_2$$

$$+W_{yx}(\theta_1,\phi_1,\theta_2,\phi_2)\cos\phi_1\sin\phi_1(\cos\theta_1-1)\cos\phi_2\sin\theta_2$$

$$\left.+W_{yy}(\theta_1,\phi_1,\theta_2,\phi_2)\cos\phi_1\sin\phi_1(\cos\theta_1-1)\sin\phi_2\sin\theta_2\right\}$$

$$\left.\times d\theta_1 d\theta_2 d\phi_1 d\phi_2\right\},$$

(79)

$$W_{yy}(r_1,\varphi_1,r_2,\varphi_2,z)=\frac{f^2 n_1}{\lambda^2}\left\{\int_0^{\theta_{max}}\int_0^{\theta_{max}}\int_0^{2\pi}\int_0^{2\pi}\sqrt{\cos\theta_1\cos\theta_2}\sin\theta_1\sin\theta_2\right.$$

$$\times\exp[-ik_1(z\cos\theta_2+r_2\sin\theta_2\cos(\phi_2-\varphi_2))]$$

$$\times\exp[ik_1(z\cos\theta_1+r_1\sin\theta_1\cos(\phi_1-\varphi_1))]$$

$$\left\{W_{xx}(\theta_1,\phi_1,\theta_2,\phi_2)\cos\phi_1\sin\phi_1(\cos\theta_1-1)\right.$$

$$\times\cos\phi_2\sin\phi_2(\cos\theta_2-1)+W_{xy}(\theta_1,\phi_1,\theta_2,\phi_2)$$

$$\times\cos\phi_1\sin\phi_1(\cos\theta_1-1)\left[\cos\theta_2+\cos^2\phi_2(1-\cos\theta_2)\right]$$

$$+W_{yx}(\theta_1,\phi_1,\theta_2,\phi_2)\left[\cos\theta_1+\cos^2\phi_1(1-\cos\theta_1)\right]$$

$$\times\cos\phi_2\sin\phi_2(\cos\theta_2-1)+W_{yy}(\theta_1,\phi_1,\theta_2,\phi_2)$$

$$\times\left[\cos\theta_1+\cos^2\phi_1(1-\cos\theta_1)\right]$$

$$\left.\times\left[\cos\theta_2+\cos^2\phi_2(1-\cos\theta_2)\right]\right\}d\theta_1 d\theta_2 d\phi_1 d\phi_2\Bigg\},$$

$$(80)$$

$$W_{yz}(r_1,\varphi_1,r_2,\varphi_2,z_2)=-\frac{f^2 n_1}{\lambda^2}\left\{\int_0^{\theta_{max}}\int_0^{\theta_{max}}\int_0^{2\pi}\int_0^{2\pi}\sqrt{\cos\theta_1\cos\theta_2}\sin\theta_1\sin\theta_2\right.$$

$$\times\exp[-ik_1(z_2\cos\theta_2+r_2\sin\theta_2\cos(\phi_2-\varphi_2))]$$

$$\times\exp[ik_1(z\cos\theta_1+r_1\sin\theta_1\cos(\phi_1-\varphi_1))]$$

$$\left\{W_{xx}(\theta_1,\phi_1,\theta_2,\phi_2)\cos\phi_1\sin\phi_1(\cos\theta_1-1)\cos\phi_2\sin\theta_2\right.$$

$$+W_{xy}(\theta_1,\phi_1,\theta_2,\phi_2)\cos\phi_1\sin\phi_1(\cos\theta_1-1)\sin\phi_2\sin\theta_2$$

$$+W_{yx}(\theta_1,\phi_1,\theta_2,\phi_2)\left[\cos\theta_1+\cos^2\phi_1(1-\cos\theta_1)\right]$$

$$\times\cos\phi_2\sin\theta_2+W_{yy}(\theta_1,\phi_1,\theta_2,\phi_2)$$

$$\times\left[\cos\theta_1+\cos^2\phi_1(1-\cos\theta_1)\right]$$

$$\left.\times\sin\phi_2\sin\theta_2\right\}d\theta_1 d\theta_2 d\phi_1 d\phi_2\Bigg\},$$

$$(81)$$

$$W_{zz}(r_1,\varphi_1,r_2,\varphi_2,z) = \frac{f^2 n_1}{\lambda^2}\left\{\int_0^{\theta_{max}}\int_0^{\theta_{max}}\int_0^{2\pi}\int_0^{2\pi}\sqrt{\cos\theta_1\cos\theta_2}\sin\theta_1\sin\theta_2\right.$$

$$\times\exp\left[-ik_1(z\cos\theta_2 + r_2\sin\theta_2\cos(\phi_2-\varphi_2))\right]$$

$$\times\exp\left[ik_1(z\cos\theta_1 + r_1\sin\theta_1\cos(\phi_1-\varphi_1))\right]$$

$$\left[W_{xx}(\theta_1,\phi_1,\theta_2,\phi_2)\cos\phi_1\sin\theta_1\cos\phi_2\sin\theta_2\right.$$

$$+W_{xy}(\theta_1,\phi_1,\theta_2,\phi_2)\cos\phi_1\sin\theta_1\sin\phi_2\sin\theta_2$$

$$+W_{yx}(\theta_1,\phi_1,\theta_2,\phi_2)\sin\phi_1\sin\theta_1\cos\phi_2\sin\theta_2$$

$$\left.+W_{yy}(\theta_1,\phi_1,\theta_2,\phi_2)\sin\phi_1\sin\theta_1\sin\phi_2\sin\theta_2\right]$$

$$\left.\times d\theta_1 d\theta_2 d\phi_1 d\phi_2\right\},$$

(82)

$$W_{yx}(r_1,\varphi_1,r_2,\varphi_2,z) = W_{xy}^*(r_2,\varphi_2,r_1,\varphi_1,z),$$

(83)

$$W_{zx}^{(ee)}(r_1,\varphi_1,r_2,\varphi_2,z) = W_{xz}^{(ee)*}(r_2,\varphi_2,r_1,\varphi_1,z),$$

(84)

$$W_{zy}(r_1,\varphi_1,r_2,\varphi_2,z) = W_{yz}^*(r_2,\varphi_2,r_1,\varphi_1,z).$$

(85)

The total intensity distribution of the tightly focused beam in the focal region is given by

$$I(r,\varphi,z) = I_x(r,\varphi,z) + I_y(r,\varphi,z) + I_z(r,\varphi,z)$$

$$= W_{xx}(r,\varphi,r,\varphi,z) + W_{yy}(r,\varphi,r,\varphi,z) + W_{zz}(r,\varphi,r,\varphi,z).$$

(86)

The element of the matrix the 2×2 cross-spectral density matrix of the partially coherent azimuthally polarized is expressed as

$$W_{xx}(x_1,y_1,x_2,y_2,0) = \frac{y_1 y_2}{w_0^2}\exp\left[-\frac{x_1^2 + y_1^2 + x_2^2 + y_2^2}{w_0^2} - \frac{(x_1-x_2)^2 + (y_1-y_2)^2}{2\sigma_{xx}^2}\right],$$

(87)

$$W_{xy}(x_1,y_1,x_2,y_2,0) = -B_{xy}\frac{y_1 x_2}{w_0^2}$$

$$\times\exp\left[-\frac{x_1^2 + y_1^2 + x_2^2 + y_2^2}{w_0^2} - \frac{(x_1-x_2)^2 + (y_1-y_2)^2}{2\sigma_{xy}^2}\right],$$

(88)

$$W_{yx}(x_1,y_1,x_2,y_2,0)=W_{xy}^*(x_2,y_2,x_1,y_1,0), \qquad (89)$$

$$W_{yy}(x_1,y_1,x_2,y_2,0)=\frac{x_1 x_2}{w_0^2}\exp\left[-\frac{x_1^2+y_1^2+x_2^2+y_2^2}{w_0^2}-\frac{(x_1-x_2)^2+(y_1-y_2)^2}{2\sigma_{yy}^2}\right].$$
$$(90)$$

Eqs. (87)-(90) can be expressed in the following alternative form

$$W_{xx}(\theta_1,\phi_1,\theta_2,\phi_2)=\frac{f^2\sin\theta_1\sin\theta_2}{w_0^2}$$
$$\times\exp\left[\begin{array}{c}-f^2\dfrac{\sin^2\theta_1+\sin^2\theta_2}{w_0^2}\\[2mm]-f^2\dfrac{\sin^2\theta_1+\sin^2\theta_2-2\sin\theta_1\sin\theta_2\cos(\phi_1-\phi_2)}{2\sigma_{xx}^2}\end{array}\right]$$
$$\times\sin\phi_1\sin\phi_2,$$
$$(91)$$

$$W_{xy}(r_1,\phi_1,r_2,\phi_2)=-B_{xy}\frac{f^2\sin\theta_1\sin\theta_2}{w_0^2}$$
$$\times\exp\left[\begin{array}{c}-f^2\dfrac{\sin^2\theta_1+\sin^2\theta_2}{w_0^2}\\[2mm]-f^2\dfrac{\sin^2\theta_1+\sin^2\theta_2-2\sin\theta_1\sin\theta_2\cos(\phi_1-\phi_2)}{2\sigma_{xy}^2}\end{array}\right]$$
$$\times\sin\phi_1\cos\phi_2,$$
$$(92)$$

$$W_{yx}(\theta_1,\phi_1,\theta_2,\phi_2)=W_{xy}^*(\theta_2,\phi_2,\theta_1,\phi_1), \qquad (93)$$

$$W_{yy}(\theta_1,\phi_1,\theta_2,\phi_2) = \frac{f^2 \sin\theta_1 \sin\theta_2}{w_0^2}$$

$$\times \exp\left[\begin{array}{l} -f^2 \dfrac{\sin^2\theta_1 + \sin^2\theta_2}{w_0^2} \\[2ex] -f^2 \dfrac{\sin^2\theta_1 + \sin^2\theta_2 - 2\sin\theta_1 \sin\theta_2 \cos(\phi_1 - \phi_2)}{2\sigma_{yy}^2} \end{array} \right]$$

$$\times \cos\phi_1 \cos\phi_2.$$

$$(94)$$

Applying Eqs. (76)-(86) and (91)-(94), one can study the tight focusing properties of a partially coherent azimuthally polarized beam.

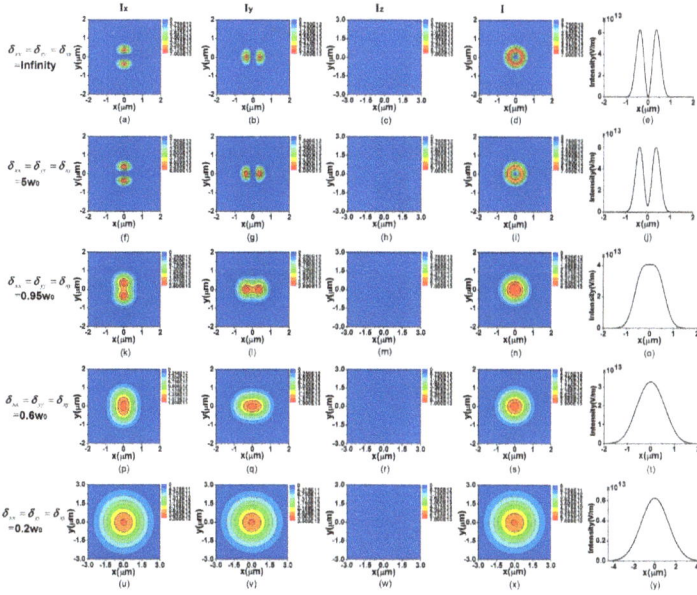

Fig. 24. Intensity distributions I_x, I_y, I_z, I and the corresponding cross line of the total intensity distribution I of a tightly focused partially coherent azimuthally polarized beam at the focal plane for different values of the correlation coefficients δ_{xx}, δ_{yy}, δ_{xy} with $f = 10$mm, $n_1 = 1.33$, $w_0 = 5$mm, and $B_{xy} = 1$.

One finds from Fig. 24 that the intensity distribution I_z of a coherent or partially coherent tightly focused azimuthally polarized beam equals to zero, while I_x, I_y and I of a tightly focused azimuthally polarized beam depend closely on the initial correlation coefficients (i.e., spatial coherence). For a completely coherent azimuthally polarized beam ($\sigma_{xx} = \sigma_{xy} = \sigma_{yy}$ = Infinity), I_x and I_y have two beamlets along y-and x-directions, respectively, and I has a circular dark hollow beam profile. With the decrease of correlation coefficients, the two beamlets structure of the intensity distributions I_x and I_y gradually disappear, and the dark hollow beam profile of I also disappears gradually. For certain values of correlation coefficients ($\delta_{xx} = \delta_{xy} = \delta_{yy} = 0.95w_0$), I_x and I_y have beam spots with "dumbbell" structure, and I have a circular flat-topped beam profile. For certain values of correlation coefficients ($\delta_{xx} = \delta_{xy} = \delta_{yy} = 0.6w_0$), I_x and I_y have elliptical beam spots, and I has a Gaussian beam spot. When the correlation coefficients are very small ($\delta_{xx} = \delta_{xy} = \delta_{yy} = 0.2w_0$), I_x, I_y and I all have circular Gaussian beam spots.

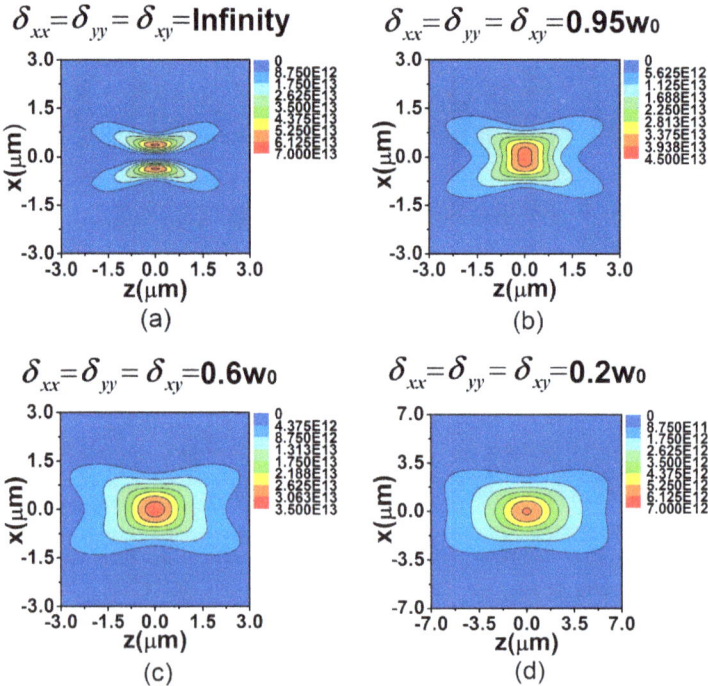

Fig. 25. Intensity distribution I of a tightly focused partially coherent azimuthally polarized beam in the xz plane near focus for different values of the correlation coefficients δ_{xx}, δ_{yy}, δ_{xy}.

One finds from Fig. 25 that a dark channel is formed in the xz plane for a coherent tightly focused azimuthally polarized beam. With the decrease of the correlation coefficients, the dark channel disappears gradually, and a bright beam spot in xz plane is formed. The bright beam spot is of rectangular symmetry when the correlation coefficients are not very small, and the bright beam spot is of elliptical symmetry when the correlation coefficients are very small. Thus, decreasing the spatial coherence of an azimuthally polarized beam provides a convenient way for shaping its tightly focused beam spot. Our results will be useful for material thermal processing, where a tightly focused flat-topped beam spot is required,[136] and for particle trapping, where a tightly focused dark hollow beam spot or bright beam spot is required.[3] More information about the radiation forces of a tightly focused partially coherent azimuthally polarized beam on a Rayleigh particle can be found in Ref. 11. Spectral stokes singularities of tightly focused partially coherent radially polarized beam were explored in Ref. 137. The propagation properties of partially coherent radially polarized beam in turbulent atmosphere and in uniaxial crystals can be found in Ref. 138 and Ref. 139, respectively.

6. Partially coherent vector beams with non-uniform state of polarization: experiment

In this section, we will introduce experimental generation of a partially coherent azimuthally or radially polarized beam with controllable spatial coherence.

Part 1 of Fig. 26 shows the experimental setup for generating a partially coherent azimuthally polarized beam with controllable spatial coherence. A linearly polarized beam generated by a He-Ne laser is focused by a thin lens L_1, then it illuminates a rotating ground-glass disk (RGGD), producing a partially coherent beam with Gaussian statistics. After passing through a collimation lens L_2 and a Gaussian amplitude filter (GAF), the generated partially coherent beam becomes a linearly polarized GSM beam. After passing through a liquid crystal polarization converter (Acroptix, Switzerland), the generated GSM beam becomes a partially coherent azimuthally polarized beam. In the experiment, the liquid crystal polarization located just behind the GAF is used to convert the generated GSM beam into a partially coherent azimuthally polarized beam, it just modulate the state of polarization of the GSM beam, while it doesn't alter its spatial coherence, thus the correlation coefficients of the generated partially coherent azimuthally polarized beam are approximated as $\delta_{xx} = \delta_{yy} = \delta_{xy} = \delta_0$ with δ_0 being the coherence width of the generated GSM beam . The parameter w_0 is determined by the GAF. δ_0 can be measured by the procedure described in 4.1.

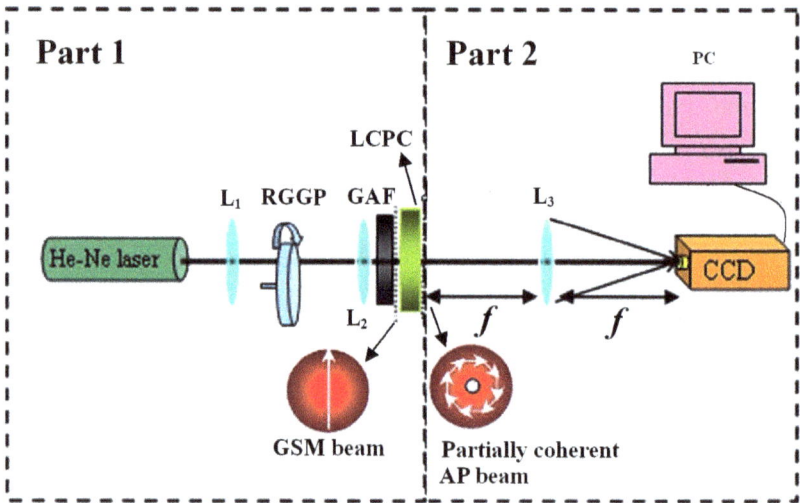

Fig. 26. Experimental setup for generating a partially coherent azimuthally polarized beam and measuring its intensity distribution after passing through a thin lens. L_1, L_2, L_3, thin lenses; RGGP, rotating ground-glass plate; GAF, Gaussian amplitude filter; LCPC, liquid crystal polarization converter; CCD, charge-coupled device; PC, personal computer.

Part 2 of Fig. 26 shows the setup for measuring the intensity of a partially coherent azimuthally polarized beam at the focal plane after passing through a thin lens L_3 with focal length $f = 40$ cm. The distance from liquid crystal polarization converter to L_3 and the distance from L_3 to CCD both equal to f. With the measured beam parameters and the propagation formulae,[11] we can simulate the focusing properties of a partially coherent azimuthally polarized beam, and carry out comparison with experimental results. In a similar way, we can generate a partially coherent radially polarized beam by adjusting the liquid crystal polarization converter and measure its focused intensity.

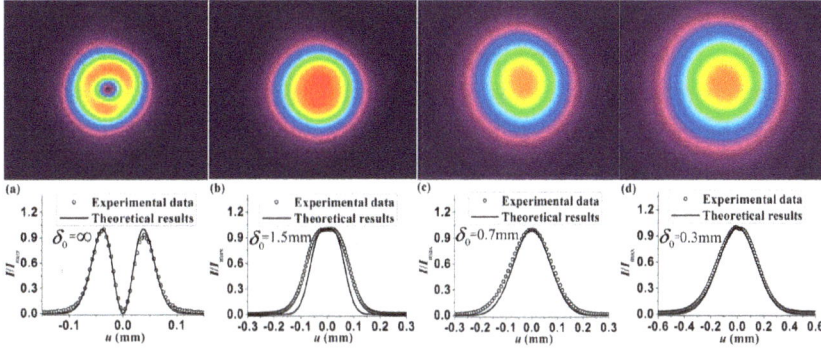

Fig. 27. Experimental results of the intensity distribution of the focused partially coherent azimuthally polarized beam and the corresponding cross line (dotted curve) at the focal plane for different values of δ_0. The solid curves are calculated by the theoretical formulae.

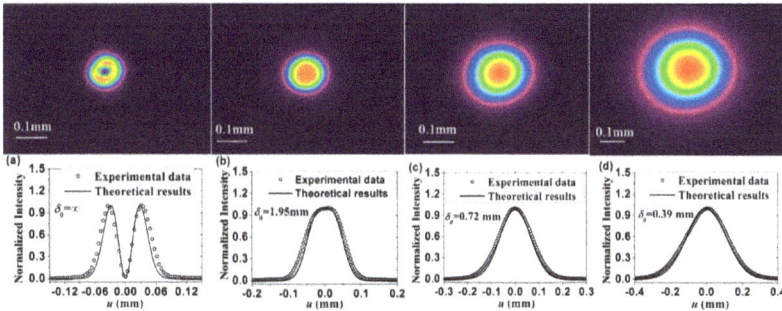

Fig. 28. Experimental results of the intensity distribution of the focused partially coherent radially polarized beam and the corresponding cross line (dotted curve) at the focal plane for different values of δ_0. The solid curves are calculated by the theoretical formulae.

Figures 27 and 28 show the experimental results of the intensity distributions of the focused partially coherent azimuthally and radially polarized beams and corresponding cross lines, respectively. The results calculated by theoretical formulae are also shown. One finds from

Figs. 27 and 28 that the focused beam spots of coherent azimuthally and radially polarized beams have dark hollow beam profiles as expected. With the decrease of δ_0 (i.e., spatial coherence), the focused beam spots gradually transform from dark hollow beam profiles into Gaussian beam profiles. For suitable value of δ_0, the focused beam spots with flat-topped beam profiles can be formed. The experimental results agree well with the theoretical predictions. Thus, decreasing the spatial coherence of an azimuthally or raidally polarized beam indeed provides a convenient way for shaping its beam profile. More detailed information about generation and measurement of partially coherent radially and azimuthally polarized beams can be found in Refs. 10, 11 and 140. More recently, experimental study of the scintillation index of a partially coherent RP beam propagating through thermally induced turbulence was reported,[141] and it was found that a partially coherent RP beam has advantage over a linearly polarized partially coherent beam for reducing turbulence-induced scintillation, which will be useful in free-space optical communications.

7. Summary

Recent developments on partially coherent vector beams with uniform or non-uniform state of polarization are reviewed. Partially coherent vector beams have displayed many unique interesting properties, which are useful in many applications, such as free-space optical communication, material thermal processing, remote sensing, particle trapping and optical imaging. For partially coherent vector beam with uniform state of polarization, numerous theoretical papers have been published on such beam, while only few experimental papers were reported on such beams. For partially coherent vector beam with non-uniform state of polarization, only few theoretical and experimental papers have been published. Further studies in this area may focus on exploring new properties and applications of partially coherent vector beams both theoretically and experimentally.

Acknowledgments

This work is supported by the National Natural Science Foundation of China under Grant Nos. 11274005, 0904102, 61008009, and 11104195, the Foundation for the Author of National Excellent Doctoral Dissertation of PR China under Grant No. 200928, the Huo Ying Dong Education Foundation of China under Grant No. 121009, the Key Project of Chinese Ministry of Education under Grant No. 210081, the Universities Natural Science Research Project of Jiangsu Province Grant Nos. 10KJB140011 and 11KJB140007, the Project Funded by the Priority Academic Program Development of Jiangsu Higher Education Institutions, the support by the Innovation Plan for Graduate Students in the Universities of Jiangsu Province under Grant Nos. CXLX11_0064 and CXLX12_0780, and the Project Sponsored by the Scientific Research Foundation for the Returned Overseas Chinese Scholars, State Education Ministry.

References

1. L. Mandel and E. Wolf, *Optical Coherence and Quantum Optics* (Cambridge U. Press,1995).
2. C. Brosseau, Fundamentals of polarized light-a statistical approach (Wiley, New York, 1998).
3. Q. Zhan, *Adv. Opt. Photon.* **1**, 1 (2009).
4. D. James, *J. Opt. Soc. Am. A* **11**, 1641 (1994).
5. E. Wolf, *Phys. Lett. A* **312**, 263 (2003).
6. E. Wolf, *Opt. Lett.* **28**, 1078 (2003).
7. E. Wolf, *Introduction to the theory of coherence and polarization of light* (Cambridge U. Press, 2007).
8. F. Gori, *Opt. Lett.* **23**, 241 (1998).
9. Y. Dong, Y. Cai, C. Zhao, and M. Yao, *Opt. Express* **19**, 5979 (2011).
10. F. Wang, Y. Cai, Y. Dong, and O. Korotkova, *Appl. Phys. Lett.* **100**, 051108 (2012).
11. Y. Dong, F. Wang, C. Zhao, and Y. Cai, *Phys. Rev. A* **86**, 013840 (2012).
12. S. N. Volkov, D. James, T. Shirai, and E. Wolf, *J. Opt. A: Pure Appl. Opt.* **10**, 055001 (2008).
13. O. Korotkova, *J. Opt. A: Pure Appl. Opt.* **8**, 30 (2006).
14. T. Shirai and E. Wolf, *Opt. Commun.* **272**, 289 (2007).
15. O. Korotkova and E. Wolf, *Opt. Comm.* **246**, 35 2005).
16. O. Korotkova and E. Wolf, *Opt. Lett.* **30**, 198 (2005).
17. T. Setälä, M. Kaivola, and A. T. Friberg, *Phys. Rev. Lett.* **88**, 123902 (2002).

18. T. Setälä, A. Shevchenko, M. Kaivola, and A. T. Friberg, *Phys. Rev. E* **66**, 016615 (2002).
19. K. Lindfors, A. Priimagi, T. Setälä, A. Shevchenko, and A. T. Friberg, *Nat. Photonics* **1**, 228 (2007).
20. K. Lindfors, T. Setälä, M. Kaivola, and A. T. Friberg, *J. Opt. Soc. Am. A* **22**, 561 (2005).
21. T. Setälä, K. Lindfors, and A. T. Friberg, *Opt. Lett.* **34**, 3394 (2009).
22. J. Ellis, A. Dogariu, S. Ponomarenko, and E. Wolf, *Opt. Commun.* **248**, 333 (2005).
23. O. Korotkova and E Wolf, *J. Opt. Soc. Am. A* **21**, 2382 (2004).
24. A. Luis, *Phys. Rev. A* **71**, 063815 (2005).
25. K. Duan and B. Lu, *J. Opt. Soc. Am. A* **21**, 1924 (2004).
26. L. Zhang and Y. Cai, *Opt. Express* **19**, 13312 (2011).
27. Y. Dong, F. Feng, Y. Chen, C. Zhao, and Y. Cai, *Opt. Express* **20**, 15908 (2012).
28. Z. Mei, *Opt. Express* **18**, 22826 (2010).
29. F. Gori, M. Santarsiero, G. Piquero, R. Borghi, A. Mondello, and R. Simon, *J. Opt. A: Pure Appl. Opt.* **3**, 1 (2001).
30. O. Korotkova, M. Salem, and E. Wolf, *Opt. Lett.* **29**, 1173 (2004).
31. H. Wang, X. Wang, A. Zeng, and K. Yang, *Opt. Lett.* **32**, 2215 (2007).
32. Y. Cai and O. Korotkova, *Appl. Phys. B* **96**, 499 (2009).
33. H. Roychowdhury and O. Korotkova, *Opt. Commun.* **249**, 379 (2005).
34. F. Gori, M. Santarsiero, R. Borghi, and V. Ramírez-Sánchez, *J. Opt. Soc. Am. A* **25**, 1016 (2008).
35. J. A. Arnaud, "Hamiltonian theory of beam mode propagation in *Progress in Optics,* Vol 11, E. Wolf, ed. (North-Holland, 1973), pp. 247–304.
36. J. A. Arnaud, "Nonorthogonal optical waveguides and resonators," Bell Syst. Tech. J. **49**, 2311 (1970).
37. Q. Lin and Y. Cai, *Opt. Lett.* **27**, 216 (2002).
38. Q. Lin and Y. Cai, *Opt. Lett.* **27**, 1672 (2002).
39. Y. Cai and L. Hu, *Opt. Lett.* **31**, 685 (2006).
40. Y. Cai and S. He, *Appl. Phys. Lett.* **89**, 041117 (2006).
41. Y. Cai, Q. Lin, H. T. Eyyuboğlu, and Y. Baykal, *Opt. Commun.* **278,** 157 (2007).
42. Y. Cai and U. Peschel, *Opt. Express* **15**, 15480 (2007).
43. Y. Cai, Q. Lin, and O. Korotkova, *Opt. Express* **17**, 2450 (2009).
44. Y. Cai, D. Ge, and Q. Lin, *J. Opt. A: Pure Appl. Opt.* **5**, 453 (2003).
45. D. Ge, Y. Cai, and Q. Lin, *Chin. Phys.* **14**, 128 (2005).
46. M. Yao, Y. Cai, H. T. Eyyuboğlu, Y. Baykal, and O. Korotkova, *Opt. Lett.* **33**, 2266 (2008).
47. O. Korotkova, M. Yao, Y. Cai, H. T. Eyyuboğlu, and Y. Baykal, *J. Opt. Soc. Am. A* **25**, 2710 (2008).
48. Y. Cai, O. Korotkova, H. T. Eyyuboğlu, and Y. Baykal, *Opt. Express* **16**, 15834 (2008).
49. C. Zhao, Y. Cai, and O. Korotkova, *Opt. Express* **17**, 21472 (2009).

50. O. Korotkova, Y. Cai, and E. Watson, *Appl. Phys. B* **94**, 681 (2009)
51. M. Yao, Y. Cai, O. Korotkova, Q. Lin, and Z. Wang, *Opt. Express* **18**, 22503 (2010).
52. Z. Tong, Y. Cai, and O. Korotkova, *Opt. Commun.* **283**, 3838 (2010).
53. T. Shirai, *Opt. Commun.* **256**, 197 (2005).
54. O. Korotkova, B. G. Hoover, V. L. Gamiz, and E. Wolf, *J. Opt. Soc. Am. A* **22**, 2547 (2005).
55. M. Salem and E. Wolf, *Opt. Lett.* **33**, 1180 (2008).
56. O. Korotkova, T. D. Visser, and E. Wolf, *Opt. Commun.* **281**, 515 (2008).
57. G. Wu and Y. Cai, *Opt. Express* **19**, 8700 (2011).
58. A. G. Fox and T. Li, *Bell. Syst. Tech. J.* **40**, 453 (1961).
59. E. Wolf, *Phys. Lett.* **3**, 166 (1963).
60. E. Wolf and G. S. Agarwal, *J. Opt. Soc. Am. A* **1**, 541 (1984).
61. F. Gori, *Atti Fond. Giorgio Ronchi* **35**, 434 (1980).
62. P. DeSantis, A. Mascello, C. Palma, and M. R. Perrone, *IEEE J. Quantum Electron.* **32**, 802 (1996).
63. C. Palma, G. Cardone, and G. Cincotti, *IEEE J. Quantum Electron.* **34**, 1082 (1998).
64. E. Wolf, *Opt. Commun.* **265**, 60-62 (2006).
65. T. Saastamoinen, J. Turunen, J. Tervo, T. Setala, and A. T. Friberg, *J. Opt. Soc. Am. A* **22**, 103-108 (2005).
66. Z. Tong, O. Korotkova, Y. Cai, H. T. Eyyuboğlu, and Y. Baykal, *Appl. Phys. B* **97**, 849-857 (2009).
67. S. Zhu and Y. Cai, *Opt. Express* **18**, 27567 (2010).
68. M. Yao, Y. Cai, and O. Korotkova, *Opt. Commun.* **283**, 4505 (2010)
69. S. Zhu, F. Zhou, Y. Cai, and L. Zhang, *Appl. Phys. B* **102**, 953 (2011).
70. H. Kogelnik and T. Li, *Appl. Opt.* **5**, 1550 (1966).
71. A. Ishimaru, *Wave propagation and scattering in random media*, (Academic Press, New York, 1978) Vol. 2.
72. L. C. Andrews, R. L. Phillips, and C. Y. Hopen, *Laser Beam Scintillation with Applications* (SPIE Press, Washington, 2001).
73. Y. Cai and S. He, *Opt. Lett.* **31**, 568 (2006).
74. Y. Cai and S. He, *Opt. Express* **14**, 1353 (2006).
75. Y. Cai, Y. Chen, H. T. Eyyuboğlu, and Y. Baykal, *Opt. Lett.* **32**, 2405 (2007).
76. Y. Cai, Q. Lin, H. T. Eyyuboğlu, and Y. Baykal, *Opt. Express* **16**, 7665 (2008).
77. Y. Yuan, Y. Cai, J. Qu, H. T. Eyyuboğlu, Y. Baykal, and O. Korotkova, *Opt. Express* **17**, 17344 (2009).
78. F. Wang and Y. Cai, *Opt. Express* **18**, 24661 (2010).
79. G. Wu and Y. Cai, *Opt. Lett.* **36**, 1939 (2011).
80. F. Wang, Y. Cai, H. T. Eyyuboğlu, and Y. Baykal, *Opt. Lett.* **37**, 184 (2012).
81. O. Korotkova, M. Salem, and E. Wolf, *Opt. Commun.* **233**, 225 (2004).
82. M. Salem, O. Korotkova, A. Dogariu, and E. Wolf, *Waves in Random Media* **14**, 513 (2004).
83. O. Korotkova, M. Salem, A. Dogariu, and E. Wolf, *Waves Random Complex Media* **15**, 353 (2005).

84. H. Roychowdhury, S. A. Ponomarenko, and E. Wolf, *J. Mod. Opt.* **52**, 1611 (2005).

85. O. Korotkova, *Opt. Commun.* **281**, 2342 (2008).

86. X. Du, D. Zhao, and O. Korotkova, *Opt. Express* **15**, 16909 (2007).

87. W. Lu, L. Liu, J. Sun, Q. Yang, and Y. Zhu, *Opt. Commun.* **271**, 1 (2007).

88. S. Zhu, Y. Cai, and O. Korotkova, *Opt. Express* **18**, 12587 (2010).

89. E. Shchepakina and O. Korotkova, *Opt. Express* **18**, 10650 (2010).

90. S. Zhu and Y. Cai, *Appl. Phys. B* **103**, 971 (2011).

91. H. T. Yura and S. G. Hanson, *J. Opt. Soc. Am A* **4**, 1931 (1987).

92. H. T. Yura and S. G. Hanson, *J. Opt. Soc. Am A* **6**, 564 (1989).

93. S. Sahin, Z. Tong, and O. Korotkova, *Opt. Commun.* **283**, 4512 (2010).

94. V. Namias, *J. Inst. Math. Appl.* **25**, 241 (1973).

95. A. W. Lohmann, *J. Opt. Soc. Am. A* **10**, 2181 (1993).

96. D. Mendlovic and H. M. Ozaktas, *J. Opt. Soc. Am. A* **10**, 1875 (1993).

97. A. Torre, in *Progress in Optics Vol. XLIII*, E. Wolf, ed. (Elsevier, Amsterdam, 2002).

98. Y. Cai, Q. Lin, and S. Zhu, *Appl. Phy. Lett.* **86**, 021112 (2005).

99. Y. Cai and S. Zhu, *J. Opt. Soc. Am. A* **22**, 1798 (2005).

100. T. B. Pittman, Y. H. Shih, D. V. Strekalov, and A. V. Sergienko, *Phys. Rev. A* **52**, R3429 (1995).

101. D. V. Strekalov, A. V. Sergienko, D. N. Klyshko, and Y. H. Shih, *Phys. Rev. Lett.* **74**, 3600 (1995).

102. A. Gatti, E. Brambilla, M. Bache, and L. A. Lugiato, *Phys. Rev. Lett.* **93**, 093602 (2004).

103. J. Cheng and S. Han, *Phys. Rev. Lett.* **92**, 093903 (2004).

104. Y. Cai and S. Zhu, *Opt. Lett.* **29**, 2716 (2004).

105. Y. Cai and S. Zhu, *Phys. Rev. E* **71**, 056607 (2005).

106. A. Valencia, G. Scarcelli, M. D'Angelo, and Y. Shih, *Phys. Rev. Lett.* **94**, 063601 (2005).

107. F. Ferri, D. Magatti, A. Gatti, M. Bache, E. Brambilla, and L. A. Lugiato, *Phys. Rev. Lett.* **94**, 183602 (2005).

108. Y. Cai, Q. Lin, and S. Zhu, *J. Opt. Soc. Am. A* **23**, 835 (2006).

109. Y. Cai and F. Wang, *Opt. Lett.* **31**, 2278 (2006).

110. F. Wang, Y. Cai, and S. He, *Opt. Express* **14**, 6999 (2006).

111. F. Wang, S. Zhu, X. Hu, and Y. Cai, *Opt. Commun.* **284**, 5275 (2011).

112. O. Gawhary and S. Severini, *Opt. Lett.* **33**, 1360 (2008).

113. O. Gawhary and S. Severini, *Opt. Lett.* **33**, 1866 (2008).

114. O. Gawhary and S. Severini, *Opt. Commun.* **283**, 2481, (2010).

115. F. Wang, Y. Cai, and O. Korotkova, *J. Opt. Soc. Am. A* **27**, 1120 (2010).

116. L. Zhang, F. Wang, Y. Cai, and O. Korotkova, *Opt. Commun.* **284**, 1111 (2011).

117. G. Piquero, F. Gori, P. Romanini, M. Santarsiero, R. Borghi, and A. Mondello, *Opt. Commun.* **208** 9 (2002).

118. T. Shirai, O. Korotkova, and E. Wolf, *J. Opt. A, Pure Appl. Opt.* **7**, 232 (2005).

119. M. Santarsiero, R. Borghi, and V. Ramirez-Sanchez, *J. Opt. Soc. Am. A* **26**, 1437 (2009).
120. A. S. Ostrovsky, G. Rodríguez-Zurita, C. Meneses-Fabián, M. Á. Olvera-Santamaría, and C. Rickenstorff-Parrao, *Opt. Express* **18**, 12864 (2010).
121. A. S. Ostrovsky, M. A. Olvera, C. Richenstorff, G. Martinez-Niconoff, and V. Arrizon, *Opt. Commun.* **283**, 4490 (2010).
122. B. Kanseri, S. Rath, and H. C. Kandpal, *IEEE J. Quantum Electron.* **45**, 1163 (2009).
123. B. Kanseri, S. Rath, and H. C. Kandpal, *Opt. Commun.* **282**, 3059 (2009).
124. B. Kanseri and H. C. Kandpal, *Opt. Lett.* **33**, 2410 (2008).
125. I. Vidal, E. J. S. Fonseca, and J. M. Hickmann, *Phys. Rev. A* **84**, 033836 (2011).
126. F. Wang, G. Wu, X. Liu, S. Zhu, and Y. Cai, *Opt. Lett.* **36**, 2722 (2011).
127. F. Wang and Y. Cai, *J. Opt. Soc. Am. A* **24**, 1937 (2007).
128. L. C. Cohen, *Bell Syst. Tech. J.* **51**, 573 (1972).
129. P. Winzer and W. Leeb, *Opt. Lett.* **23**, 986 (1998).
130. D. J. Wheeler and J. D. Schmidt, *J. Opt. Soc. Am. A* **28**, 1224 (2011).
131. M. Salem and G. P. Agrawal, *Opt. Lett.* **34**, 2829 (2009).
132. M. Salem and G. P. Agrawal, *J. Opt. Soc. Am. A* **26**, 2452 (2009).
133. M. Salem and G. P. Agrawal, *J. Opt. Soc. Am. A* **28**, 307 (2011).
134. C. Zhao, Y. Dong, G. Wu, F. Wang, Y. Cai, and O. Korotkova, *Appl. Phys. B* **108**, 891 (2012).
135. B. Richards and E. Wolf, *Proc. R. Soc. A* **253**, 358 (1959).
136. D. W. Coutts, *IEEE J. Quantum Electron.* **38**, 1217 (2002).
137. Y. Luo and B. Lu, *J. Opt.* **12**, 115703 (2010).
138. H. Wang, D. Liu, and Z. Zhou, *Appl. Phys. B* **101**, 361 (2010).
139. L. Guo, Z. Tang, C. Liang, and Z. Tan, *Eur. Phys. J. Appl. Phys.* **52**, 31301 (2010).
140. G. Wu, F. Wang, and Y. Cai, *Opt. Express* **20**, 28301 (2012).
141. F. Wang, X. Liu, Y. Yuan, and Y. Cai, *Appl. Phys. Lett.* **103**, 091102 (2013).

INDEX

3D polarization 134, 139–141, 145

Archimedes spiral slot.....................183
Atomic Spin Analyzer.....................217
azimuthal polarization........................4
azimuthally polarized doughnut
    modes (APDM) 74

beam purity 148
Berry's phase 9
Bessel-Gauss.......................................4
bull's eye................................. 172, 193
Bull's eye structures........................169

cascaded $\lambda/2$-plates................. 203, 204
CdSe/ZnS quantum dots..................108
circular polarization extinction ratio188
circular slot antenna 182
collection efficiency function (CEF).81
cross-spectral density matrix ... 223, 224
cylindrical vector beams (CVBs) .. 1, 74

diffractive optical element (DOE)... 125
dipole antenna.................................142

electric dipole array.........................147
electric dipole radiation............. 22, 134
electromagnetic Gaussian Schell-model
    (EGSM) 227
evanescent Bessel beam 168, 169
excited-state tautomerization 99
extraordinary optical transmission
    (EOT) 177

few-mode fiber..................................11
flattop focus 127
flattop focusing 125, 129, 130–132
four-segment mode converter 76

geometric phase 9, 189
GNRs 112, 113
gold nanorods (GNR).......................111
Gouy phase 3, 22

Helmholtz condition objective 19
Helmholtz equation.............................2
Hermite-Gauss mode...........................2
Herschel condition 18
Hybrid spiral plasmonic lens...........189

Kretschmann configuration.............163

Laguerre-Gauss mode 3
LHS plasmonic lens 184, 187
LHS plasmonic lens array 187

magnetic dipole array.............. 146, 153
miniature circular polarization
    analyzers 186
Mueller matrix polarimetry (MMP) 216

nanoantenna 178
near-field scanning optical microscope
    (NSOM).....................................167
non-mechanical polarization
    rotator 204, 205
notational symmetry........................207
nulling detection......................211, 213
nulling microellipsometer 211–213
numerical aperture (NA) 2, 16, 17

optical antenna 178
optical bubble...................................132
optical cage 45, 46, 48, 50
optical chain............................155, 157
optical needle.......... 147, 148, 149, 151
optical tube......................................153

orbital angular momentum
(OAM) 63–65

Pancharatnam's phase 9
parabolic mirror (PM) 81
Partially Coherent Vector Beams 227
plasmonic lens 163, 169
Poincaré sphere 30, 31
polarization mode matching 182
polarization rotator (PR) ... 11, 203–205
pupil apodization function 18

radial polarization 6
radial polarization interferometer
(RPI) 214
ray projection function 18
Richards-Wolf vector diffraction
theory 16, 46, 77, 130, 168
rotational symmetry 206, 207, 211
rotationally symmetric
microellipsometer 207–210
Radially Polarized Donut Modes
(RPDM) 75

scalar optical field 28
scanning near-field optical microscopy
(SNOM) 111
segmented λ/2-plate 8

segments of λ/2 plates 76
Si nanocrystals 108
silicon nanocrystals 106
sine condition objective 18
single quantum absorber 82
single quantum system 81, 84
SiO$_2$ nanoparticles (NPs) 103, 104
spatial light modulators (SLM) . 11, 133
spectral degree of coherence 223
spectral degree of cross-polarization
.... 224
spectral degree of polarization 223
spherical spot 141, 142, 144–146
spin angular momentum (SAM) 29
spin-orbit interaction 185
spiral plasmonic lens 182–184, 186,
189, 190
Spiral plasmonic lens array 187
Stokes parameters 30
Surface plasmon polaritons (SPP) ... 161

Three-dimensional Polarization 133
transition dipole moment (TDM) 74

vector optical field 28

Young's two-slit interference 58–61

www.ingramcontent.com/pod-product-compliance
Lightning Source LLC
Chambersburg PA
CBHW050546190326
41458CB00007B/1938